DISCOVERING ALGEBRA

Examples with Keystrokes on the TI-82 / TI-83 and TI-85 / TI-86

A Laboratory Approach

Donna Marie Pirich, Ph. D.
New York Institute of Technology

Patricia A. Bigliani
New York Institute of Technology
SUNY Farmingdale

PRENTICE HALL Upper Saddle River, NJ 07458

Senior Editor: *Ann Marie Jones*
Assistant Editor: *Audra Walsh*
Production Editor: *Maria T. Molinari*
Special Projects Manager: *Barbara A. Murray*
Manufacturing Buyer: *Alan Fischer*
Supplement Cover Manager: *Paul Gourhan*
Supplement Cover Designer: *PM Workshop Inc.*

© 1997 by **PRENTICE-HALL, INC.**
Simon & Schuster/A Viacom Company
Upper Saddle River, NJ 07458

All rights reserved. No part of this book may be
reproduced, in any form or by any means,
without permission in writing from the publisher.

Printed in the United States of America

10 9 8 7 6 5 4 3 2

ISBN 0-13-649203-7

Prentice-Hall International (UK) Limited, *London*
Prentice-Hall of Australia Pty. Limited, *Sydney*
Prentice-Hall Canada, Inc., *Toronto*
Prentice-Hall Hispanoamericana, S.A., *Mexico*
Prentice-Hall of India Private Limited, *New Delhi*
Prentice-Hall of Japan, Inc., *Tokyo*
Simon & Schuster Asia Pte. Ltd., *Singapore*
Editora Prentice-Hall do Brasil, Ltda., *Rio de Janeiro*

PREFACE

The graphing calculator is a powerful tool which can be used not only to analyze data, but also to recognize patterns, test hypotheses, and discover axioms. Instructors and students alike have found that integration of the graphing calculator within the classroom setting adds a new dimension to standard material. The graphing calculator takes students beyond basic mathematical manipulation and enables them to draw upon their visual and tactile senses in a discovery process that makes mathematics enjoyable.

This book is intended as a supplement to any standard intermediate algebra text. Material is provided to serve the needs of a standard fifteen week course. Laboratory experiences can readily be incorporated in the classroom presentation and provide students with hands-on experience and step-by-step instruction on how to make the most of their graphing calculator. No prior knowledge of the graphing calculator is required. The intent of this book is to provide beginners with the tools necessary to succeed in algebra without anxiety.

Each laboratory contains a statement of its purpose, a brief analytic background, several example problems with step-by-step solutions, and exercises which can be assigned for additional practice. Many of the exercises lend themselves to group activities which can be readily presented in the classroom by students. Instructors need not assign all laboratories. Involvement in the laboratory experience encourages students to take a more active role in class and to think, question, and analyze with confidence.

No special equipment is needed other than a graphing calculator such as the TI-82, TI-83, TI-85, or TI-86. The TI-82 and TI-83 are the calculators most frequently used for business applications, while the TI-85 and TI-86 are used for scientific applications. Calculator screens and instructions for the TI-83 and TI-85 will be shown side-by-side for the example problems within this text. The TI-82 and TI-86 instructions will be given when they differ from those of their counterparts.

The authors and their students have found the laboratory approach to be a positive experience which has enabled them to focus their attention on analysis of data and applications to real world problems. This is the direction in which we must proceed in order to develop a true appreciation for mathematics and its relevance in our world.

Acknowledgments

The authors are most grateful to Texas-Instruments and Prentice-Hall for their generous support of the classroom testing of this material and to Mr. Cleo Demesier for his work as a teaching assistant in the laboratory.

Dedication

To our families, the Piriches: Ron, Christopher, Michael, Connie and Mike Sassano, and the late Josephine DeVito, whose memory inspired this work; and the Biglianis: Ray, Tommy, Kristi, Patti, and Kelly; your support and understanding made the completion of this project possible.

Problem Solving with Technology
10 Helpful Suggestions

The authors and their students have found the following suggestions to be helpful to the new graphing calculator user. These hints are offered to help the student to integrate problem solving and technology with as little anxiety as possible.

- If the screen is blank when you turn your calculator on try adjusting the contrast. In general a blank screen indicates that the contrast needs to be adjusted or the batteries need to be changed.

- It is helpful to have paper and pencil handy when working with the graphing calculator. Problems may not always be set up in a convenient form for entry into the calculator and it may be necessary to simplify a problem before referring to the calculator.

- It is a good idea to check the **MODE** screen of your graphing calculator before starting a problem to make sure that the current settings are consistent with the requirements of the problem to be solved.

- If you are having trouble getting the calculator to come up with the correct answer try clearing memory; but remember to be careful not to clear any stored programs you might need.

- It is helpful to remember to clear all graphs, drawings, matrices, and lists before starting a problem which makes use of these features.

- The calculator keys are color-coded and it is helpful to familiarize yourself with the keyboard layout and menu maps. The top-down order of the arithmetic operator keys is a useful reminder of the order of operations.

- Care should be taken in the correct use of the negation key and the subtraction key. It is helpful to note that the negation key is the same color as the number keys and the subtraction key is the same color as the other arithmetic operator keys.

- It is helpful to enclose fractions inside parentheses to avoid errors in calculations. Although the TI calculators will adhere to the order of operations, the TI-82 and TI-85 will treat an expression such as $1/2x$ as if the $2x$ is in the denominator. The TI-83 and TI-86 will treat the expression as $.5x$. It is also helpful to place parentheses around expressions which are in either the numerator or the denominator.

- The graphing calculator has many different features which can be helpful in solving a particular problem. It is a good idea to solve the problem in more than one way and to compare results. The last step in solving any problem is to be sure that the answer "makes sense."

- Try not to leave your calculator in a cold or hot place. Failure to do so could cause damage.

TABLE OF CONTENTS

Preface

Problem Solving with Technology
10 Helpful Suggestions

Laboratory 1:   Basic Concepts and Introduction to Graphing Calculators. 1

Laboratory 2:   Solution of Linear Equations. 13

Laboratory 3:   Solution of Linear Inequalities. 23

Laboratory 4:   Introduction to Graphing: The Equation of a Line. 35

Laboratory 5:   Data Analysis and Statistical Approach to the Equation of a Line. 47

Laboratory 6:   Systems of Linear Equations. 59

Laboratory 7:   Advanced Graphing Techniques: The Quadratic Equation. 73

Laboratory 8:   Functions. 89

Laboratory 9:   Polynomials. 101

Laboratory 10: Rational Expressions. 117

Laboratory 11: Rational Equations. 129

Laboratory 12: Radicals and Exponents. 141

Laboratory 13: Inverse Functions. 157

Laboratory 14: Exponential and Logarithmic Functions. 167

Laboratory 15: Conic Sections. 183

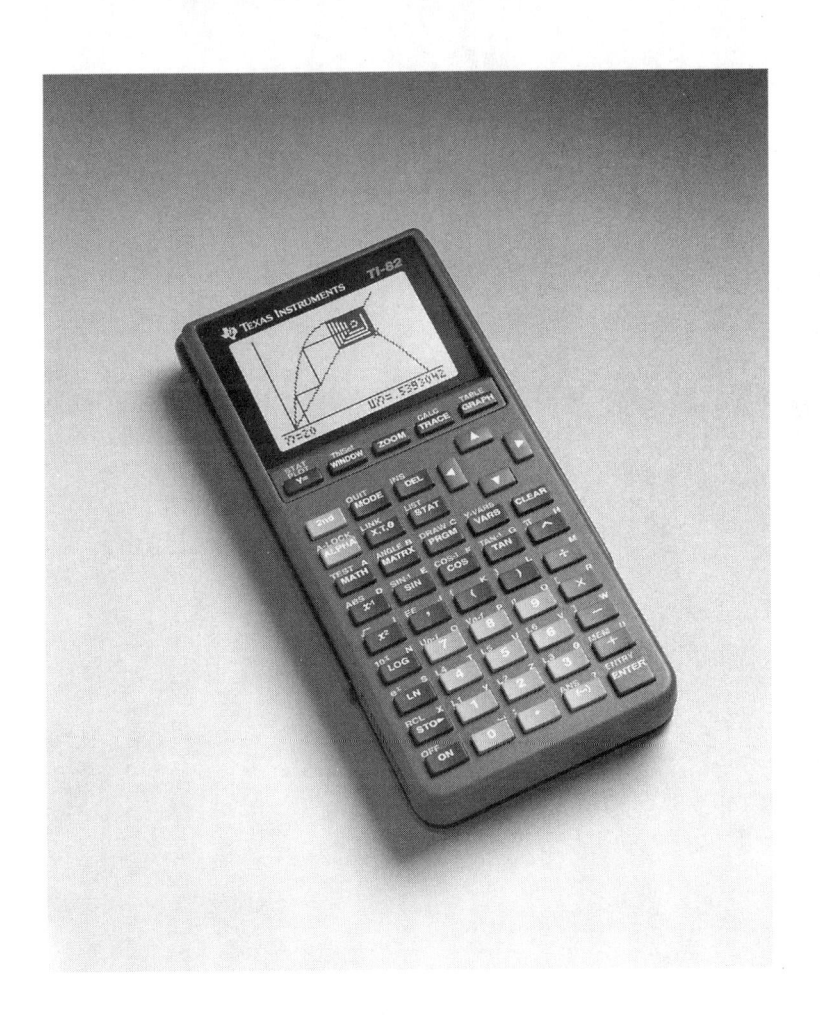

Texas Instruments
TI-82 Graphing Calculator

Texas Instruments
TI-83 Graphing Calculator

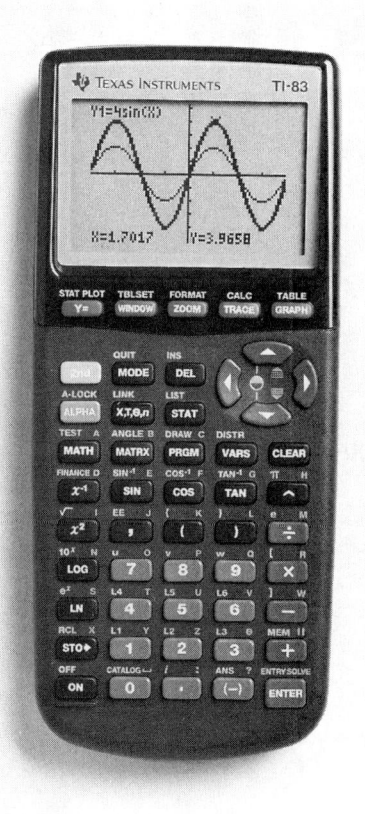

Photographs of Texas Instruments calculators appear
here courtesy of Texas Instruments

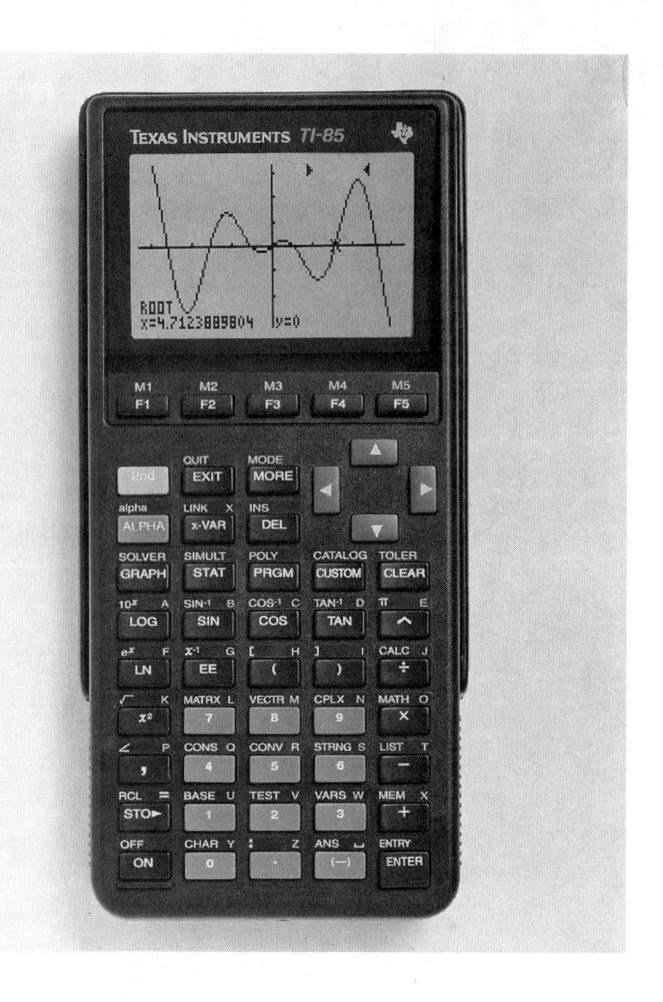

Texas Instruments
TI-85 Graphing Calculator

Texas Instruments
TI-86 Graphing Calculator

Photographs of Texas Instruments calculators appear here courtesy of Texas Instruments

LABORATORY 1: BASIC CONCEPTS AND INTRODUCTION TO GRAPHING CALCULATORS

Purpose:

The purpose of this laboratory is to provide an introduction to the TI-83/TI-82 and TI-85/TI-86 graphing calculators as they relate to the order of operations, the real number system, fractions, and absolute value. Most of the arithmetic operations are executed identically on all four calculators. Therefore, only one set of keystrokes and screens will be presented for each example unless the operation is calculator dependent.

Analytic Approach:

A. Getting Started on the Graphing Calculator

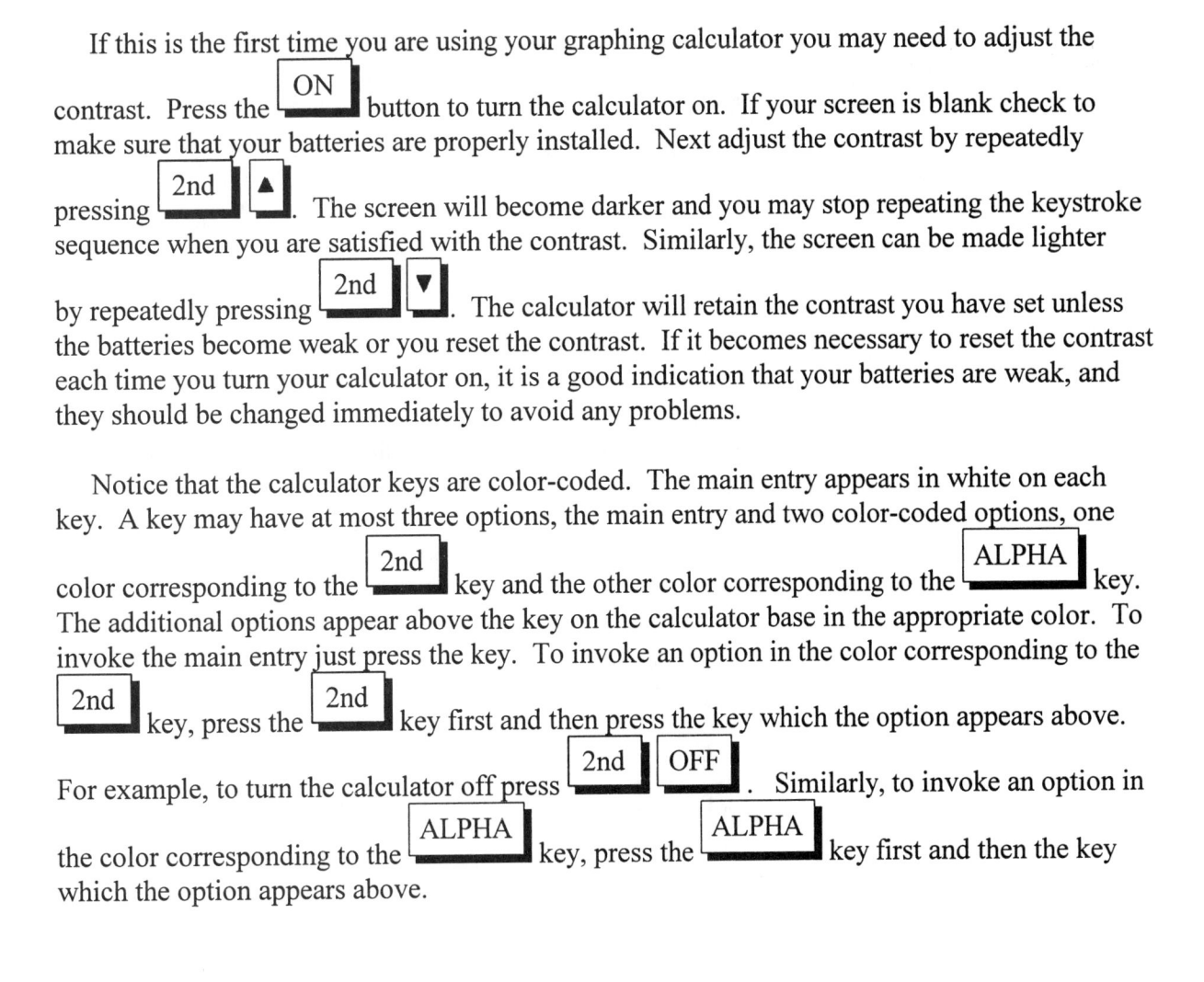

If this is the first time you are using your graphing calculator you may need to adjust the contrast. Press the $\boxed{\text{ON}}$ button to turn the calculator on. If your screen is blank check to make sure that your batteries are properly installed. Next adjust the contrast by repeatedly pressing $\boxed{\text{2nd}}$ $\boxed{\blacktriangle}$. The screen will become darker and you may stop repeating the keystroke sequence when you are satisfied with the contrast. Similarly, the screen can be made lighter by repeatedly pressing $\boxed{\text{2nd}}$ $\boxed{\blacktriangledown}$. The calculator will retain the contrast you have set unless the batteries become weak or you reset the contrast. If it becomes necessary to reset the contrast each time you turn your calculator on, it is a good indication that your batteries are weak, and they should be changed immediately to avoid any problems.

Notice that the calculator keys are color-coded. The main entry appears in white on each key. A key may have at most three options, the main entry and two color-coded options, one color corresponding to the $\boxed{\text{2nd}}$ key and the other color corresponding to the $\boxed{\text{ALPHA}}$ key. The additional options appear above the key on the calculator base in the appropriate color. To invoke the main entry just press the key. To invoke an option in the color corresponding to the $\boxed{\text{2nd}}$ key, press the $\boxed{\text{2nd}}$ key first and then press the key which the option appears above. For example, to turn the calculator off press $\boxed{\text{2nd}}$ $\boxed{\text{OFF}}$. Similarly, to invoke an option in the color corresponding to the $\boxed{\text{ALPHA}}$ key, press the $\boxed{\text{ALPHA}}$ key first and then the key which the option appears above.

LABORATORY 1: BASIC CONCEPTS AND INTRODUCTION TO GRAPHING CALCULATORS

B. Order of Operations

In algebra there is a particular order in which calculations must be performed. It is referred to as the order of operations. The order of operations is often remembered by using the acronym PEMDAS or "Please Excuse My Dear Aunt Sally." The letters in the acronym stand for parentheses, exponentiation, multiplication, division, addition, and subtraction, respectively. Moving from left to right, all calculations inside parentheses must be performed first. In the case of nested parentheses (parentheses within parentheses) work from the innermost parentheses out. Perform all exponentiations and take any indicated roots. Next multiply and divide numbers in the order in which they appear from left to right. Finally, add and subtract numbers in the order in which they appear from left to right.

The keyboard layout on the TI graphing calculators provides another means of remembering the order of operations:

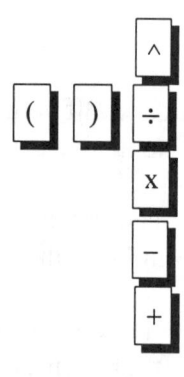

Notice that the exponentiation, division, multiplication, subtraction, and addition operators are located in the rightmost column of the calculator keyboard. They are placed in "top-down" order of operations; exponentiation first, followed by division and multiplication (which are of equal importance), and finally subtraction and addition (which are of equal importance). The parentheses are located to the left of the operator column on the calculator. These operator keys are different in color from the numeric or number keys on the calculator:

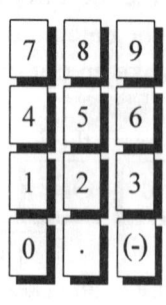

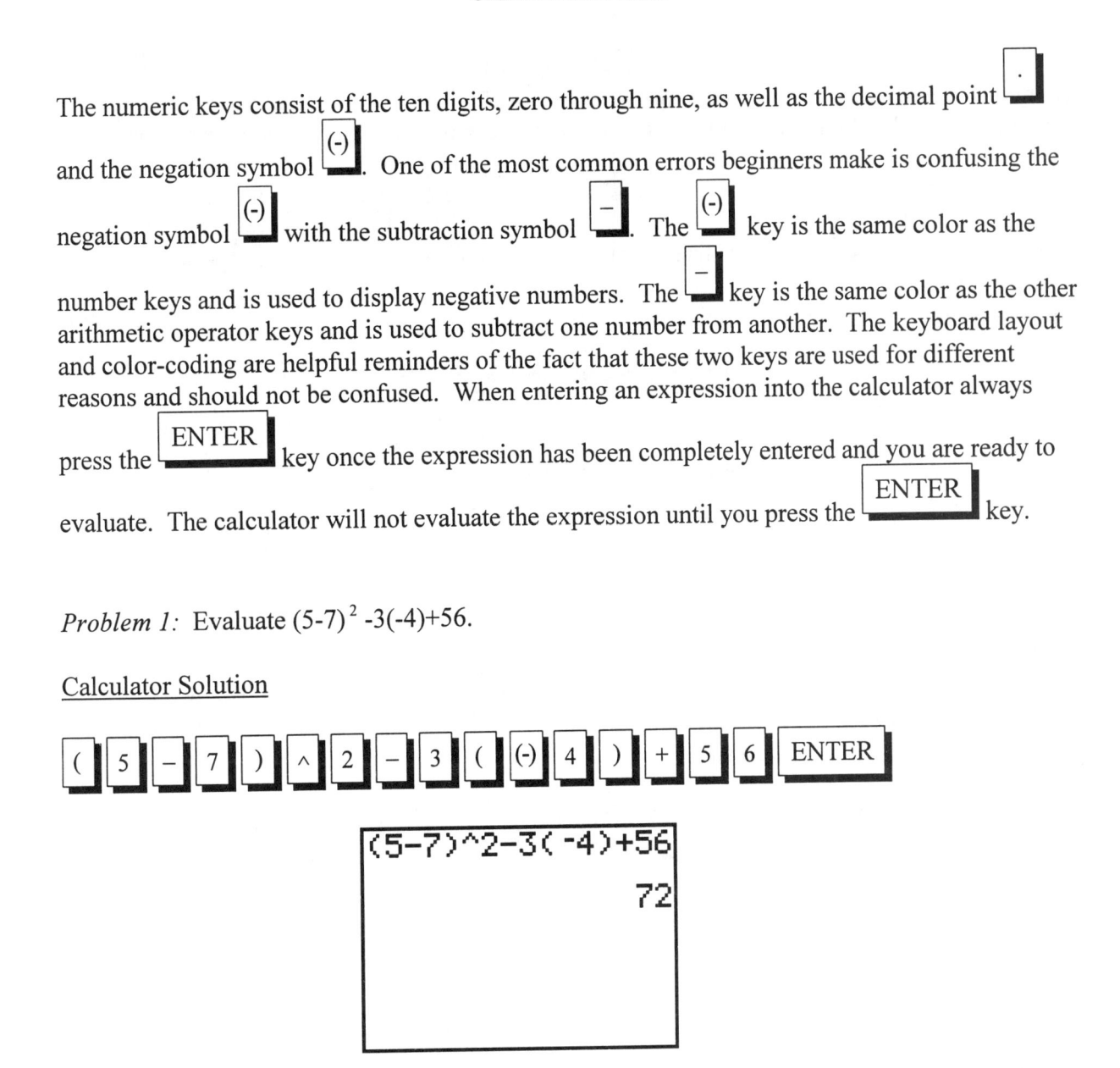

LABORATORY 1: BASIC CONCEPTS AND INTRODUCTION TO GRAPHING CALCULATORS

The numeric keys consist of the ten digits, zero through nine, as well as the decimal point [.] and the negation symbol [(-)]. One of the most common errors beginners make is confusing the negation symbol [(-)] with the subtraction symbol [−]. The [(-)] key is the same color as the number keys and is used to display negative numbers. The [−] key is the same color as the other arithmetic operator keys and is used to subtract one number from another. The keyboard layout and color-coding are helpful reminders of the fact that these two keys are used for different reasons and should not be confused. When entering an expression into the calculator always press the [ENTER] key once the expression has been completely entered and you are ready to evaluate. The calculator will not evaluate the expression until you press the [ENTER] key.

Problem 1: Evaluate $(5-7)^2 -3(-4)+56$.

Calculator Solution

[(] [5] [−] [7] [)] [^] [2] [−] [3] [(] [(-)] [4] [)] [+] [5] [6] [ENTER]

```
(5-7)^2-3(-4)+56
                72
```

Notice the difference in the way the subtraction and negation symbols appear on the calculator screen. The difference in size and location of the symbols is helpful in locating data entry errors.

Alternate Solution

The [x^2] key can be used to square a number. It is equivalent to the keystroke sequence [^] [2]. It is not necessary to retype the entire keystroke sequence in order to test the alternate solution. The TI calculators are equipped with an edit feature. To recall the last expression

LABORATORY 1: BASIC CONCEPTS AND INTRODUCTION TO GRAPHING CALCULATORS

entered to the calculator screen press [2nd] [ENTRY]. The last entry will now appear on the screen. The blinking box that moves across the screen as you type is called the cursor. Use the [◄] to move the cursor over the [^] and press [DEL] to delete the exponentiation symbol. Press [DEL] again to delete the number 2. Now press [2nd] [INS] [x^2] [ENTER] to edit the line and evaluate.

```
(5-7)^2-3(-4)+56
                72
(5-7)²-3(-4)+56
                72
```

Clear the screen when you are done by pressing [CLEAR]. Try entering the following keystroke sequence to verify that the same results are achieved if the original entry is not edited and a new string is typed in instead:

[( ] [5] [-] [7] [ )] [x^2] [-] [3] [( ] [(-)] [4] [ )] [+] [5] [6] [ENTER]

Problem 2: Multiply 32 by -24.

Calculator Solution

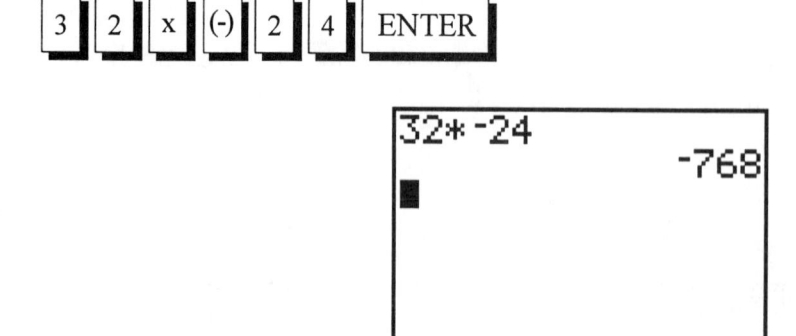

[3] [2] [x] [(-)] [2] [4] [ENTER]

```
32*-24
         -768
```

Notice that the symbol which appears on the calculator screen for multiplication is the asterisk symbol. Parentheses may be used to imply multiplication.

LABORATORY 1: BASIC CONCEPTS AND INTRODUCTION TO GRAPHING CALCULATORS

Alternate Solution

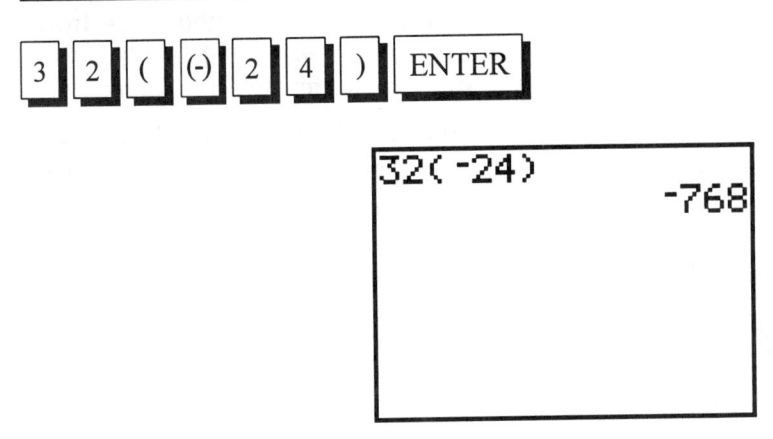

[3] [2] [(] [(-)] [2] [4] [)] [ENTER]

```
32(-24)
           -768
```

An interesting difference between the TI-83/TI-82 and TI-85/TI-86 calculators is illustrated if we type in the keystroke sequence:

[3] [2] [(-)] [2] [4] [ENTER]

TI-83 Graphing Calculator Solution

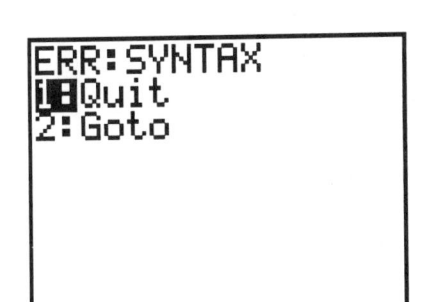

```
ERR:SYNTAX
1:Quit
2:Goto
```

TI-85 Graphing Calculator Solution

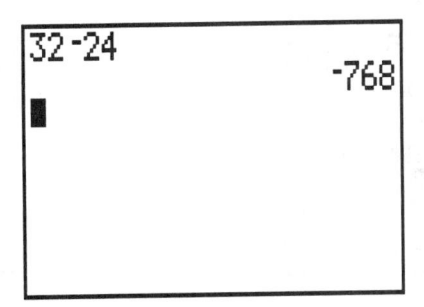

```
32-24
          -768
■
```

The TI-83 displays a syntax error. By pressing [▼] [ENTER] or [2] the calculator screen will indicate where the error occurred and it can be corrected. In this case it would be necessary to insert a multiplication symbol in order for the TI-83 to calculate the expression. Similarly, the TI-82 will display a syntax error. However, the "Quit" and "Goto" items are in the reverse order on the screen, so it would be necessary to press [ENTER] or [1] to see where the mistake was made.

The TI-85 and TI-86 calculators do not display a syntax error. Multiplication is implicit on these calculators and it is performed even though the multiplication symbol is missing. These calculators recognize that the only way the expression can be evaluated properly is as a

LABORATORY 1: BASIC CONCEPTS AND INTRODUCTION TO GRAPHING CALCULATORS

multiplication. Implicit multiplication can lead to problems if care is not taken in typing expressions into the calculator. For example, suppose we had really wanted to subtract 24 from 32 and had inadvertently used the negation symbol instead of the subtraction symbol. The keystroke sequence illustrated above could correspond to such an error. The TI-83 and TI-82 calculators would recognize a syntax error and it could be corrected. However, the TI-85 and TI-86 calculators would instead assume multiplication, thus giving an incorrect result. It is very important to check the calculator screen to make sure that expressions have been correctly entered. Failure to do so can result in calculation errors. Always remember that the calculator will do exactly what you tell it to do, even if it is not necessarily what you *want* it to do!

Problem 3: Divide 372 by 654.

It is important to note that the TI-83 and TI-82 calculators are capable of displaying results accurate to ten significant digits while the TI-85 and TI-86 calculators can display up to twelve significant digits. Check the current settings on your calculator.

<u>TI-83 Graphing Calculator Solution</u>   <u>TI-85 Graphing Calculator Solution</u>

[MODE]   [2nd] [MODE]

```
Normal Sci Eng              Normal Sci Eng
Float 0123456789            Float 012345678901
Radian Degree               Radian Degree
Func Par Pol Seq            RectC PolarC
Connected Dot               Func Pol Param DifEq
Sequential Simul            Dec Bin Oct Hex
Real a+bi re^θi             RectV CylV SphereV
Full Horiz G-T              dxDer1 dxNDer
```

If your calculator screen does not look like the screen above use the arrow keys to highlight and [ENTER] the first entry in each line of the [MODE] screen. The second line of the [MODE] screen determines the number of decimal places of accuracy displayed on the calculator screen. The **Float** mode will display the maximum number of decimal places of accuracy on your calculator. Should you choose to display anything less you can do so by using the arrow keys to highlight and [ENTER] the appropriate number on the **Float** line of the [MODE] screen. For example, to display two decimal places of accuracy simply highlight

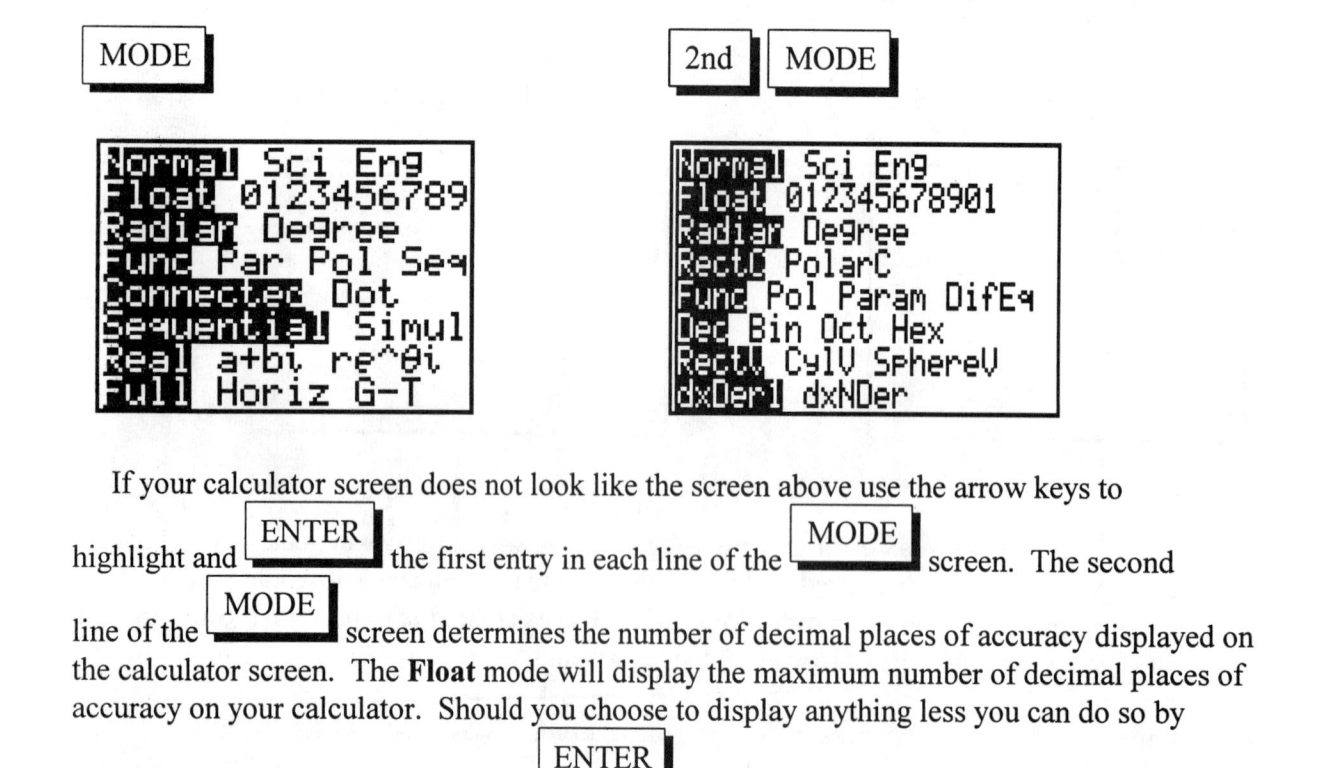

LABORATORY 1: BASIC CONCEPTS AND INTRODUCTION TO GRAPHING CALCULATORS

the 2 on the **Float** line of the [MODE] screen and then press [ENTER]. We will be using the calculator's full capability, so make sure that you are in **Float** mode before clearing the screen.

To calculate 372/654 simply press [3][7][2][÷][6][5][4][ENTER].

TI-83 Graphing Calculator Solution

```
372/654
      .5688073394
```

TI-85 Graphing Calculator Solution

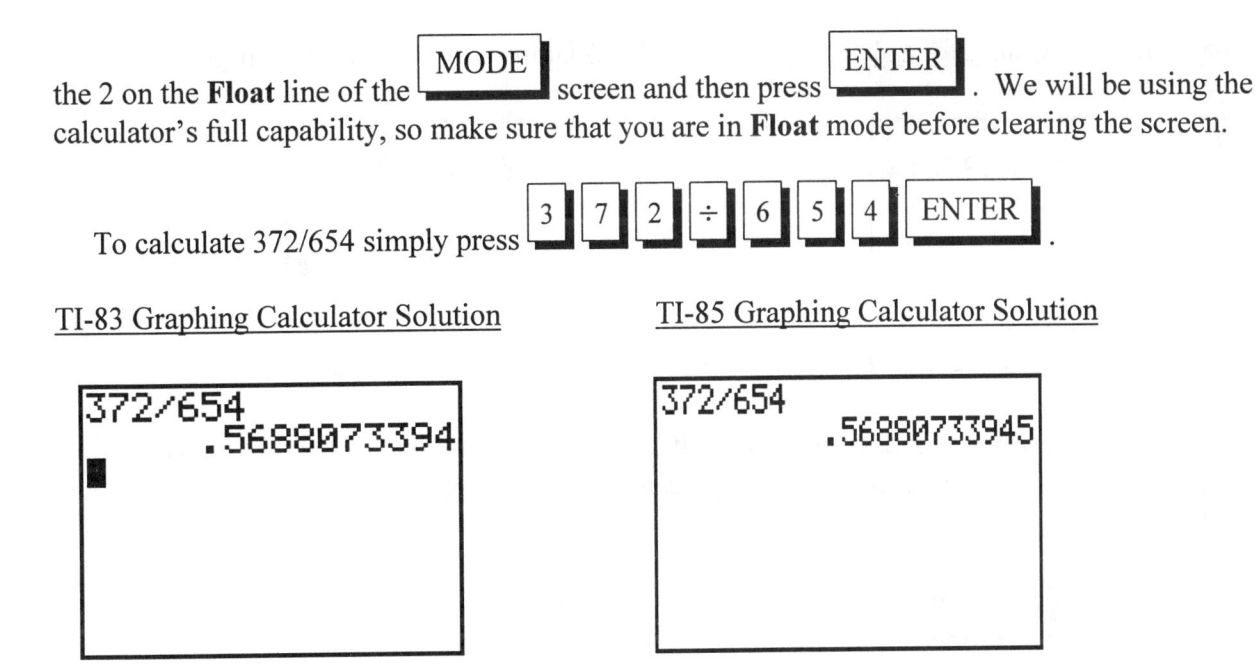

```
372/654
      .56880733945
```

C. The Real Number System

The set of natural or counting numbers can be stated as the set N, where N = { 1, 2, 3, ... }. The set of whole numbers, W, where W = { 0, 1, 2, 3, ... }, contains the set of natural numbers. Similarly, the set of whole numbers is contained in the set I, of integers, where I = { ... -3, -2, -1, 0, 1, 2, 3, ...}. The set of rational numbers, Q, include the integers and all quotients of integers (where the divisor is non-zero). Therefore, we may express the set as Q = {a/b where a and b are integers and b ≠ 0}. All rational numbers have decimal portions which are either terminating or repeating. Irrational numbers, on the other hand, have decimal portions which are nonterminating and nonrepeating. The set of real numbers, R, is the set of all numbers that are either rational or irrational.

The calculator makes classification of numbers a simple task. By examining the decimal portion of a real number we can determine if it is rational or irrational.

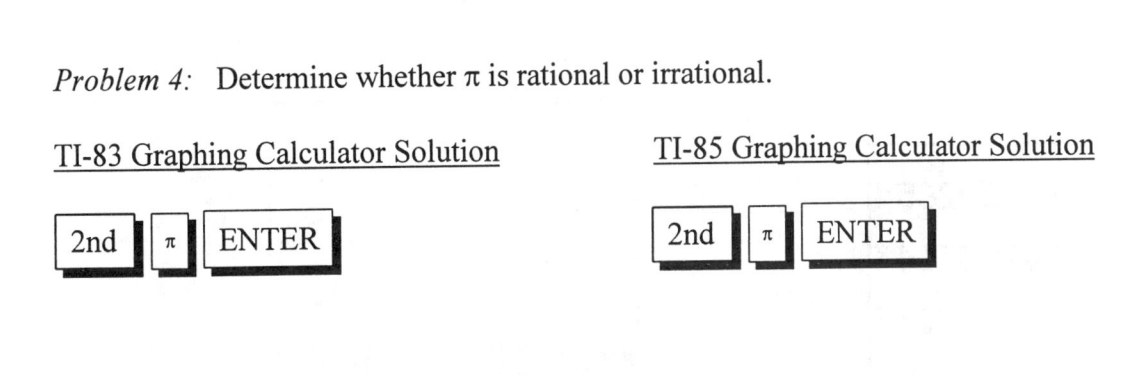

Problem 4: Determine whether π is rational or irrational.

TI-83 Graphing Calculator Solution

[2nd][π][ENTER]

TI-85 Graphing Calculator Solution

[2nd][π][ENTER]

LABORATORY 1: BASIC CONCEPTS AND INTRODUCTION TO GRAPHING CALCULATORS

TI-83 Graphing Calculator Solution

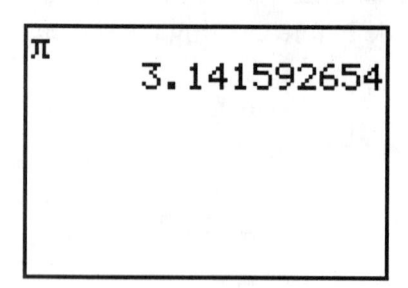

TI-85 Graphing Calculator Solution

```
π
            3.141592654
```

```
π
            3.14159265359
```

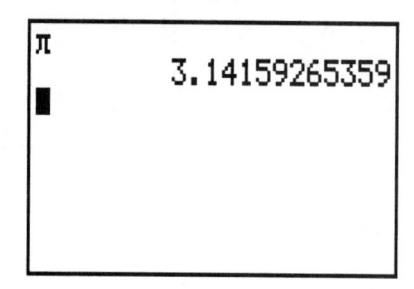

The decimal portion of π does not repeat. It is also nonterminating. Therefore, π is an irrational number. Compare the decimal representation for π to the decimal representation for 22/7.

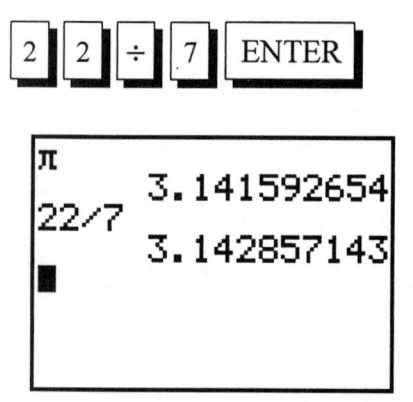

| 2 | 2 | ÷ | 7 | ENTER |

| 2 | 2 | ÷ | 7 | ENTER |

```
π
            3.141592654
22/7
            3.142857143
```

```
π
            3.14159265359
22/7
            3.14285714286
```

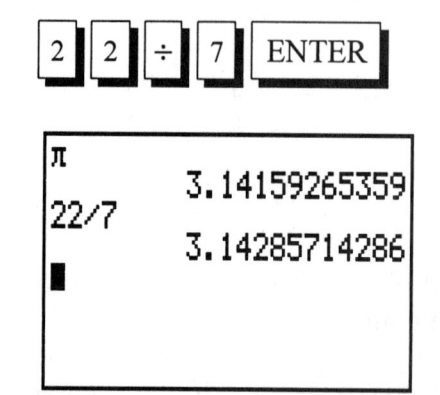

Notice that 22/7 is a rational estimate of the irrational number π correct to two decimal places. The rational number 22/7 is bigger than the irrational number π. We can convert the decimal representation for 22/7 back to fractional form by using a conversion feature on the calculator. The conversion feature reduces a rational number in decimal form to its equivalent fractional representation in lowest terms.

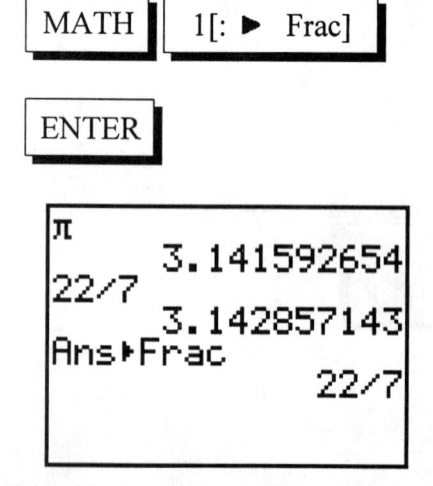

| MATH | 1[: ▶ Frac] |

| 2nd | MATH | F5[MISC] |

| ENTER |

| MORE | F1 [▶ Frac] | ENTER |

```
π
            3.141592654
22/7
            3.142857143
Ans▶Frac
                   22/7
```

```
22/7        3.14159265359
            3.14285714286
Ans▶Frac
                    22/7
NUM  PROB  ANGLE  HYP  MISC
▶Frac  %   pEval   xr   eval
```

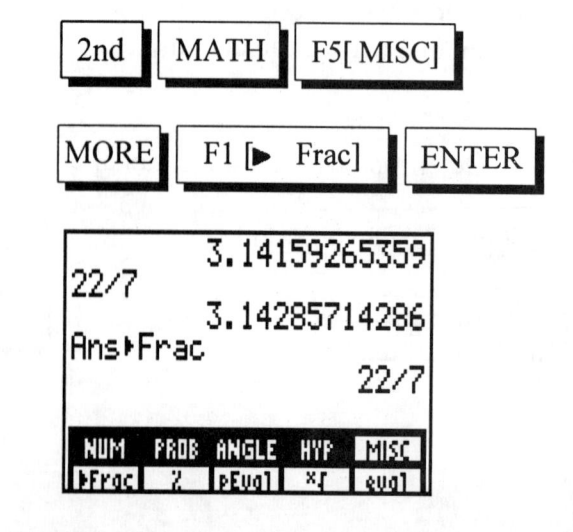

LABORATORY 1: BASIC CONCEPTS AND INTRODUCTION TO GRAPHING CALCULATORS

TI-83 Graphing Calculator Solution   TI-85 Graphing Calculator Solution

Now clear the screen.

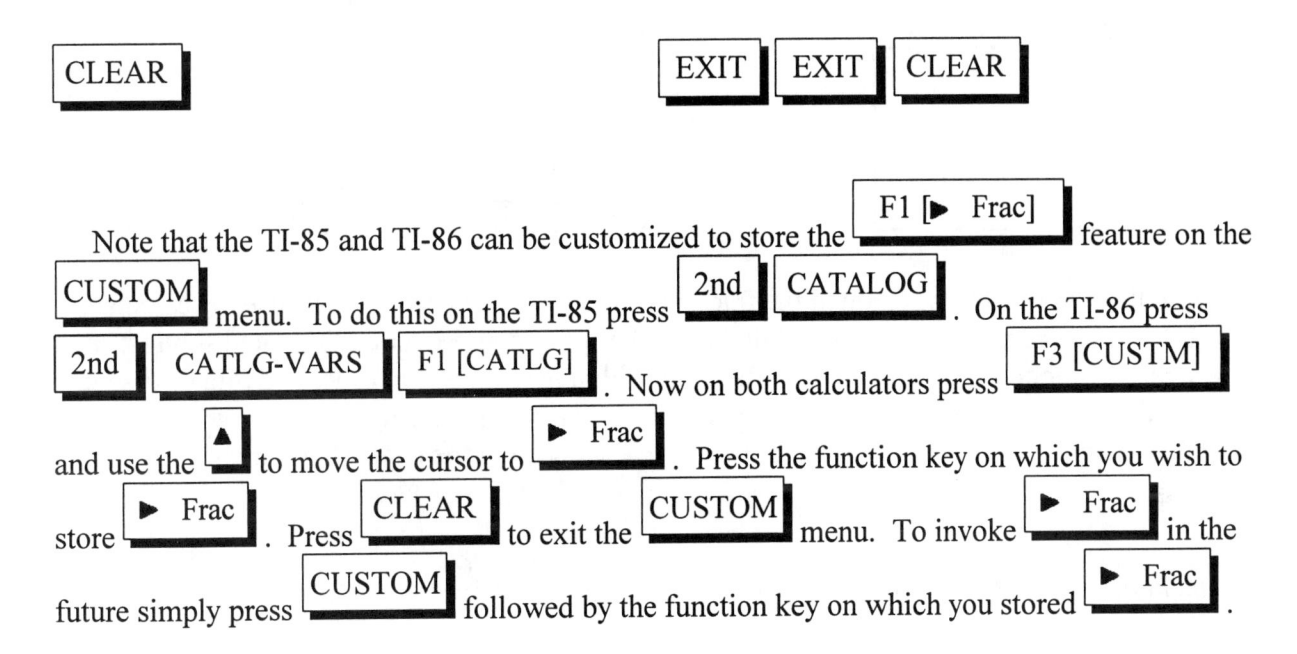

[CLEAR]   [EXIT] [EXIT] [CLEAR]

Note that the TI-85 and TI-86 can be customized to store the [F1 [▶ Frac]] feature on the [CUSTOM] menu. To do this on the TI-85 press [2nd] [CATALOG]. On the TI-86 press [2nd] [CATLG-VARS] [F1 [CATLG]]. Now on both calculators press [F3 [CUSTM]] and use the [▲] to move the cursor to [▶ Frac]. Press the function key on which you wish to store [▶ Frac]. Press [CLEAR] to exit the [CUSTOM] menu. To invoke [▶ Frac] in the future simply press [CUSTOM] followed by the function key on which you stored [▶ Frac].

Problem 5: Add 1/3 and 1/5, then subtract 2/7. Express your answer in fraction form.

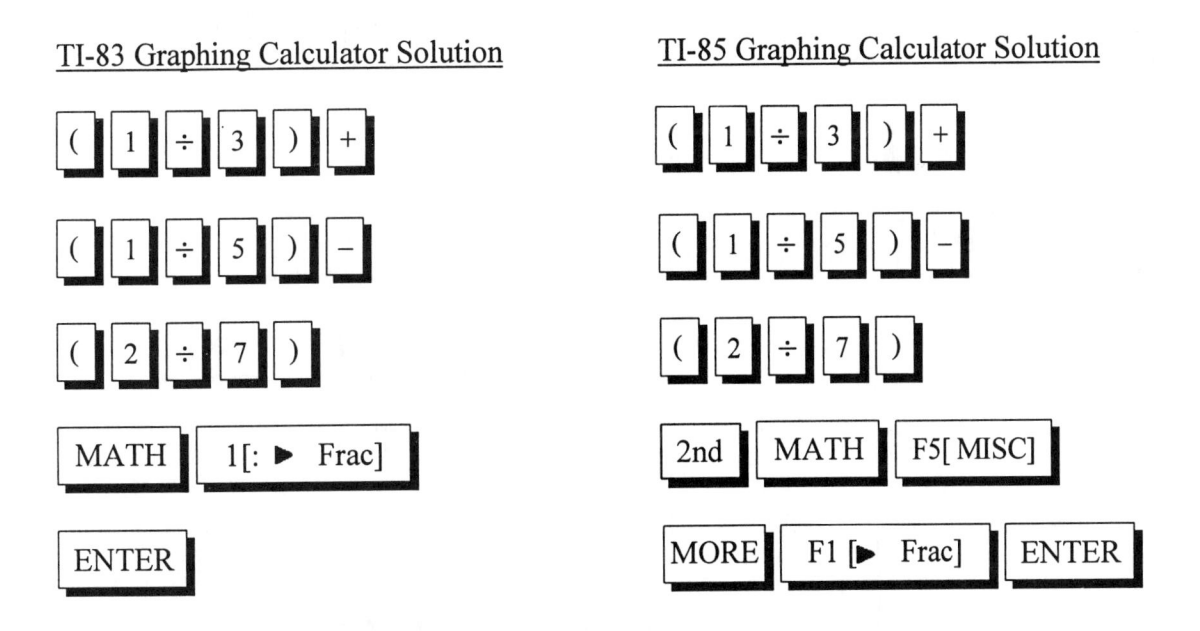

TI-83 Graphing Calculator Solution   TI-85 Graphing Calculator Solution

[(] [1] [÷] [3] [)] [+]   [(] [1] [÷] [3] [)] [+]

[(] [1] [÷] [5] [)] [−]   [(] [1] [÷] [5] [)] [−]

[(] [2] [÷] [7] [)]   [(] [2] [÷] [7] [)]

[MATH] [1[: ▶ Frac]]   [2nd] [MATH] [F5[MISC]]

[ENTER]   [MORE] [F1 [▶ Frac]] [ENTER]

LABORATORY 1: BASIC CONCEPTS AND INTRODUCTION TO GRAPHING CALCULATORS

TI-83 Graphing Calculator Solution

```
(1/3)+(1/5)-(2/7
)▶Frac
              26/105
```

TI-85 Graphing Calculator Solution

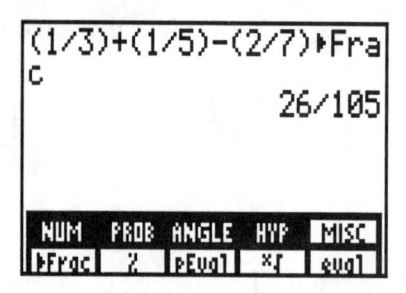

```
(1/3)+(1/5)-(2/7)▶Fra
c
              26/105
```

Note that the parentheses around the fractions are not really necessary because the calculator will follow the order of operations and division takes precedence over addition and subtraction. However, it is good practice to enclose fractions within parentheses in order to avoid errors when more involved calculations are performed. In fact, it is often necessary to enclose the numerator and the denominator in separate pairs of parentheses as seen in the example problem below.

Problem 6: Evaluate and convert to fraction form: $\dfrac{2+4(3-6)}{7-5}$.

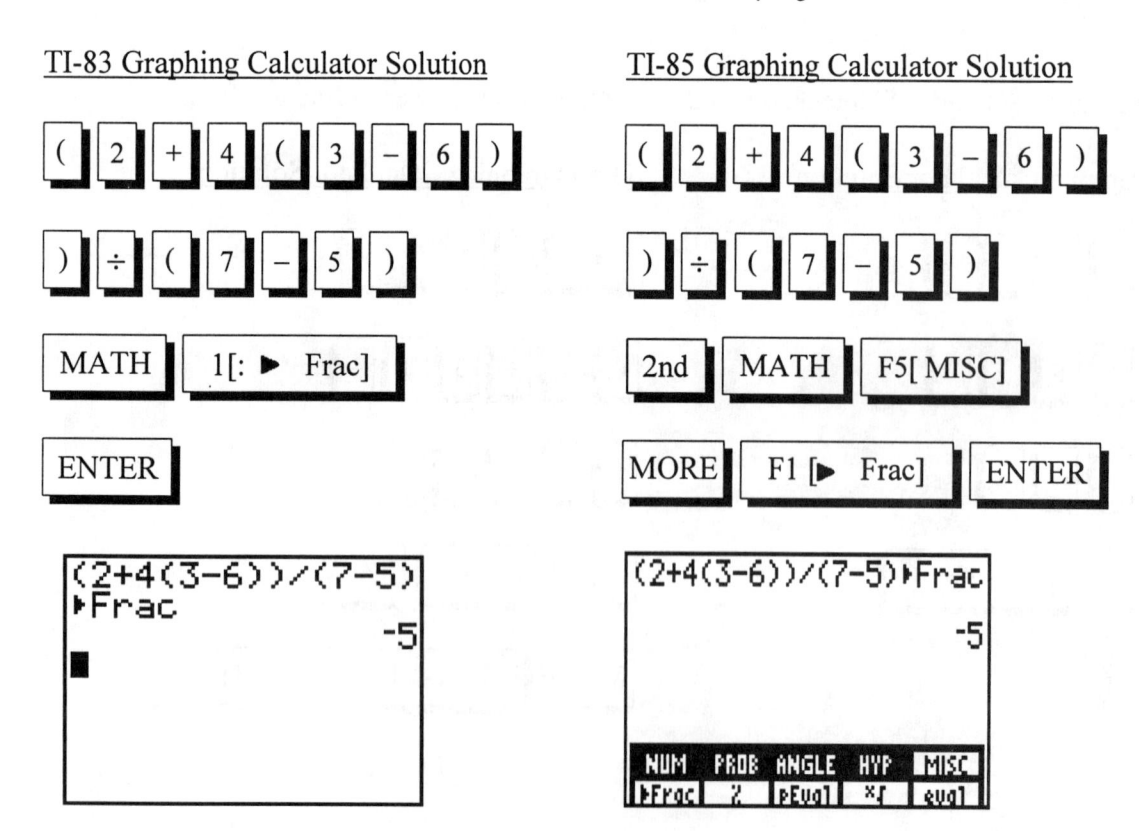

TI-83 Graphing Calculator Solution

[( ] [2] [+] [4] [( ] [3] [−] [6] [ )]

[ )] [÷] [( ] [7] [−] [5] [ )]

[MATH] [1[: ▶ Frac]]

[ENTER]

```
(2+4(3-6))/(7-5)
▶Frac
                  -5
■
```

TI-85 Graphing Calculator Solution

[( ] [2] [+] [4] [( ] [3] [−] [6] [ )]

[ )] [÷] [( ] [7] [−] [5] [ )]

[2nd] [MATH] [F5[MISC]]

[MORE] [F1 [▶ Frac]] [ENTER]

```
(2+4(3-6))/(7-5)▶Frac
                  -5
```

Repeat this problem, leaving out the parentheses which were added to the numerator and the denominator. Compare your results and explain any differences.

LABORATORY 1: BASIC CONCEPTS AND INTRODUCTION TO GRAPHING CALCULATORS

D. Absolute Value

The absolute value of a number tells us its distance from zero. The notation for the absolute value of a number x is $|x|$.

Problem 7: Evaluate the absolute value of the sum of negative three and five.

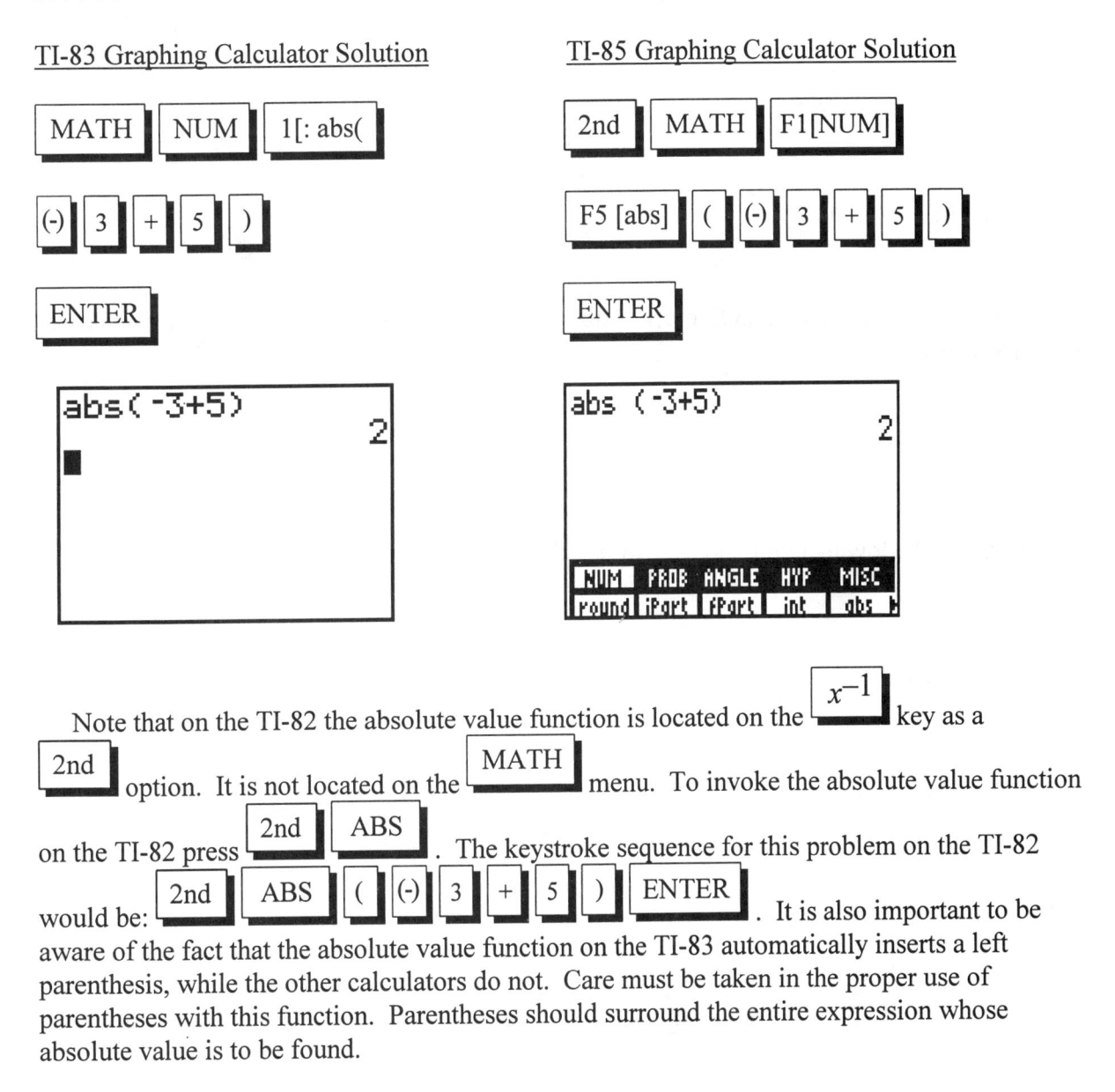

TI-83 Graphing Calculator Solution

[MATH] [NUM] [1[: abs(]

[(-)] [3] [+] [5] [)]

[ENTER]

```
abs(-3+5)
               2
■
```

TI-85 Graphing Calculator Solution

[2nd] [MATH] [F1[NUM]]

[F5 [abs]] [(] [(-)] [3] [+] [5] [)]

[ENTER]

```
abs (-3+5)
               2


NUM  PROB ANGLE HYP  MISC
round iPart fPart  int  abs
```

Note that on the TI-82 the absolute value function is located on the x^{-1} key as a [2nd] option. It is not located on the [MATH] menu. To invoke the absolute value function on the TI-82 press [2nd] [ABS]. The keystroke sequence for this problem on the TI-82 would be: [2nd] [ABS] [(] [(-)] [3] [+] [5] [)] [ENTER]. It is also important to be aware of the fact that the absolute value function on the TI-83 automatically inserts a left parenthesis, while the other calculators do not. Care must be taken in the proper use of parentheses with this function. Parentheses should surround the entire expression whose absolute value is to be found.

LABORATORY 1: BASIC CONCEPTS AND INTRODUCTION TO GRAPHING CALCULATORS

EXERCISES

1. Perform the indicated operations on your calculator:

    a. $10 + (-3) - 4 * 6 / 2$

    b. $-20 - 5 + 10 * 2$

    c. $(-3)^2 + (2 - 10 * 3)^3$

    d. $5 - (-18) + (12 / 2 - 4)$

2. Remove all the parentheses in Exercise 1 above and recalculate. Compare your answers and explain any differences.

3. Evaluate the following expressions on your calculator and express as fractions:

    a. $100/2000$

    b. $\dfrac{-20 + (-20)}{18 - 4}$

    c. $\dfrac{27 / (-3) + 4 * (-2)}{2 (6 - 4)}$

    d. $(-5)(1/5) + 2 / 3$

4. Express the answers to Exercise 3 above in decimal form, rounded to the nearest hundredth.

5. The temperature on a particular day rose from 5 degrees below zero to 17 degrees above zero. Find the net change in temperature.

LABORATORY 2: SOLUTION OF LINEAR EQUATIONS

Purpose:

The purpose of this laboratory is to outline methods used to evaluate algebraic expressions and to study the solution of linear equations using non-graphing techniques on the TI calculators. A detailed look at the graphical solution of linear equations will be covered in Laboratory 4.

Analytic Approach:

A. Algebraic Expressions

An *algebraic expression* is a collection of numbers, variables, and operators. The *terms* or parts of the expression are joined together by the arithmetic operations of addition or subtraction. A numerical *coefficient* of a term is the number which multiplies the variables in a term. For example, the algebraic expression $3x^2 + 2x - 4y + 6xy - y^2$ has five terms. The first term is $3x^2$. The variable in this term is x, the exponent of the variable is 2, and the coefficient is 3. There are two variables in this expression, x and y. The value of the algebraic expression is dependent on the values of the variables. By changing the values of the variables we can study the behavior of the expression.

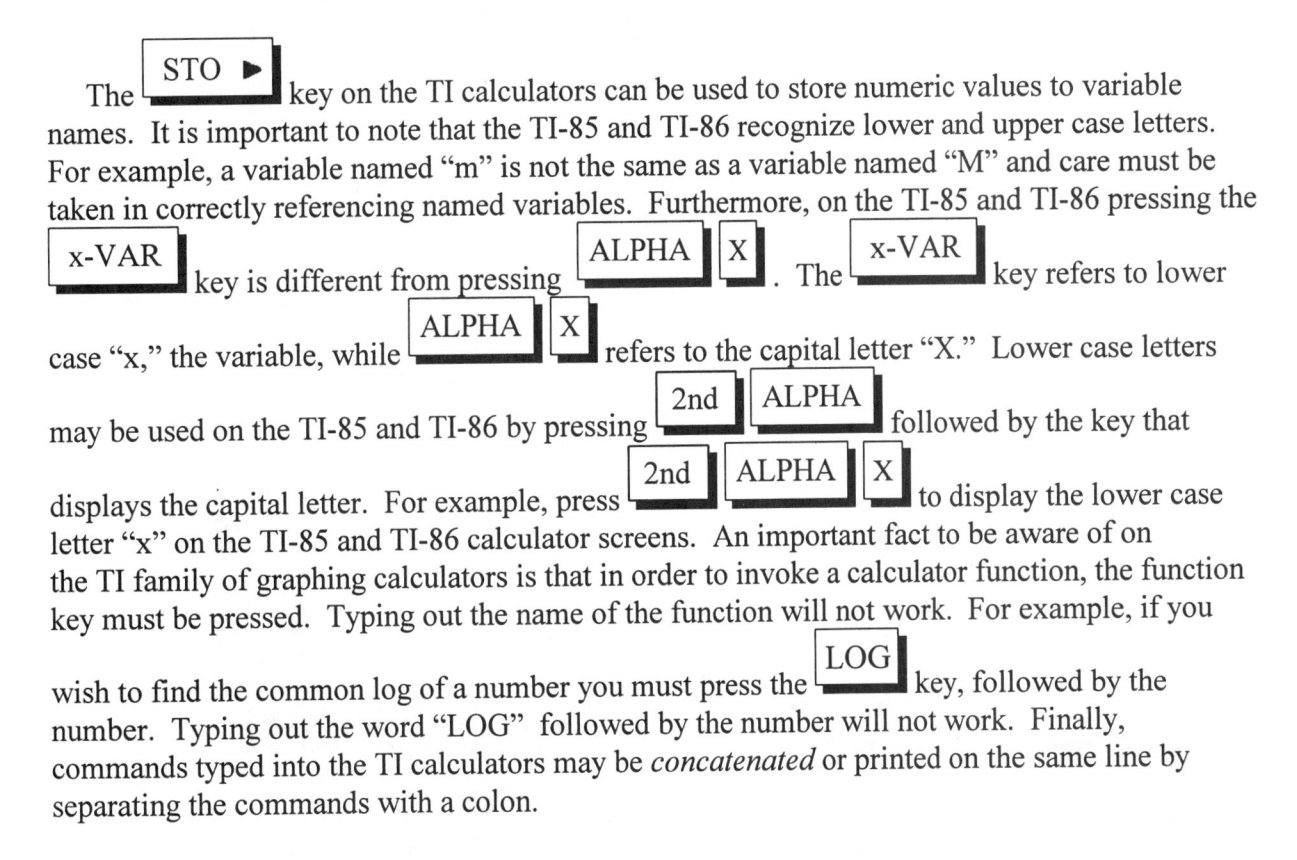

The $\boxed{\text{STO} \blacktriangleright}$ key on the TI calculators can be used to store numeric values to variable names. It is important to note that the TI-85 and TI-86 recognize lower and upper case letters. For example, a variable named "m" is not the same as a variable named "M" and care must be taken in correctly referencing named variables. Furthermore, on the TI-85 and TI-86 pressing the $\boxed{\text{x-VAR}}$ key is different from pressing $\boxed{\text{ALPHA}}\boxed{\text{X}}$. The $\boxed{\text{x-VAR}}$ key refers to lower case "x," the variable, while $\boxed{\text{ALPHA}}\boxed{\text{X}}$ refers to the capital letter "X." Lower case letters may be used on the TI-85 and TI-86 by pressing $\boxed{\text{2nd}}\boxed{\text{ALPHA}}$ followed by the key that displays the capital letter. For example, press $\boxed{\text{2nd}}\boxed{\text{ALPHA}}\boxed{\text{X}}$ to display the lower case letter "x" on the TI-85 and TI-86 calculator screens. An important fact to be aware of on the TI family of graphing calculators is that in order to invoke a calculator function, the function key must be pressed. Typing out the name of the function will not work. For example, if you wish to find the common log of a number you must press the $\boxed{\text{LOG}}$ key, followed by the number. Typing out the word "LOG" followed by the number will not work. Finally, commands typed into the TI calculators may be *concatenated* or printed on the same line by separating the commands with a colon.

LABORATORY 2: SOLUTION OF LINEAR EQUATIONS

Problem 1: Evaluate the expression $x^2 + 2x - 10$ for $x = -1, 0,$ and 1.

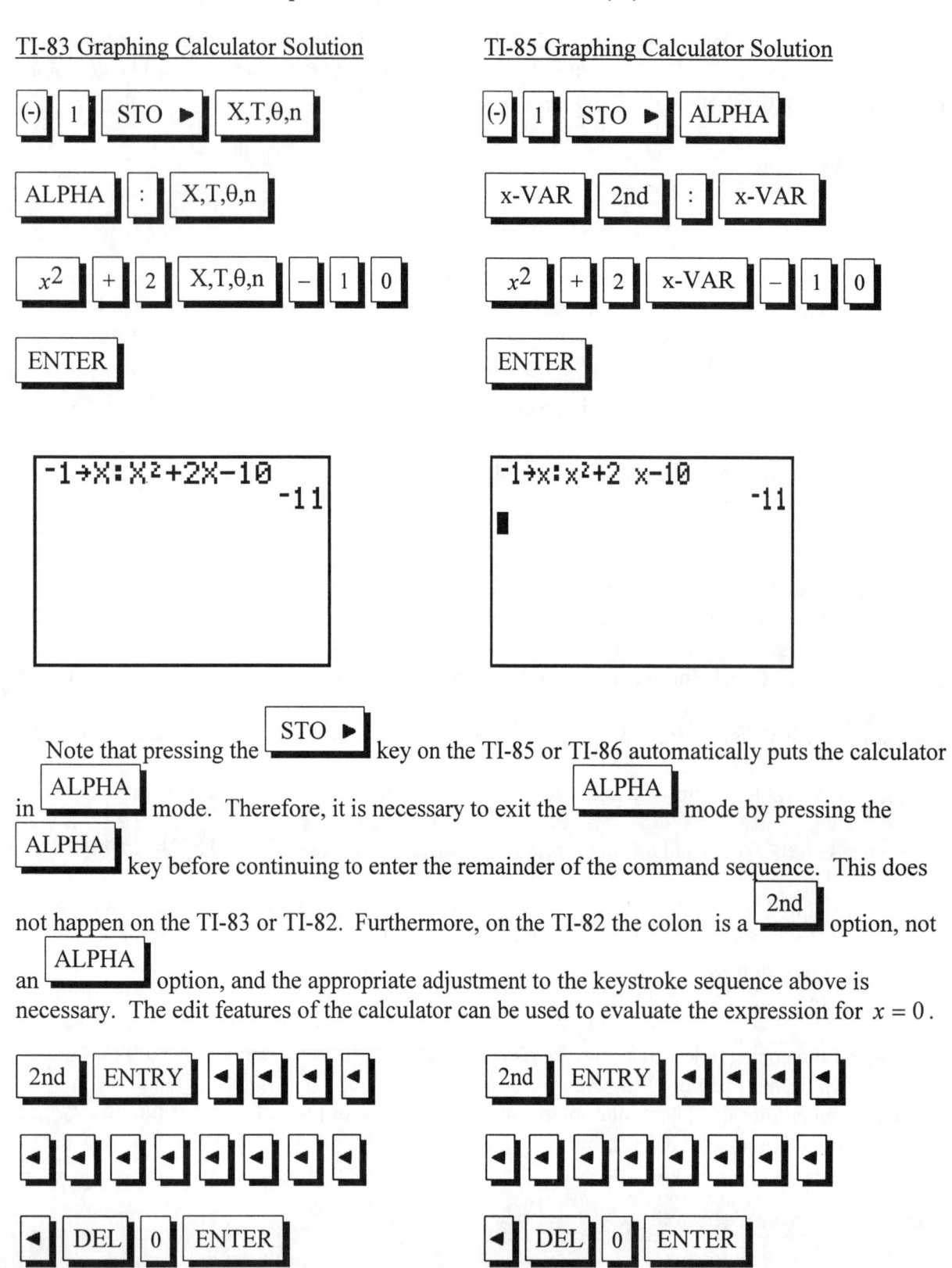

TI-83 Graphing Calculator Solution

[(-)] [1] [STO ▶] [X,T,θ,n]

[ALPHA] [:] [X,T,θ,n]

[x^2] [+] [2] [X,T,θ,n] [−] [1] [0]

[ENTER]

```
-1→X:X²+2X-10
              -11
```

TI-85 Graphing Calculator Solution

[(-)] [1] [STO ▶] [ALPHA]

[x-VAR] [2nd] [:] [x-VAR]

[x^2] [+] [2] [x-VAR] [−] [1] [0]

[ENTER]

```
-1→x:x²+2 x-10
              -11
```

Note that pressing the [STO ▶] key on the TI-85 or TI-86 automatically puts the calculator in [ALPHA] mode. Therefore, it is necessary to exit the [ALPHA] mode by pressing the [ALPHA] key before continuing to enter the remainder of the command sequence. This does not happen on the TI-83 or TI-82. Furthermore, on the TI-82 the colon is a [2nd] option, not an [ALPHA] option, and the appropriate adjustment to the keystroke sequence above is necessary. The edit features of the calculator can be used to evaluate the expression for $x = 0$.

[2nd] [ENTRY] [◄] [◄] [◄] [◄]

[◄] [◄] [◄] [◄] [◄] [◄] [◄]

[◄] [DEL] [0] [ENTER]

[2nd] [ENTRY] [◄] [◄] [◄] [◄]

[◄] [◄] [◄] [◄] [◄] [◄] [◄]

[◄] [DEL] [0] [ENTER]

LABORATORY 2: SOLUTION OF LINEAR EQUATIONS

TI-83 Graphing Calculator Solution

```
-1→X:X²+2X-10
              -11
0→X:X²+2X-10
              -10
```

TI-85 Graphing Calculator Solution

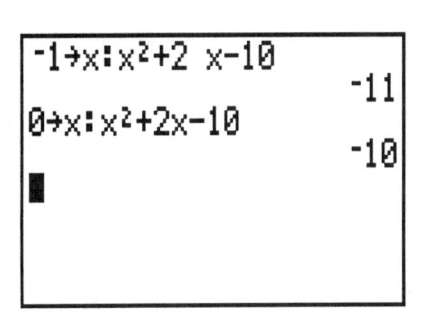

```
-1→x:x²+2 x-10
              -11
0→x:x²+2x-10
              -10
```

Finally, use the edit features again to evaluate the expression at $x = 1$.

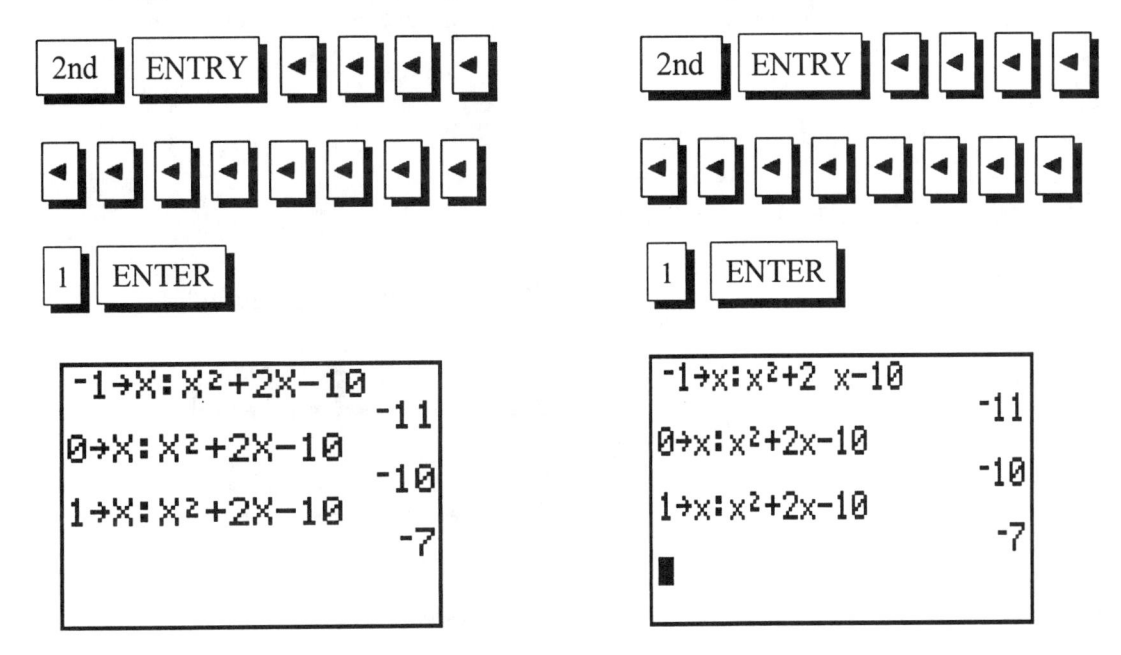

[2nd] [ENTRY] [◄] [◄] [◄] [◄]

[◄] [◄] [◄] [◄] [◄] [◄] [◄]

[1] [ENTER]

```
-1→X:X²+2X-10
              -11
0→X:X²+2X-10
              -10
1→X:X²+2X-10
              -7
```

[2nd] [ENTRY] [◄] [◄] [◄] [◄]

[◄] [◄] [◄] [◄] [◄] [◄] [◄]

[1] [ENTER]

```
-1→x:x²+2 x-10
              -11
0→x:x²+2x-10
              -10
1→x:x²+2x-10
              -7
```

When you have completed this problem, try pressing [2nd] [ENTRY] [2nd] [◄].
Where is the cursor now? How can this be used to your advantage to solve this problem?

Alternate Solution

It is not necessary to use the [STO ►] feature on the calculator to solve problems involving algebraic expressions. Another approach which can be taken is to replace all the variables in the expression by a pair of parentheses containing the value of the given variable. For example, to evaluate the algebraic expression $x^2 + 2x - 10$ at $x = -1$ type:

[(] [(-)] [1] [)] [x^2] [+] [2] [(] [(-)] [1] [)] [-] [1] [0] [ENTER]

LABORATORY 2: SOLUTION OF LINEAR EQUATIONS

Alternate Solution

```
0→x:x²+2x-10            -11
                        -10
1→x:x²+2x-10
                         -7
(-1)²+2(-1)-10
                        -11
```

The edit feature on the calculator may be used to change the number within the parentheses to evaluate the expression at $x = 0$ and $x = 1$. Try this and compare your results.

Problem 2: Evaluate the algebraic expression $Ax^2 + Bx + C$ for:
    a. $A = 10$, $B = 2$, $C = 5$, and $x = 1$.
    b. $A = -5$, $B = 1$, $C = 22$, and $x = 1$.

TI-83 Graphing Calculator Solution   TI-85 Graphing Calculator Solution

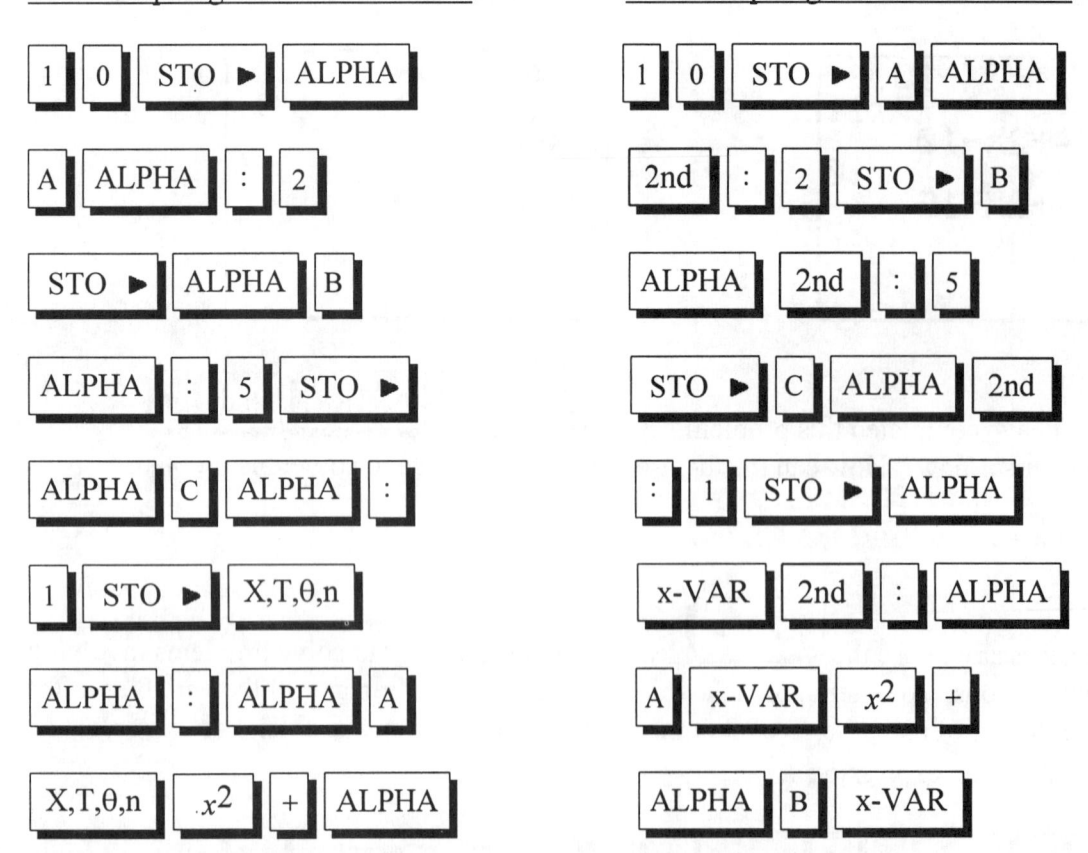

TI-83	TI-85
1 0 STO ▶ ALPHA	1 0 STO ▶ A ALPHA
A ALPHA : 2	2nd : 2 STO ▶ B
STO ▶ ALPHA B	ALPHA 2nd : 5
ALPHA : 5 STO ▶	STO ▶ C ALPHA 2nd
ALPHA C ALPHA :	: 1 STO ▶ ALPHA
1 STO ▶ X,T,θ,n	x-VAR 2nd : ALPHA
ALPHA : ALPHA A	A x-VAR x^2 +
X,T,θ,n x^2 + ALPHA	ALPHA B x-VAR

LABORATORY 2: SOLUTION OF LINEAR EQUATIONS

TI-83 Graphing Calculator Solution

[B] [X,T,θ,n] [+] [ALPHA]

[C] [ENTER]

```
10→A:2→B:5→C:1→X
:AX²+BX+C
                17
■
```

TI-85 Graphing Calculator Solution

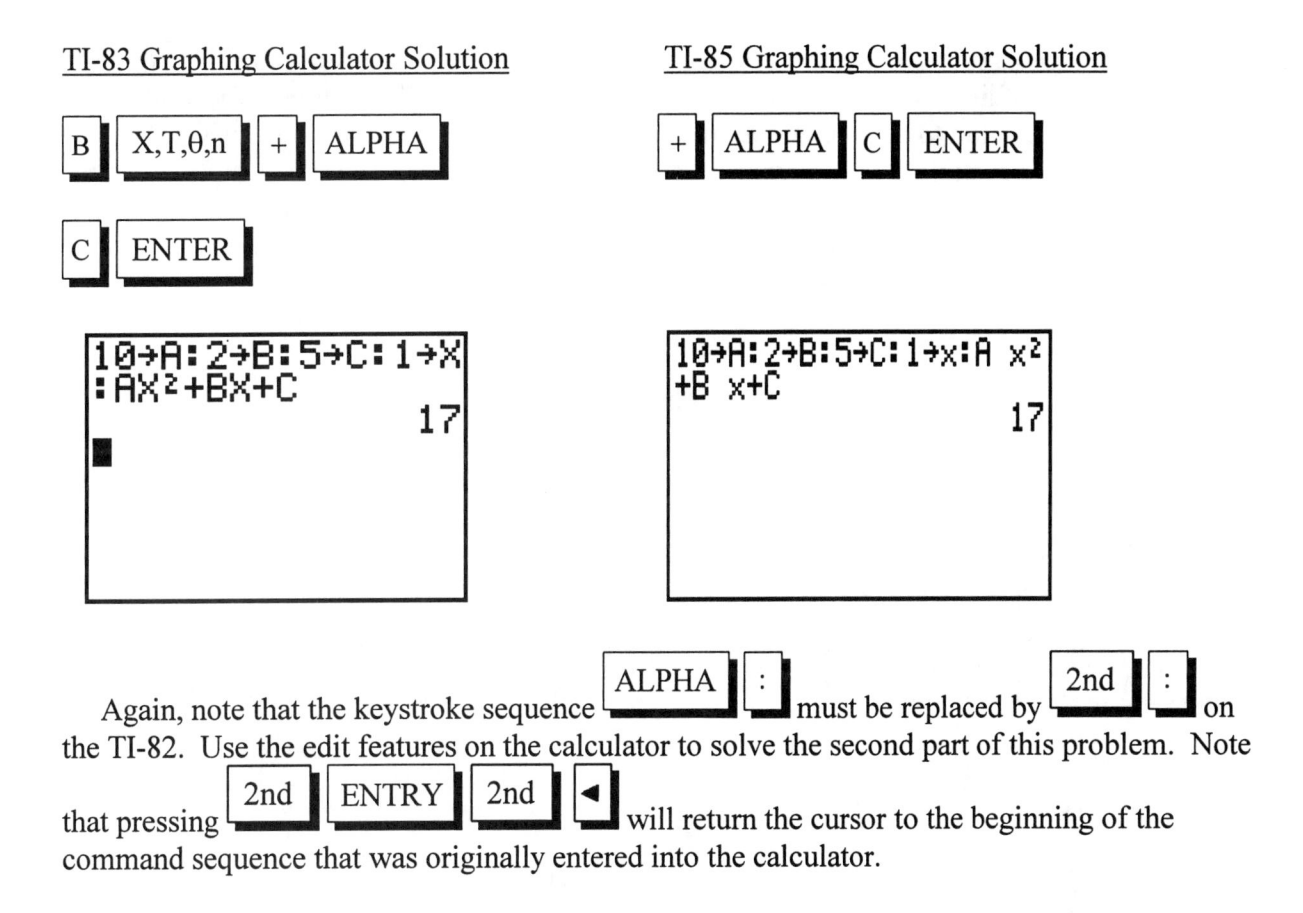

[+] [ALPHA] [C] [ENTER]

```
10→A:2→B:5→C:1→x:A x²
+B x+C
                17
```

Again, note that the keystroke sequence [ALPHA] [:] must be replaced by [2nd] [:] on the TI-82. Use the edit features on the calculator to solve the second part of this problem. Note that pressing [2nd] [ENTRY] [2nd] [◄] will return the cursor to the beginning of the command sequence that was originally entered into the calculator.

B. Solution of Linear Equations in One Unknown

An *equation* is a mathematical statement of the equality of two algebraic expressions. In particular a *linear equation* in one unknown is an equation which contains only one variable and that variable has exponent equal to one. For example, $2x + 5 = 11$ is an example of a linear equation in one unknown. The variable is x and the exponent of the variable is understood to be one. It is not necessary that the variable *appear* only once in a linear equation in one unknown, it is only necessary that there *be* only one variable. For example, $2x + 5 = 4x + 6$ is a linear equation in one unknown. Even though the variable x appears more than once, it is the *only* variable in the equation.

The solution of linear equations in one unknown involves the determination of the value of the variable that will *balance* both sides of the equation. The value of the variable that makes both sides of the equation equal is called the *solution* of the equation. A common strategy in the solution of linear equations in one unknown is to isolate the variable (get the variable "alone") on one side of the equation. The arithmetic operations of addition, subtraction, multiplication, and division may be used to isolate the variable in a linear equation in one unknown. As long as the balance of the equation is maintained by always doing the same thing to both sides of the equation (being sure never to divide by zero) the equality of the equation will be maintained and a solution will be derived.

LABORATORY 2: SOLUTION OF LINEAR EQUATIONS

The TI calculators come equipped with a **SOLVER** or **SOLVE** feature which is useful in the algebraic solution of linear equations in one unknown. The **SOLVER** feature on the TI-85 and TI-86 allows an equation to be entered into the calculator in its original form. However, the TI-83 requires the equation to be set equal to zero in order to solve it. Therefore, all terms of the equation must be brought to one side of the equation. It is also necessary to clear the last equation solved in the **SOLVER** mode before solving a new equation. The examples that follow illustrate the proper use of this feature.

Problem 3: Solve the following linear equation for x: $10x - 2 + 2x = 4x - 12$.

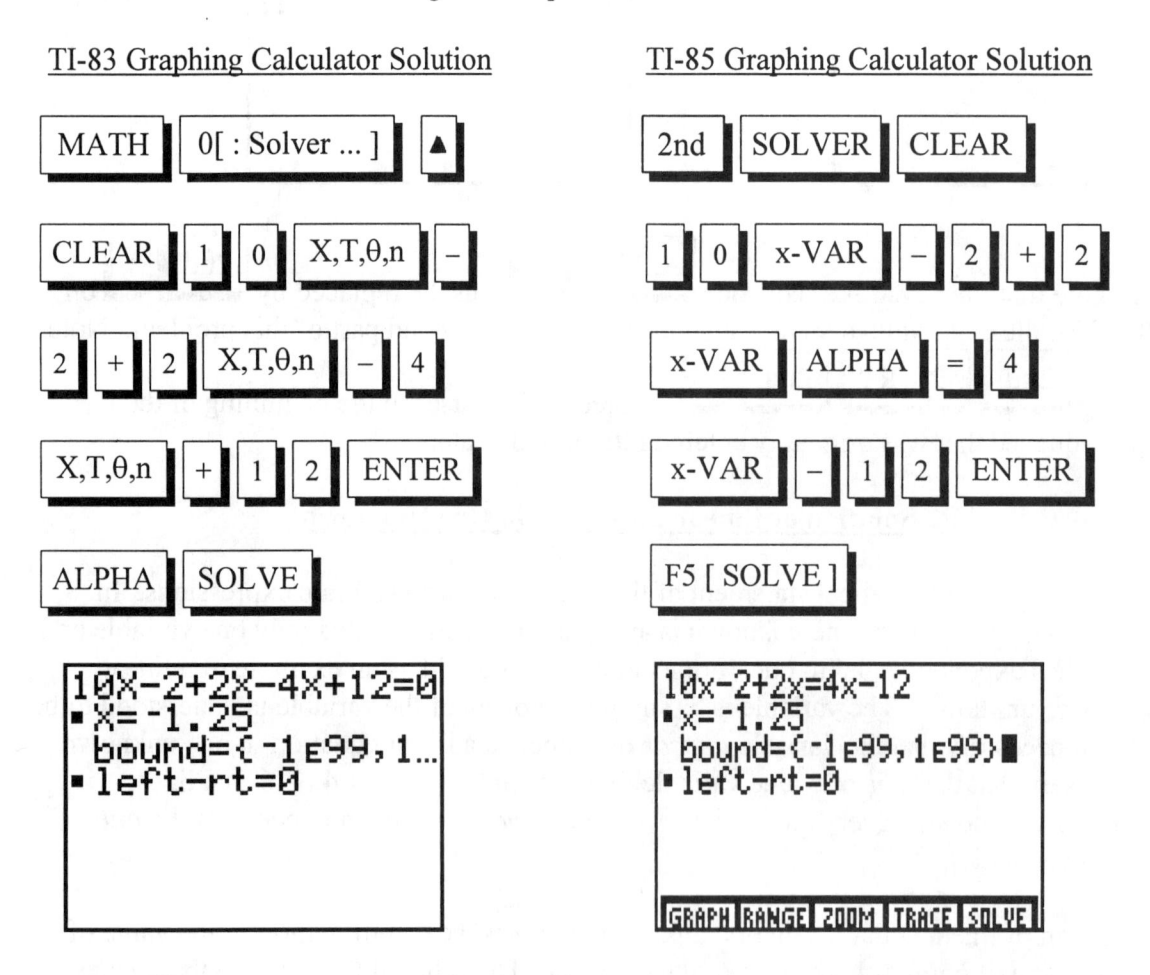

TI-83 Graphing Calculator Solution

| MATH | 0[: Solver ...] | ▲ |

| CLEAR | 1 | 0 | X,T,θ,n | − |

| 2 | + | 2 | X,T,θ,n | − | 4 |

| X,T,θ,n | + | 1 | 2 | ENTER |

| ALPHA | SOLVE |

```
10X-2+2X-4X+12=0
•X=-1.25
 bound={-1E99,1...
•left-rt=0
```

TI-85 Graphing Calculator Solution

| 2nd | SOLVER | CLEAR |

| 1 | 0 | x-VAR | − | 2 | + | 2 |

| x-VAR | ALPHA | = | 4 |

| x-VAR | − | 1 | 2 | ENTER |

| F5 [SOLVE] |

```
10x-2+2x=4x-12
•x=-1.25
 bound={-1E99,1E99}■
•left-rt=0
```

| GRAPH | RANGE | ZOOM | TRACE | SOLVE |

It is important to note that the **SOLVER** feature uses the current value of the variable x as an initial guess for the solution of the equation to be solved. An iterative procedure for the numerical approximation to the solution of the equation which is *closest to the initial guess* is executed in order to find a solution within a prescribed tolerance. Iterative procedures are often dependent on the initial guess and are subject to round-off error. This becomes even more apparent in the solution of higher order equations. It is also possible that a particularly bad guess will prevent the algorithm from converging within the allowable number of iterations. If this should happen an error message will be printed.

LABORATORY 2: SOLUTION OF LINEAR EQUATIONS

To exit the **SOLVER** mode press [2nd] [QUIT] on the TI-83, TI-85, or TI-86. Note that [EXIT] may also be used on the TI-85 or TI-86 to exit the **SOLVER** mode. The TI-82 does not have [0[: Solver ...]] on the [MATH] menu. However, a **SOLVE** feature appears on the [MATH] menu. The syntax of this feature that corresponds to the TI-83 solution outlined above is:

solve(algebraic expression after moving all terms to one side of the equation, variable, 0).

Hence the keystroke sequence on the TI-82 would be: [MATH] [0[: solve( )]] [1] [0] [X,T,θ] [−] [2] [+] [2] [X,T,θ] [−] [4] [X,T,θ] [+] [1] [2] [,] [X,T,θ] [,] [0] [)] [ENTER]. This feature may be invoked on the TI-83 by locating it in the catalog. Press [2nd] [CATALOG] and use the arrow keys to locate the feature and then press [ENTER].

Problem 4: Solve the following linear equation for x: $\dfrac{x}{2} - \dfrac{3x+1}{3} = 2(x-2)$ and convert to a fraction in lowest terms.

TI-83 Graphing Calculator Solution

[MATH] [0[: Solver ...]] [▲]

[CLEAR] [(] [X,T,θ,n] [÷]

[2] [)] [−] [(] [3] [X,T,θ,n]

[+] [1] [)] [÷] [3] [−] [2]

[(] [X,T,θ,n] [−] [2] [)]

[ENTER] [ALPHA] [SOLVE]

TI-85 Graphing Calculator Solution

[2nd] [SOLVER] [CLEAR]

[(] [x-VAR] [÷] [2] [)]

[−] [(] [3] [x-VAR]

[+] [1] [)] [÷] [3] [ALPHA]

[=] [2] [(] [x-VAR] [−] [2]

[)] [ENTER] [F5 [SOLVE]]

19

LABORATORY 2: SOLUTION OF LINEAR EQUATIONS

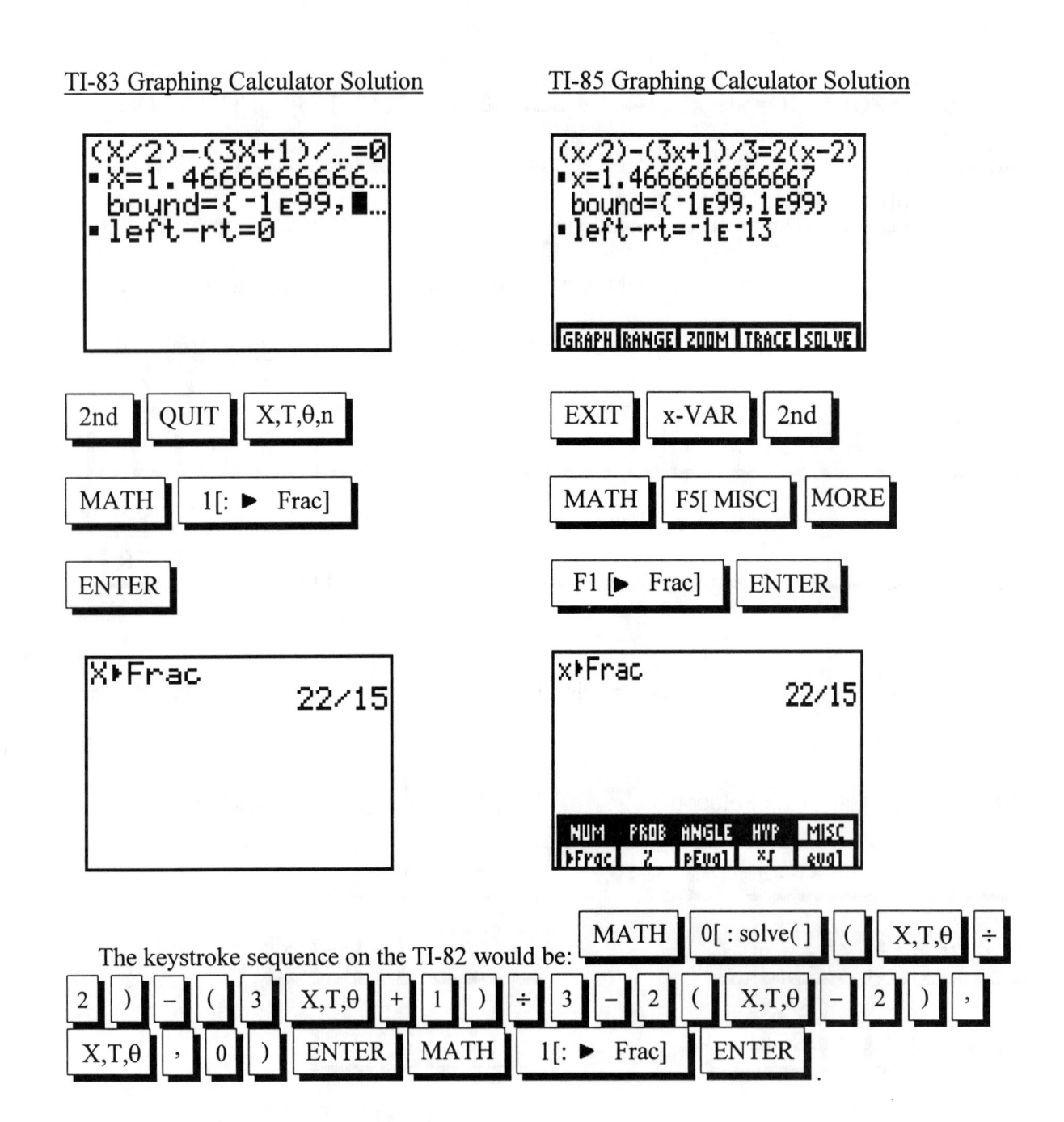

TI-83 Graphing Calculator Solution

```
(X/2)-(3X+1)/…=0
■X=1.4666666666…
 bound={-1E99,■…
■left-rt=0
```

| 2nd | QUIT | X,T,θ,n |

| MATH | 1[: ▶ Frac] |

| ENTER |

```
X▶Frac
            22/15
```

TI-85 Graphing Calculator Solution

```
(x/2)-(3x+1)/3=2(x-2)
■x=1.4666666666667
 bound={-1E99,1E99}
■left-rt=-1E-13
```
GRAPH | RANGE | ZOOM | TRACE | SOLVE

| EXIT | x-VAR | 2nd |

| MATH | F5[MISC] | MORE |

| F1 [▶ Frac] | ENTER |

```
x▶Frac
            22/15
```
NUM | PROB | ANGLE | HYP | MISC
▶Frac | % | pEval | ˣ√ | eval

The keystroke sequence on the TI-82 would be: | MATH | 0[: solve( ] | (| X,T,θ | ÷ |
| 2 |) | − | (| 3 | X,T,θ | + | 1 |) | ÷ | 3 | − | 2 | (| X,T,θ | − | 2 |) | , |
| X,T,θ | , | 0 |) | ENTER | MATH | 1[: ▶ Frac] | ENTER | .

20

LABORATORY 2: SOLUTION OF LINEAR EQUATIONS

EXERCISES

1. Evaluate the algebraic expression $10x^2 - 5x + 17$ for $x = -10$, $x = 0$, and $x = 200$.

2. Evaluate the algebraic expression $-5A^2 + 3B - \dfrac{C}{2}$ for $A = \dfrac{2}{3}$, $B = -5$, and $C = 7$. Express your answer as a fraction in lowest terms.

3. Solve the following linear equation for x: $15x - 100 + 2x = 4x + 7$. Express your answer as a fraction in lowest terms.

4. Solve the following linear equation for x: $\dfrac{3}{4}x - \dfrac{2}{3} = 2x - 1$. Express your answer as a fraction in lowest terms.

5. Solve the following linear equation for x: $3(2x - 7) + x = -5(x + 1)$. Express your answer as a fraction in lowest terms.

LABORATORY 3: SOLUTION OF LINEAR INEQUALITIES

Purpose:

The purpose of this laboratory is to study the solution of linear inequalities in one variable. Calculator features will be used to compare numbers as well as to solve and graph linear inequalities in one unknown. A preview of some of the graphing features on the TI calculators will be given in this laboratory as they relate to the graphical solution of linear inequalities in one variable. A more detailed overview of the graphing features of the TI calculators will be presented beginning in Laboratory 4 and throughout the remainder of this text.

Analytic Approach:

A. Testing the Relationship Between Two Numbers

A *number line* is a useful tool in describing the relationship between two numbers. As we move to the left on a number line, numbers get smaller in value, while as we move to the right on a number line, numbers get larger in value. The value of a number on the number line includes *magnitude* as well as *direction*. Negative numbers are to the left of zero, while positive numbers are to the right of zero. The absolute value of a number tells us its magnitude or distance from zero, while the algebraic sign of the number tells us the direction of the displacement. For example, -5 is five units to the left of zero while 5 is five units to the right of zero. Note that $|-5| = |5| = 5$, and therefore both numbers are five units from zero. However, they are five units from zero in opposite directions. Therefore, the distance between -5 and 5 is ten units. Note that in general the distance between two numbers, a and b, may be described as the absolute value of their difference, which is denoted by $|a - b|$ or equivalently by $|b - a|$. A number a is said to be *less than* a number b if it is to the left of b on the number line. The symbolic description for *a is less than b* is $a < b$. Similarly, a number a is said to be *greater than* a number b if it is to the right of b on the number line. The symbolic description for *a is greater than b* is $a > b$. Three other symbols are used to describe the relationship between two numbers or algebraic expressions. These symbols are $=$, $\leq$, and $\geq$ and refer to the relationships *equal to*, *less than or equal to*, and *greater than or equal to*, respectively. Notice that the *point of the arrow* in each of the inequality symbols always points to the smaller number. This is often helpful in recalling the verbal definition of the inequality symbols.

The TI calculators are equipped with a **TEST** feature to test the relationship between two numbers or algebraic expressions. The *relational operators* described above may be used to test the *truth value* of a statement. The **TEST** feature will return a value of 0 if the statement is false and a value of 1 if the statement is true.

LABORATORY 3: SOLUTION OF LINEAR INEQUALITIES

Problem 1: Is $\frac{1}{7} < \frac{1}{9}$?

TI-83 Graphing Calculator Solution

[( ] [1] [÷] [7] [ )] [2nd] [TEST]

[5[: <]] [( ] [1] [÷] [9] [ )]

[ENTER]

```
(1/7)<(1/9)
                    0
■
```

TI-85 Graphing Calculator Solution

[( ] [1] [÷] [7] [ )] [2nd] [TEST]

[F2 [<]] [( ] [1] [÷] [9] [ )]

[ENTER]

```
(1/7)<(1/9)
                    0
■
 ══  <  >  ≤  ≥
```

The calculator returns a zero, and we see that $\frac{1}{7}$ is not less than $\frac{1}{9}$. In fact $\frac{1}{7}$ is greater than $\frac{1}{9}$. Use the edit features on the calculator to verify this fact. Remember to clear the screen when you are done. Note that on the TI-85 and TI-86 it will be necessary to clear the [TEST] menu from the screen by pressing [EXIT].

Problem 2: Is the distance between -2 and 7 greater than the distance between 1 and 8?

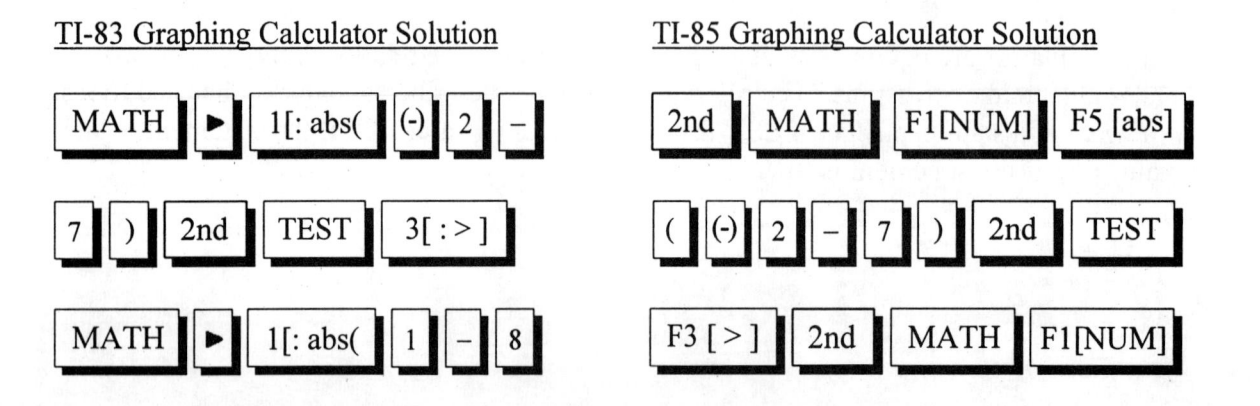

TI-83 Graphing Calculator Solution

[MATH] [▶] [1[: abs( ] [(-)] [2] [−]

[7] [ )] [2nd] [TEST] [3[: >]]

[MATH] [▶] [1[: abs( ] [1] [−] [8]

TI-85 Graphing Calculator Solution

[2nd] [MATH] [F1[NUM]] [F5 [abs]]

[( ] [(-)] [2] [−] [7] [ )] [2nd] [TEST]

[F3 [>]] [2nd] [MATH] [F1[NUM]]

LABORATORY 3: SOLUTION OF LINEAR INEQUALITIES

TI-83 Graphing Calculator Solution   TI-85 Graphing Calculator Solution

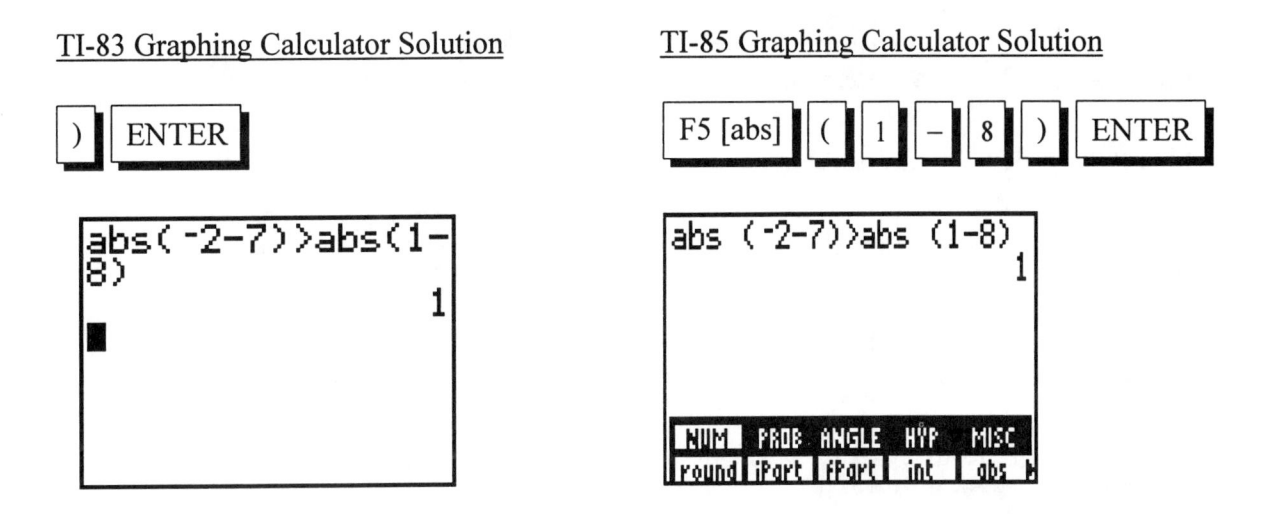

|) | ENTER | | F5 [abs] | (| 1 | − | 8 |) | ENTER |

```
abs(-2-7)>abs(1-
8)
                 1
■
```

```
abs (-2-7)>abs (1-8)
                     1



NUM   PROB  ANGLE  HYP   MISC
round iPart fPart  int   abs
```

The calculator returns a one. Therefore, we have verified that the distance between -2 and 7 is greater than the distance between 1 and 8. What are the individual distances?

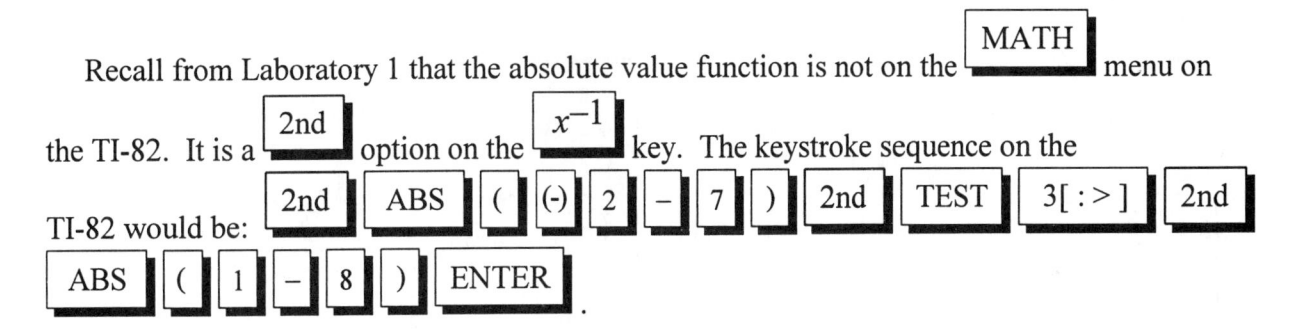

Recall from Laboratory 1 that the absolute value function is not on the [MATH] menu on the TI-82. It is a [2nd] option on the x^{-1} key. The keystroke sequence on the TI-82 would be: [2nd] [ABS] [(] [(-)] [2] [−] [7] [)] [2nd] [TEST] [3[: >]] [2nd] [ABS] [(] [1] [−] [8] [)] [ENTER].

B. Solution of Linear Inequalities in One Unknown

The solution of linear *inequalities* in one unknown may be approached in much the same way that the solution of linear *equalities* in one unknown is approached. Whatever is done to one side of the inequality must also be done to the other side. However, when the operation being performed on both sides of the inequality is multiplication or division by a negative number, the direction of the inequality must be reversed. For example, if we divide both sides of the inequality $-2x \leq 4$ by -2, the result is $x \geq -2$. To verify the necessity of reversing the inequality in this problem, rework the problem by first adding $2x$ to both sides of the original inequality. This results in the equivalent inequality $0 \leq 2x + 4$. Then subtract 4 from both sides so that $-4 \leq 2x$. Finally, divide both sides by 2 so that $-2 \leq x$. The two inequalities, $x \geq -2$ and $-2 \leq x$, are two different ways of saying exactly the same thing.

The solution of linear inequalities in one unknown can be performed on the TI calculators by using the graphing feature in addition to the **SOLVER** or **SOLVE** feature. The graphing feature can be used to provide a graph of the solution set, while the **SOLVER** or **SOLVE** feature can be used to find the endpoints of the solution intervals.

LABORATORY 3: SOLUTION OF LINEAR INEQUALITIES

Problem 3: Solve the linear inequality $2x - 10 \leq 4x + 2$. Consider first the solution to the equality $2x - 10 = 4x + 2$ and then discuss the solution to the given inequality.

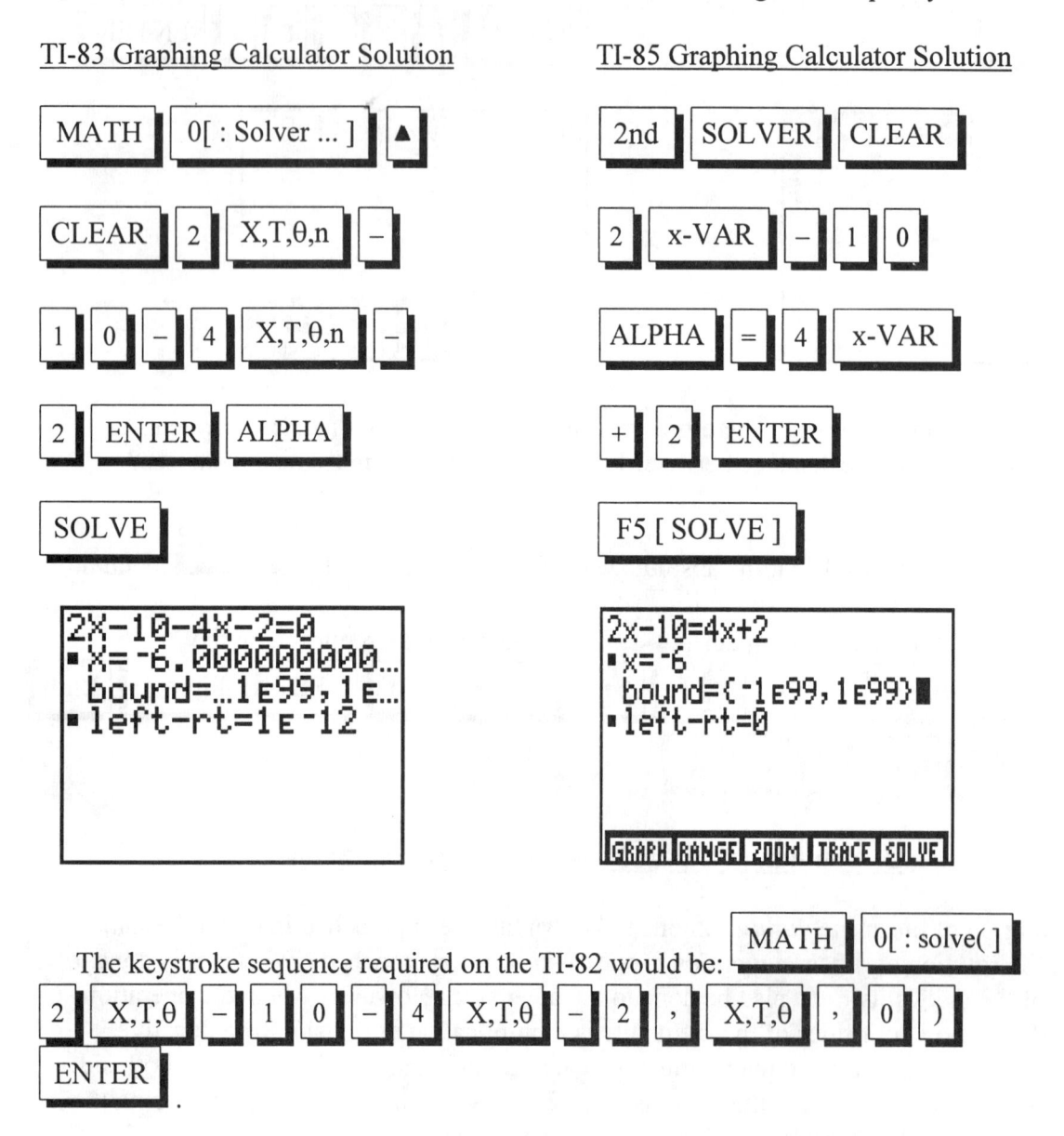

TI-83 Graphing Calculator Solution   TI-85 Graphing Calculator Solution

| MATH | 0[: Solver ...] | ▲ |

| 2nd | SOLVER | CLEAR |

| CLEAR | 2 | X,T,θ,n | − |

| 2 | x-VAR | − | 1 | 0 |

| 1 | 0 | − | 4 | X,T,θ,n | − |

| ALPHA | = | 4 | x-VAR |

| 2 | ENTER | ALPHA |

| + | 2 | ENTER |

| SOLVE |

| F5 [SOLVE] |

```
2X-10-4X-2=0
·X=-6.000000000...
 bound=...1E99,1E...
·left-rt=1E-12
```

```
2x-10=4x+2
·x=-6
 bound={-1E99,1E99}
·left-rt=0
```

GRAPH RANGE ZOOM TRACE SOLVE

The keystroke sequence required on the TI-82 would be: | MATH | 0[: solve( ] |
| 2 | X,T,θ | − | 1 | 0 | − | 4 | X,T,θ | − | 2 | , | X,T,θ | , | 0 |) |
| ENTER |.

We see that the solution to the equality is $x = -6$. We will now turn to the graphing feature on the calculator to determine the solution interval for the original inequality. In order to use the graphing feature of the calculator to complete the solution of this problem it will be necessary to provide a brief summary of the graphing menu. Notice that the graphing menu will be displayed on the TI-85 and TI-86 by pressing the | GRAPH | key. The graphing menu keys on the TI-83 and TI-82 are located directly below the display screen. The basic menu consists of five keys, a key to enter a function in the form *y* as an expression represented in terms of the variable *x*, a

26

LABORATORY 3: SOLUTION OF LINEAR INEQUALITIES

key to adjust the graphing window or range on the x and y axes, a key to zoom in or out on the graph, a key to trace along the graph of the curve being analyzed, and a key to simply graph the given expression on the current graphing window or range. The standard graphing range on the TI calculators is $-10 \le x \le 10$, $-10 \le y \le 10$, with a step size of one unit between the tic marks on the x and y axes. The graphing menu includes other options which will be discussed in the laboratories that follow. It will be sufficient to graph this problem on the standard axes since the solution of the equality lies within the standard viewing window.

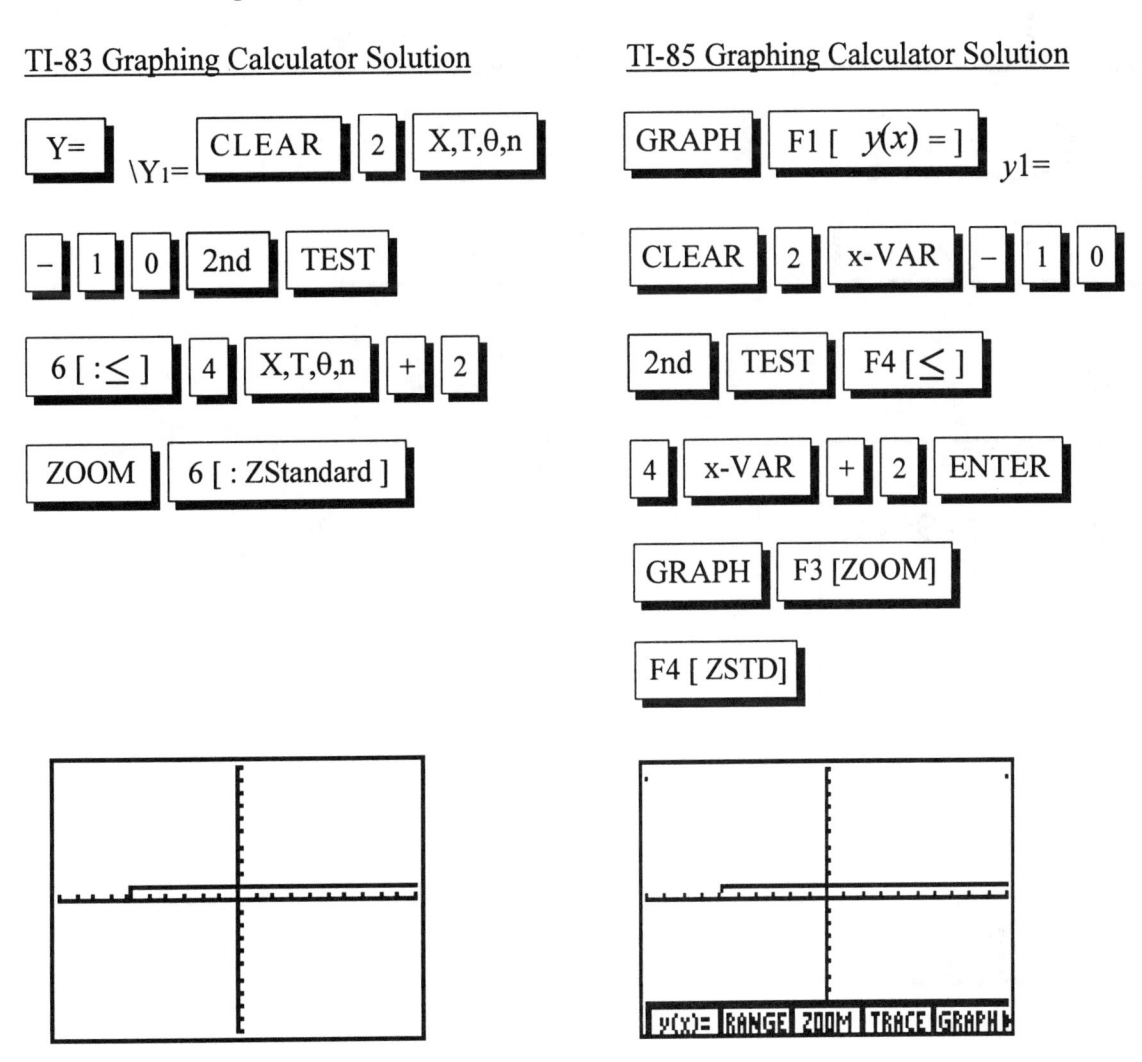

TI-83 Graphing Calculator Solution

[Y=] \Y₁= [CLEAR] [2] [X,T,θ,n]

[−] [1] [0] [2nd] [TEST]

[6 [:≤]] [4] [X,T,θ,n] [+] [2]

[ZOOM] [6 [: ZStandard]]

TI-85 Graphing Calculator Solution

[GRAPH] [F1 [y(x) =]] y1=

[CLEAR] [2] [x-VAR] [−] [1] [0]

[2nd] [TEST] [F4 [≤]]

[4] [x-VAR] [+] [2] [ENTER]

[GRAPH] [F3 [ZOOM]]

[F4 [ZSTD]]

We see that the solution to this problem is $x \ge -6$. Remember to clear the screen when you are done. This may be accomplished on the TI-83 and TI-82 by pressing [CLEAR] and on the TI-85 and TI-86 by pressing [EXIT].

27

LABORATORY 3: SOLUTION OF LINEAR INEQUALITIES

Problem 4: Solve the linear inequality $5x + 10 > x - 50$ by first solving the equality $5x + 10 = x - 50$ and then graphing the given inequality.

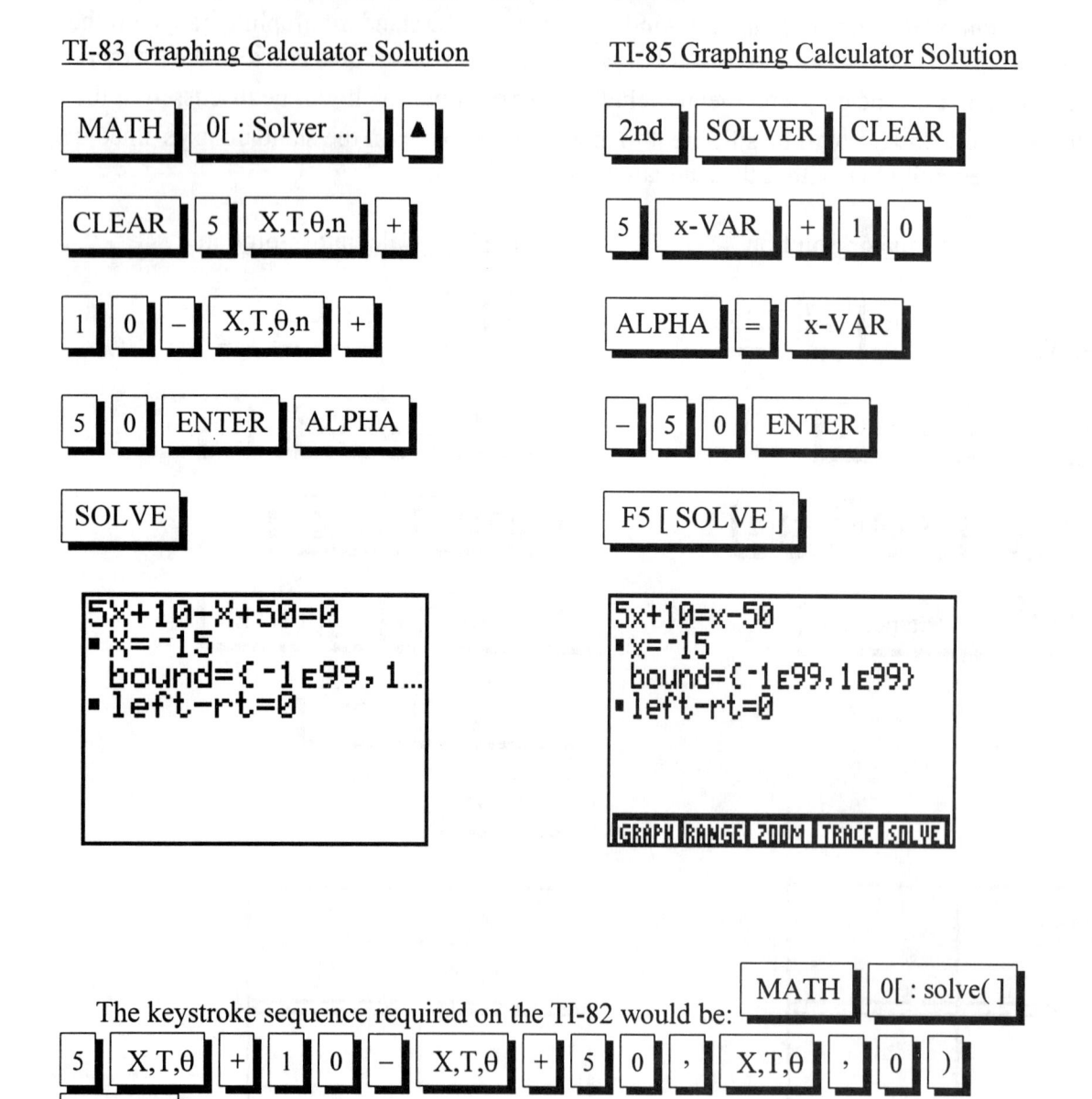

TI-83 Graphing Calculator Solution   TI-85 Graphing Calculator Solution

| MATH | 0[: Solver ...] | ▲ | | 2nd | SOLVER | CLEAR |

| CLEAR | 5 | X,T,θ,n | + | | 5 | x-VAR | + | 1 | 0 |

| 1 | 0 | – | X,T,θ,n | + | | ALPHA | = | x-VAR |

| 5 | 0 | ENTER | ALPHA | | – | 5 | 0 | ENTER |

| SOLVE | | F5 [SOLVE] |

```
5X+10-X+50=0              5x+10=x-50
•X= -15                   •x= -15
 bound={-1E99,1...         bound={-1E99,1E99}
•left-rt=0                •left-rt=0
```
                                          GRAPH RANGE ZOOM TRACE SOLVE

The keystroke sequence required on the TI-82 would be: | MATH | 0[: solve( ] |

| 5 | X,T,θ | + | 1 | 0 | – | X,T,θ | + | 5 | 0 | , | X,T,θ | , | 0 |) |

| ENTER |

The solution to $5x + 10 = x - 50$ is $x = -15$. Since -15 is outside the standard viewing window, we must adjust the range so that the minimum value on the x-axis is less than -15. In this problem we will adjust the minimum value on the x-axis to be -20 before graphing the solution of the inequality.

LABORATORY 3: SOLUTION OF LINEAR INEQUALITIES

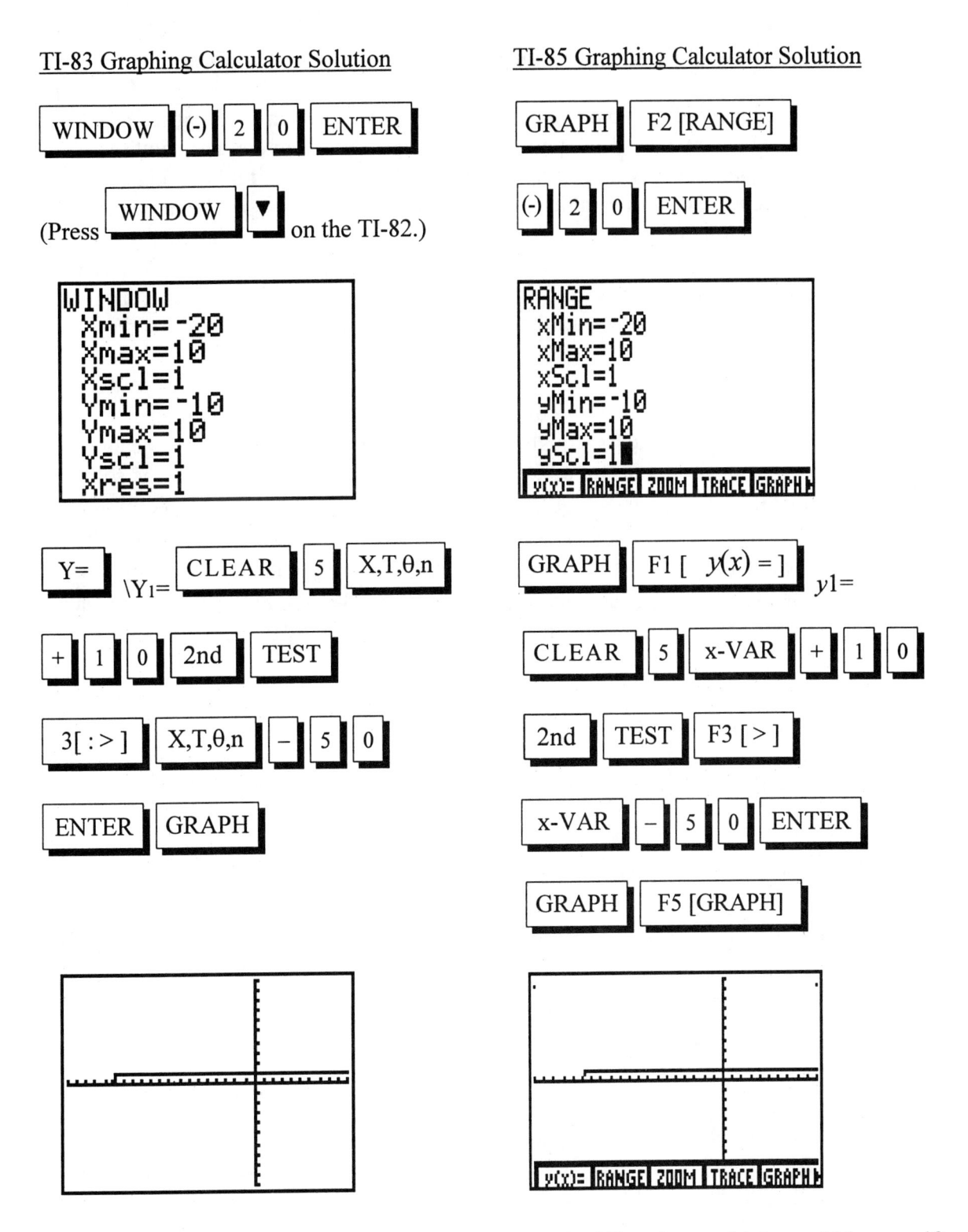

TI-83 Graphing Calculator Solution

| WINDOW | (-) | 2 | 0 | ENTER |

(Press | WINDOW | ▼ | on the TI-82.)

```
WINDOW
 Xmin=-20
 Xmax=10
 Xscl=1
 Ymin=-10
 Ymax=10
 Yscl=1
 Xres=1
```

| Y= | \Y₁= | CLEAR | 5 | X,T,θ,n |

| + | 1 | 0 | 2nd | TEST |

| 3 [:>] | X,T,θ,n | − | 5 | 0 |

| ENTER | GRAPH |

TI-85 Graphing Calculator Solution

| GRAPH | F2 [RANGE] |

| (-) | 2 | 0 | ENTER |

```
RANGE
 xMin=-20
 xMax=10
 xScl=1
 yMin=-10
 yMax=10
 yScl=1
y(x)= RANGE ZOOM TRACE GRAPH▶
```

| GRAPH | F1 [y(x) =] | y1=

| CLEAR | 5 | x-VAR | + | 1 | 0 |

| 2nd | TEST | F3 [>] |

| x-VAR | − | 5 | 0 | ENTER |

| GRAPH | F5 [GRAPH] |

We see that the solution to this problem is $x > -15$. What do you think would happen if we reversed the inequality in the original statement of the problem? Try it and see what happens to the solution. Remember to clear the screen when you are done.

LABORATORY 3: SOLUTION OF LINEAR INEQUALITIES

C. Solution of Compound Linear Inequalities in One Unknown

A *compound linear inequality* in one unknown is an inequality which is composed of two inequalities connected by the logical *and* operator or the logical *or* operator. The solution of a compound linear inequality involving the logical *and* operator consists of all numbers which satisfy both parts of the compound inequality. Thus, the solution to this type of compound inequality is the *intersection* of the solution sets of the individual parts of the inequality. The solution of a compound linear inequality involving the logical *or* operator consists of all numbers that solve at least one of the parts of the compound inequality. Hence, the solution to this type of compound inequality is the *union* of the solution sets of the individual parts of the inequality.

The TI calculators may be used to solve compound inequalities involving the logical *or* operator by solving and graphing each inequality separately on the same axes. This means that we must enter two separate algebraic expressions, $y1$ and $y2$, into the calculator. The solution of compound inequalities involving the logical *and* statement requires that each expression be solved individually in order to determine the endpoints. However, the expressions should be entered into the calculator for graphing as a single expression, $y1$, by enclosing the individual parts of the inequality in parentheses and placing them side by side. For example, the inequality $-5 < 3x - 2 < 2$ would be entered through the graphing menu as $y1 = (3x - 2 > -5)(3x - 2 < 2)$.

Problem 5: Graph the region $x < -5$ or $x > 4$.

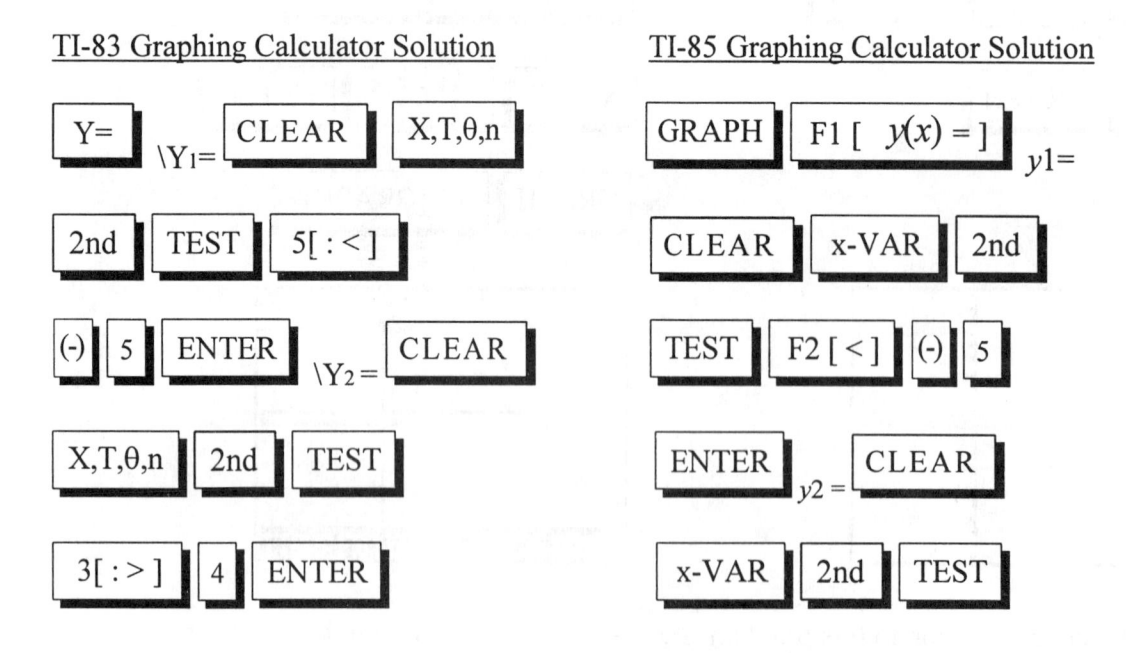

TI-83 Graphing Calculator Solution

Y=	\Y₁= CLEAR	X,T,θ,n
2nd	TEST	5[: <]
(-)	5	ENTER	\Y₂= CLEAR
X,T,θ,n	2nd	TEST
3[: >]	4	ENTER

TI-85 Graphing Calculator Solution

GRAPH	F1 [y(x) =]	y1=
CLEAR	x-VAR	2nd
TEST	F2 [<]	(-)	5
ENTER	y2= CLEAR
x-VAR	2nd	TEST

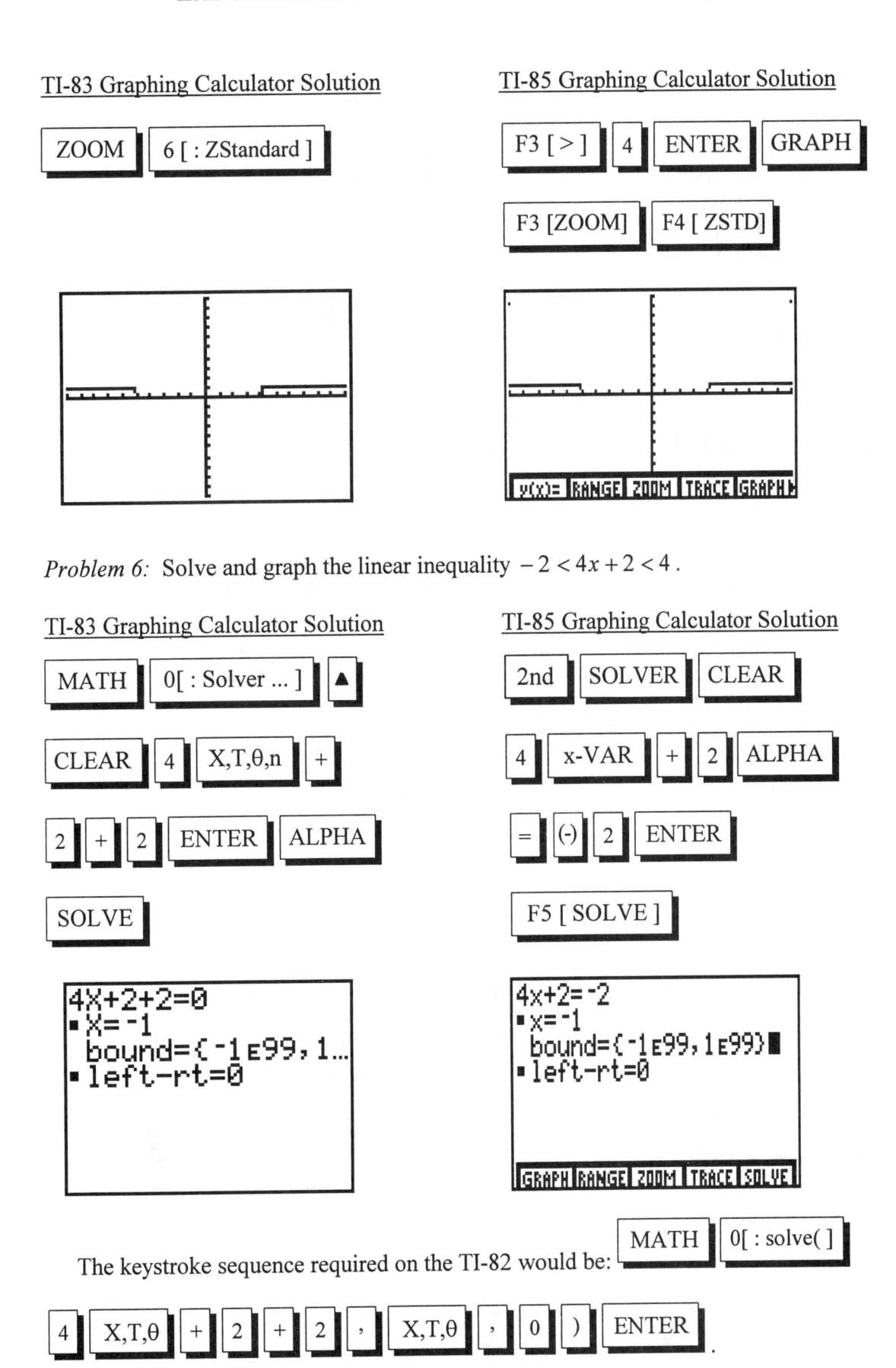

LABORATORY 3: SOLUTION OF LINEAR INEQUALITIES

TI-83 Graphing Calculator Solution

| ZOOM | 6 [: ZStandard] |

TI-85 Graphing Calculator Solution

| F3 [>] | 4 | ENTER | GRAPH |

| F3 [ZOOM] | F4 [ZSTD] |

Problem 6: Solve and graph the linear inequality $-2 < 4x + 2 < 4$.

TI-83 Graphing Calculator Solution

| MATH | 0[: Solver ...] | ▲ |

| CLEAR | 4 | X,T,θ,n | + |

| 2 | + | 2 | ENTER | ALPHA |

| SOLVE |

```
4X+2+2=0
•X=-1
 bound={-1E99,1...
•left-rt=0
```

TI-85 Graphing Calculator Solution

| 2nd | SOLVER | CLEAR |

| 4 | x-VAR | + | 2 | ALPHA |

| = | (-) | 2 | ENTER |

| F5 [SOLVE] |

```
4x+2=-2
•x=-1
 bound={-1E99,1E99}
•left-rt=0
```

The keystroke sequence required on the TI-82 would be: | MATH | 0[: solve( ] |

| 4 | X,T,θ | + | 2 | + | 2 | , | X,T,θ | , | 0 |) | ENTER |.

LABORATORY 3: SOLUTION OF LINEAR INEQUALITIES

TI-83 Graphing Calculator Solution

[MATH] [0[: Solver ...]] [▲]

[▶] [▶] [▶] [▶] [–] [4] [ENTER]

[ALPHA] [SOLVE]

```
4X+2-4=0
■X=.5
 bound={-1E99,1…
■left-rt=0
```

TI-85 Graphing Calculator Solution

[2nd] [SOLVER] [▶] [▶] [▶] [▶]

[▶] [4] [DEL] [ENTER]

[F5 [SOLVE]]

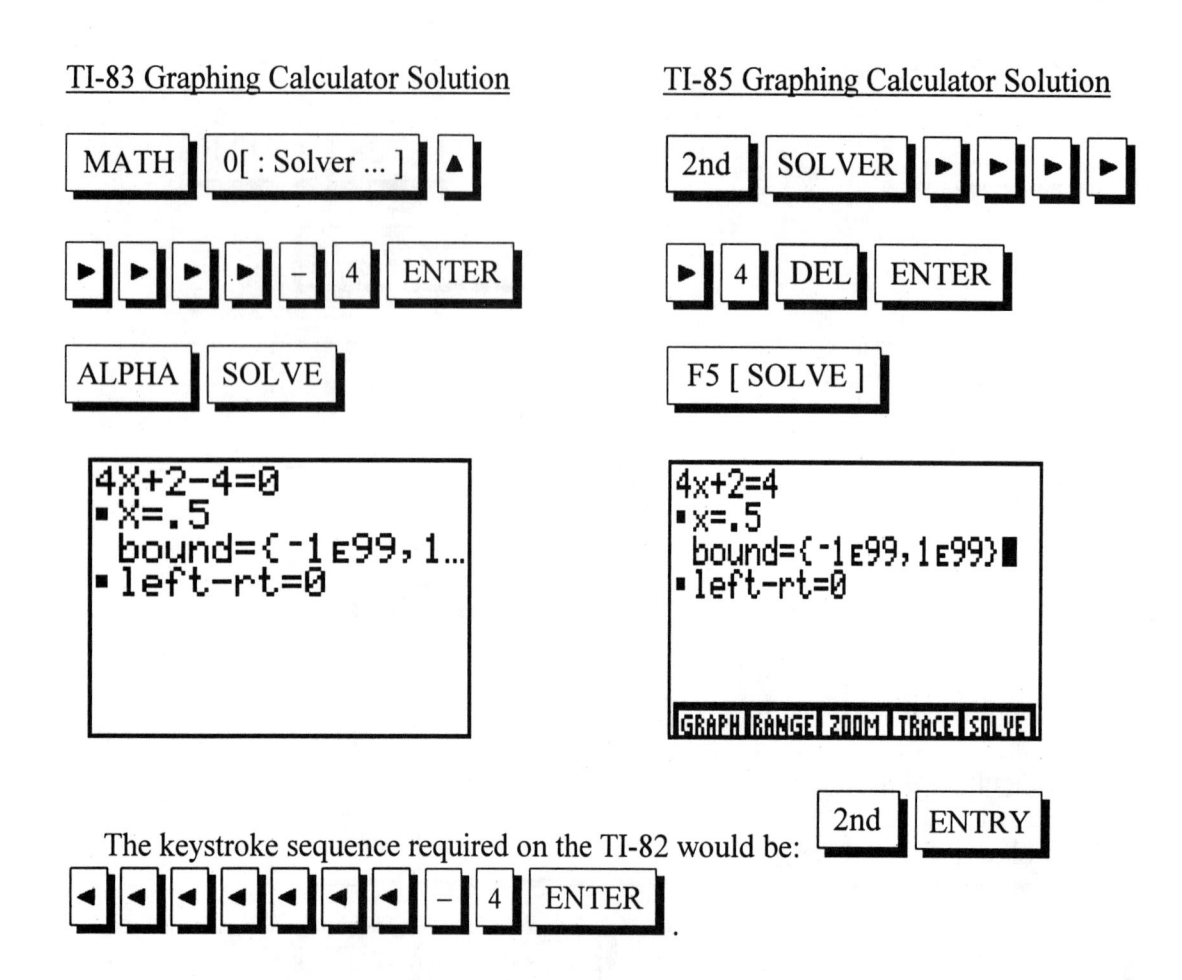

```
4x+2=4
■x=.5
 bound={-1E99,1E99}■
■left-rt=0
```
GRAPH RANGE ZOOM TRACE SOLVE

The keystroke sequence required on the TI-82 would be: [2nd] [ENTRY]

[◀] [◀] [◀] [◀] [◀] [◀] [◀] [–] [4] [ENTER] .

This tells us that the endpoints of the solution interval are −1 and .5. We will now graph the solution. Remember to clear all equations from the graph menu prior to graphing the solution to this problem.

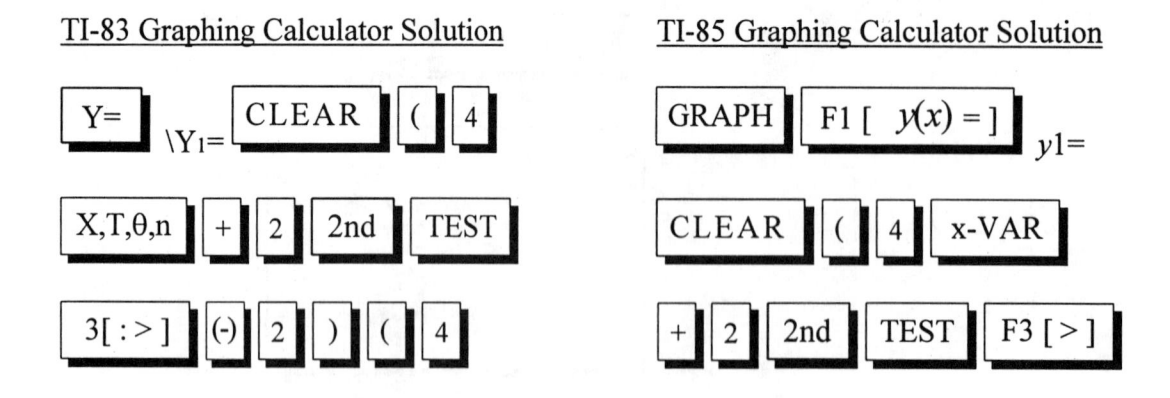

TI-83 Graphing Calculator Solution

[Y=] \Y₁= [CLEAR] [(] [4]

[X,T,θ,n] [+] [2] [2nd] [TEST]

[3[: >]] [(-)] [2] [)] [(] [4]

TI-85 Graphing Calculator Solution

[GRAPH] [F1 [y(x) =]] y1=

[CLEAR] [(] [4] [x-VAR]

[+] [2] [2nd] [TEST] [F3 [>]]

32

LABORATORY 3: SOLUTION OF LINEAR INEQUALITIES

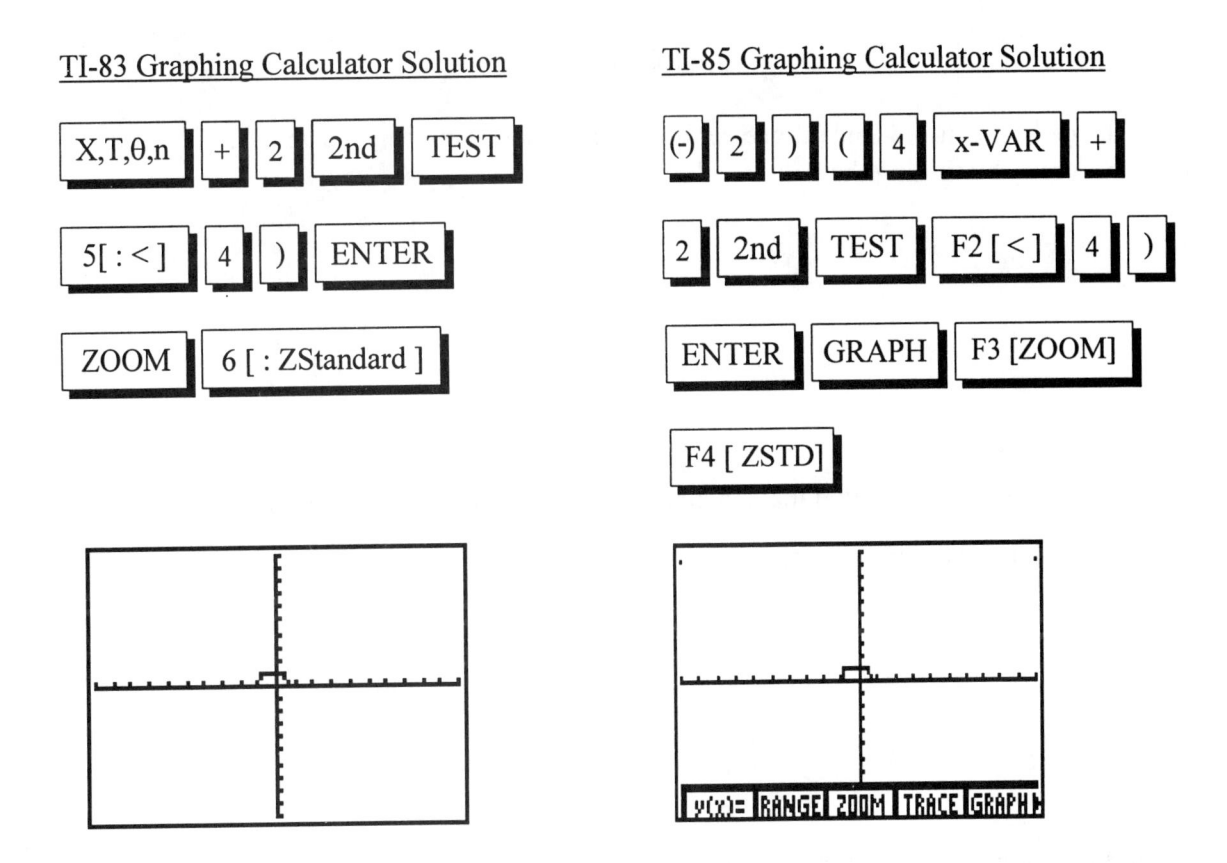

TI-83 Graphing Calculator Solution

| X,T,θ,n | + | 2 | 2nd | TEST |

| 5[: <] | 4 |) | ENTER |

| ZOOM | 6 [: ZStandard] |

TI-85 Graphing Calculator Solution

| (-) | 2 |) | (| 4 | x-VAR | + |

| 2 | 2nd | TEST | F2 [<] | 4 |) |

| ENTER | GRAPH | F3 [ZOOM] |

| F4 [ZSTD] |

We see that the solution to this problem is $-1 < x < \frac{1}{2}$. Adjust the viewing window so that the minimum value on the x-axis is -3 and the maximum value on the x-axis is 3. How does this change the way the graph looks?

LABORATORY 3: SOLUTION OF LINEAR INEQUALITIES

EXERCISES

1. Is the distance between -4 and 12 greater than the distance between -5 and 6? Test the inequality and verify by finding the individual distances.

2. Is $\dfrac{10}{3} > \dfrac{9}{2}$? Is $\dfrac{5}{9} > \dfrac{1}{2}$? Can you make a general statement about the relationship between two fractions that have the same numerator but different denominators?

3. Solve and graph the linear inequality $2x + 7 > \frac{2}{3}x - 4$.

4. Solve and graph the compound linear inequality $3x + 4 < 5$ or $1 - x < -7$.

5. Solve and graph the compound linear inequality $-10 < 2x - \frac{1}{2} < 6$.

LABORATORY 4: INTRODUCTION TO GRAPHING: THE EQUATION OF A LINE

Purpose:

The purpose of this laboratory is to study basic graphing techniques on the TI calculators through an exploration of graphing the equation of a given line. A brief discussion of the concept of slope and the slope-intercept form of the equation of a line will be included. Methods for determining the equation of a line based on given data will be covered in Laboratory 5.

Analytic Approach:

We have seen that the number line is a useful tool in analyzing the relationship between real numbers. Consider now *two* number lines, one *horizontal* (with numbers increasing from left to right) and one *vertical* (with numbers increasing from bottom to top) that intersect at right angles at a point called the *origin* (zero on both number lines) to form a rectangular coordinate system for graphing in two dimensions. The horizontal number line is called the *x-axis*, the vertical number line is called the *y-axis*. Points in the coordinate system are defined by an *x-coordinate* and a *y-coordinate*, (x,y). To graph a point (x,y) in the rectangular coordinate system, simply mark the point at which an imaginary vertical line passing through *x* on the *x-axis* intersects with an imaginary horizontal line passing through *y* on the *y-axis*. The axes divide the rectangular coordinate system into four *quadrants*. The quadrants are numbered *I, II, III,* and *IV,* starting with the upper rightmost quadrant, in which the *x*-coordinate and the *y*-coordinate are both positive, and moving in a counter-clockwise direction for the remaining three quadrants. The rectangular coordinate system makes it possible to graph algebraic equalities and inequalities in two unknowns. The solution set of an algebraic equality or inequality in two unknowns consists of all points, (x,y), that *satisfy* or make the statement true.

Consider two points, (x_1, y_1) and (x_2, y_2). We say that *two points determine a line*, since given any two points in the rectangular coordinate system, we may connect them by drawing a straight line through them. The *slope*, *m*, of a line is a measure of the change in *y* versus the change in *x* as we move from left to right along the line. We have the following formula for the slope of a line passing through the points (x_1, y_1) and (x_2, y_2):

$$m = \frac{y_2 - y_1}{x_2 - x_1}$$

where $x_1 \neq x_2$. A line which slopes upward as we move from left to right on the line is said to have positive slope, while a line which slopes downward as we move from left to right is said to have negative slope. Furthermore, a line which is parallel to the *x*-axis is said to have slope equal to zero, while a line which is parallel to the *y*-axis is said to have no slope, undefined slope, or nonexistent slope. (It is important to note that *having slope equal to zero* is not the same as *having no slope*.) If two lines are parallel then their slopes are equal ($m_1 = m_2$). Finally, if two lines are perpendicular then their slopes are negative reciprocals ($m_1 = -\frac{1}{m_2}$).

LABORATORY 4: INTRODUCTION TO GRAPHING: THE EQUATION OF A LINE

Other points of interest in the graph of an equation include the *x-intercept* and the *y-intercept*. An *x-intercept* is a point at which the graph of the equation passes through the *x-axis*, while a *y-intercept* is a point at which the graph passes through the *y-axis*. An *x-intercept* is also called a *root* or a *zero* of the equation.

The equation of a line may be studied in various forms. The *standard form* for the equation of a straight line is $Ax + By = C$, where A, B, and C, are real numbers, and A and B are not both equal to zero. The *slope-intercept form* of the equation of a line is $y = mx + b$, where m is the slope and b is the *y*-intercept. Finally, the *point-slope form* of the equation of a line is $y - y_1 = m(x - x_1)$, where m is the slope and (x_1, y_1) is a point known to be on the line.

The graphing menus of the TI calculators expect to have the equation of a line entered in a form which isolates y (i.e. $y =$ an expression in terms of x). If the equation of the line to be graphed is *not* already in that form it is necessary to *put* it in that form. Note that vertical lines will be of the form $x = a$, where a is a real number, and do not contain the variable y. The TI calculators handle the special case of the graphing of vertical lines via a separate **Vert** or **Vertical** feature.

Problem 1: Graph the line $y = 2x + 5$ on the standard axes and find the *x-intercept* and *y-intercept*. Remember to clear all graphs before starting this problem.

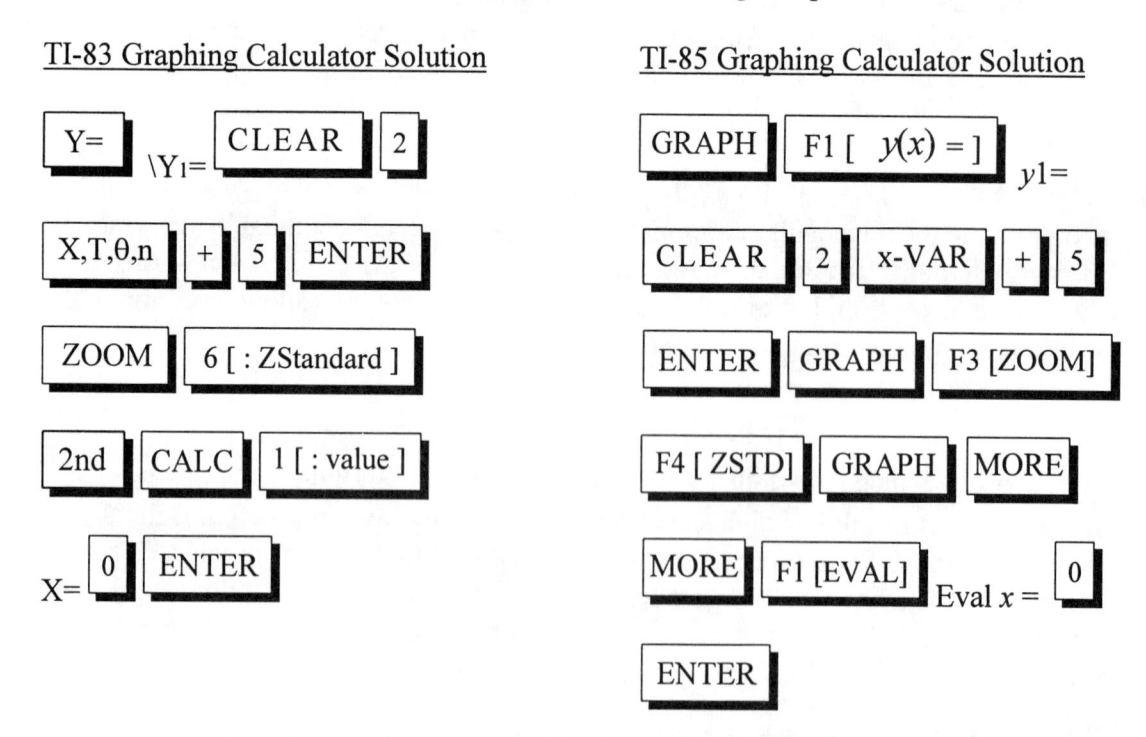

TI-83 Graphing Calculator Solution

Y=	\Y₁=	CLEAR	2
X,T,θ,n	+	5	ENTER
ZOOM	6 [: ZStandard]
2nd	CALC	1 [: value]
X=	0	ENTER

TI-85 Graphing Calculator Solution

GRAPH	F1 [y(x) =]	y1=
CLEAR	2	x-VAR	+	5
ENTER	GRAPH	F3 [ZOOM]
F4 [ZSTD]	GRAPH	MORE
MORE	F1 [EVAL]	Eval x =	0
ENTER

LABORATORY 4: INTRODUCTION TO GRAPHING: THE EQUATION OF A LINE

TI-83 Graphing Calculator Solution

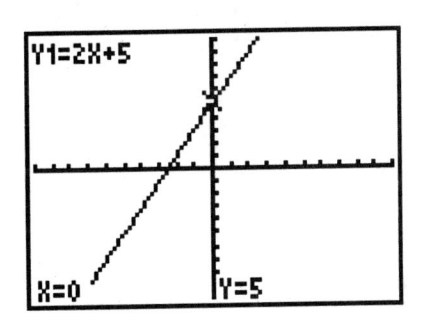

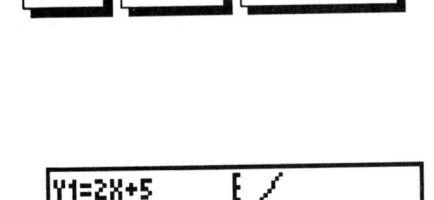

TI-85 Graphing Calculator Solution

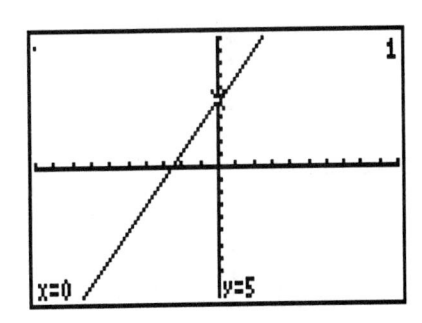

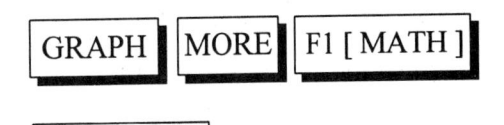

Note that the *x*-value of the *y*-intercept of a line is zero. Therefore, we may use the **EVAL** or **value** function on the calculator to evaluate the formula for the equation of the line at $x = 0$ in order to find the *y*-intercept. The TI-85 and TI-86 calculators also have a *y*-intercept option. To invoke this option on the TI-85 press: GRAPH MORE F1 [MATH] MORE F4 [YICPT] ENTER . Note that the TI-86 has the **YICPT** function located on the **F2** key of the corresponding **MATH** submenu. We will now find the *x*-intercept (root or zero) of the line.

TI-83 Graphing Calculator Solution

2nd CALC 2 [: zero]

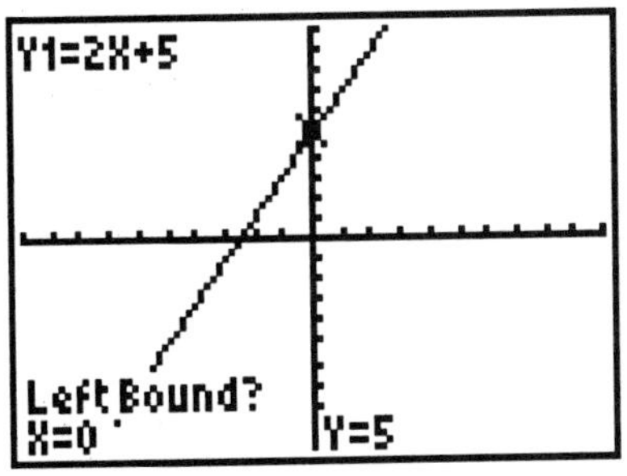

TI-85 Graphing Calculator Solution

GRAPH MORE F1 [MATH]

F3 [ROOT]

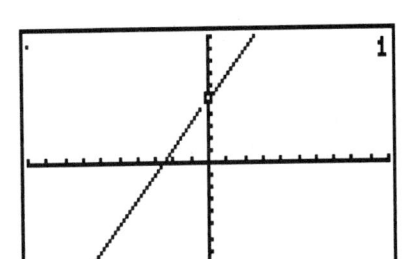

(Move the cursor to the left of the root.)

(There is only one root to this equation.)

ENTER

ENTER

LABORATORY 4: INTRODUCTION TO GRAPHING: THE EQUATION OF A LINE

TI-83 Graphing Calculator Solution   TI-85 Graphing Calculator Solution

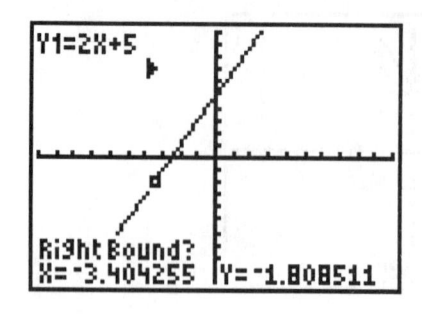

(Move the cursor to the right of the root.)

ENTER

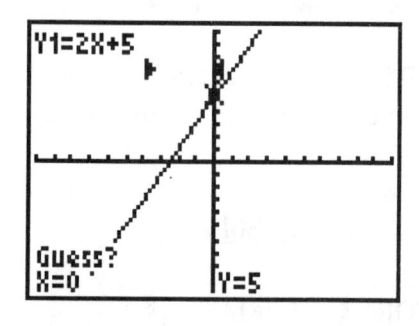

ENTER

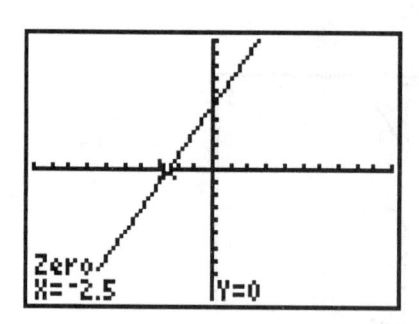

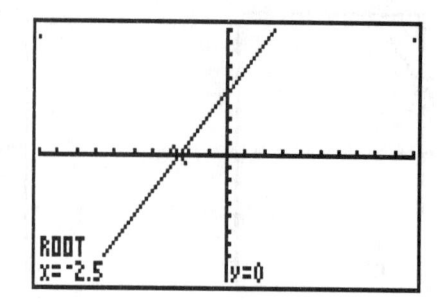

Note that on the TI-86 the **ROOT** function is located on the **F1** key of the corresponding **MATH** submenu. Furthermore, on the TI-86, it is necessary to enter the left bound, right bound, and guess, as in the TI-83 keystroke sequence illustrated above.

The equation of a line to be graphed may not always be given in a form in which y is isolated. It will be necessary to convert such equations to this form before graphing them on the TI calculators.

38

LABORATORY 4: INTRODUCTION TO GRAPHING: THE EQUATION OF A LINE

Problem 2: Graph the lines $3x + 2y = 10$ and $6x + 4y = 8$ on the same set of axes and evaluate each equation at $x = 3$. Remember to clear all graphs and write the equation of each line in *slope-intercept form* before starting the problem.

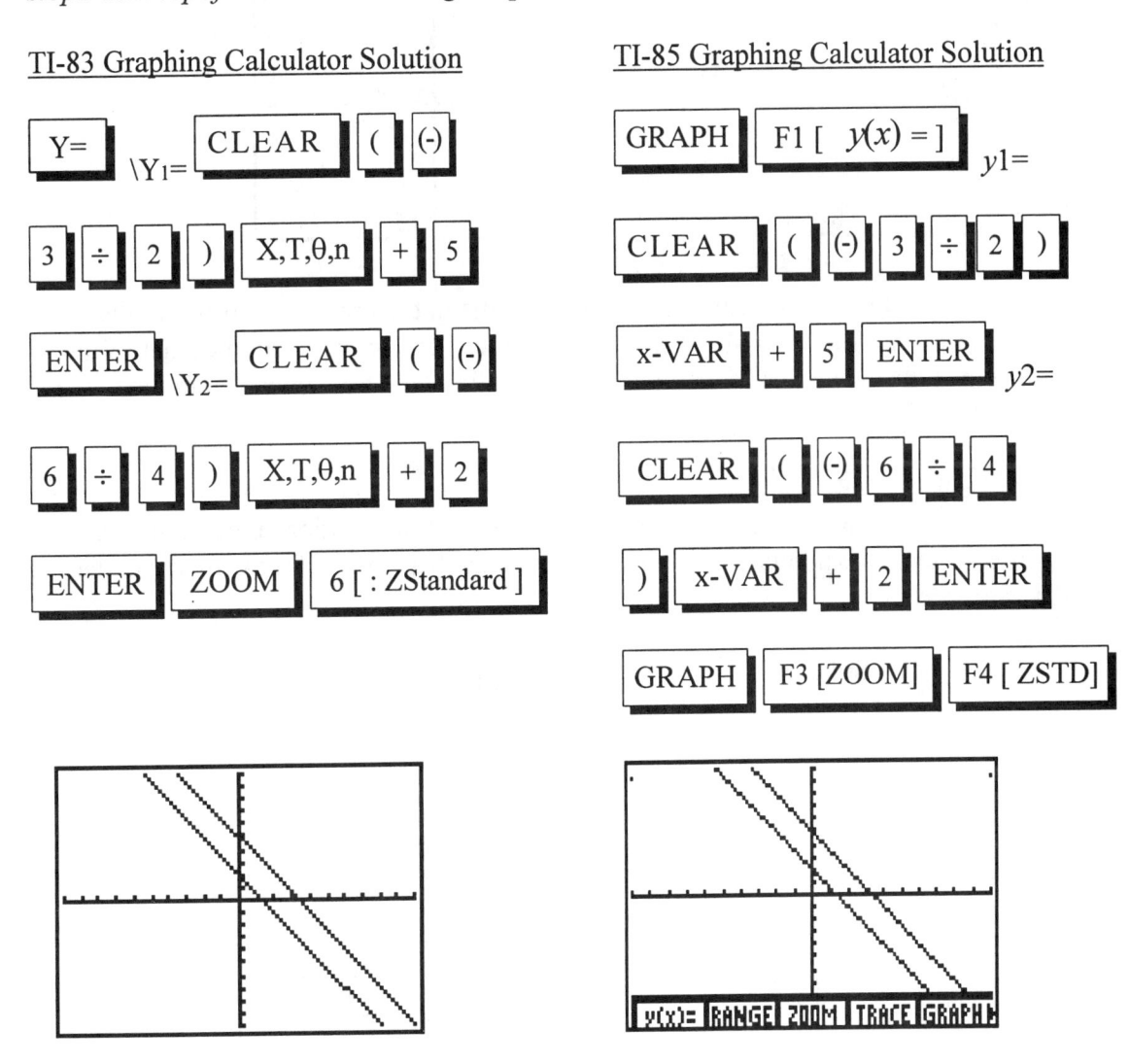

TI-83 Graphing Calculator Solution

[Y=] \Y₁= [CLEAR] [(] [(-)]
[3] [÷] [2] [)] [X,T,θ,n] [+] [5]
[ENTER] \Y₂= [CLEAR] [(] [(-)]
[6] [÷] [4] [)] [X,T,θ,n] [+] [2]
[ENTER] [ZOOM] [6 [: ZStandard]]

TI-85 Graphing Calculator Solution

[GRAPH] [F1 [y(x) =]] y1=
[CLEAR] [(] [(-)] [3] [÷] [2] [)]
[x-VAR] [+] [5] [ENTER] y2=
[CLEAR] [(] [(-)] [6] [÷] [4]
[)] [x-VAR] [+] [2] [ENTER]
[GRAPH] [F3 [ZOOM]] [F4 [ZSTD]]

The calculator will display the graph of both lines on the same axes. Now evaluate the first equation at $x = 3$.

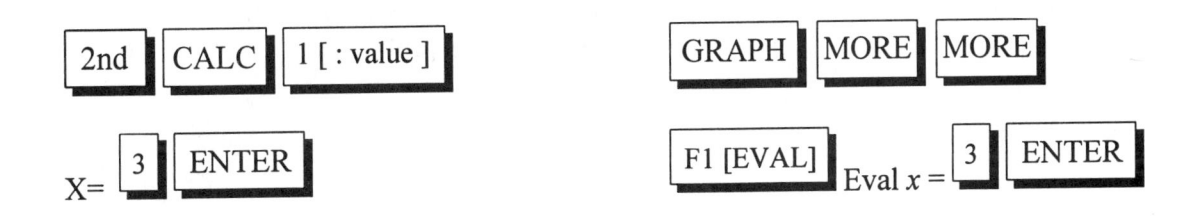

[2nd] [CALC] [1 [: value]]
X= [3] [ENTER]

[GRAPH] [MORE] [MORE]
[F1 [EVAL]] Eval x = [3] [ENTER]

39

LABORATORY 4: INTRODUCTION TO GRAPHING: THE EQUATION OF A LINE

TI-83 Graphing Calculator Solution

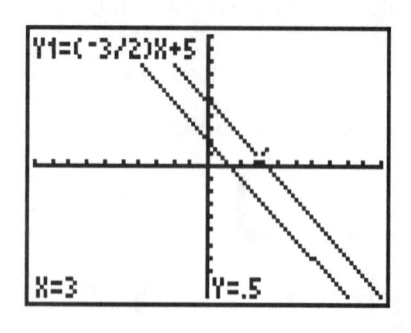

TI-85 Graphing Calculator Solution

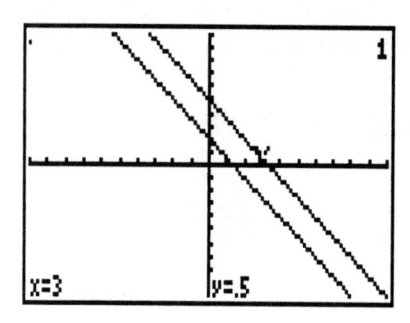

Notice that the TI-83 displays the equation of the first line in the upper left corner of the screen, while the TI-85, TI-86, and TI-82 display the number 1 in the upper right corner of the screen to indicate that the evaluation is taking place on the first equation entered into the calculator.

The ▲ and ▼ keys may be used to move through the list of equations entered into the calculator in order to evaluate points on other equations in the list. We will now evaluate the second equation at $x = 3$.

TI-83 Graphing Calculator Solution

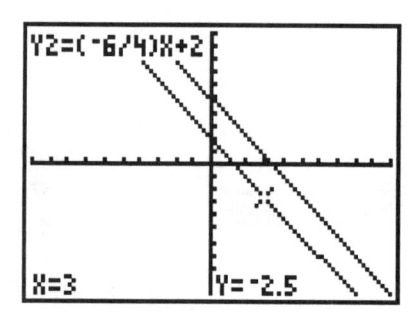

TI-85 Graphing Calculator Solution

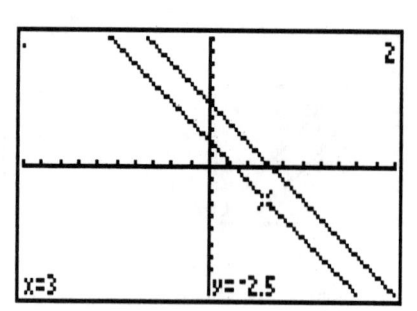

The calculator screen now shows that the evaluation is taking place using the second equation in the list. Make a statement about the slopes of the two equations. Are the lines parallel? Try evaluating each of the equations in this problem at $x = -5$ and $x = \dfrac{3}{2}$.

LABORATORY 4: INTRODUCTION TO GRAPHING: THE EQUATION OF A LINE

Problem 3: Graph the line $x = 20$ on an appropriate viewing window. Be sure to clear all graphs before starting this problem.

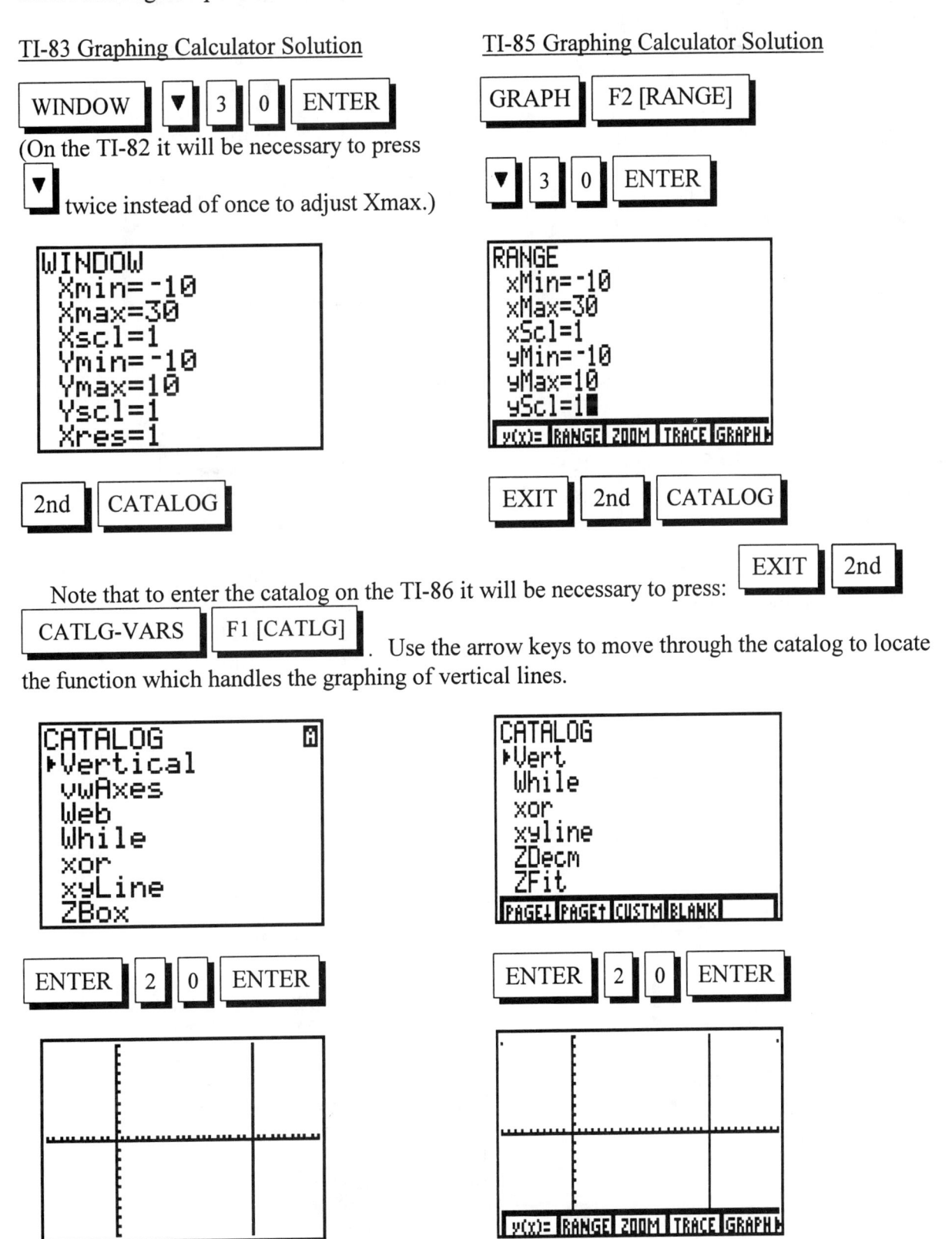

TI-83 Graphing Calculator Solution

[WINDOW] [▼] [3] [0] [ENTER]

(On the TI-82 it will be necessary to press [▼] twice instead of once to adjust Xmax.)

```
WINDOW
 Xmin=-10
 Xmax=30
 Xscl=1
 Ymin=-10
 Ymax=10
 Yscl=1
 Xres=1
```

[2nd] [CATALOG]

TI-85 Graphing Calculator Solution

[GRAPH] [F2 [RANGE]]

[▼] [3] [0] [ENTER]

```
RANGE
 xMin=-10
 xMax=30
 xScl=1
 yMin=-10
 yMax=10
 yScl=1
```

[EXIT] [2nd] [CATALOG]

Note that to enter the catalog on the TI-86 it will be necessary to press: [EXIT] [2nd] [CATLG-VARS] [F1 [CATLG]]. Use the arrow keys to move through the catalog to locate the function which handles the graphing of vertical lines.

```
CATALOG
▶Vertical
 vwAxes
 Web
 While
 xor
 xyLine
 ZBox
```

[ENTER] [2] [0] [ENTER]

```
CATALOG
▶Vert
 While
 xor
 xyline
 ZDecm
 ZFit
```

[ENTER] [2] [0] [ENTER]

LABORATORY 4: INTRODUCTION TO GRAPHING: THE EQUATION OF A LINE

The keystroke sequence to draw this vertical line on the TI-82 is: [2nd] [DRAW] [4 [: Vertical]] [2] [0] [ENTER]. The drawing must be cleared from the screen by following the procedure outlined below:

TI-83 Graphing Calculator Solution

[2nd] [DRAW] [1 [: ClrDraw]]

[2nd] [QUIT]

TI-85 Graphing Calculator Solution

[GRAPH] [MORE] [F2 [DRAW]]

[MORE] [F5 [CLDRW]] [EXIT]

[EXIT]

Note that on the TI-86 the keystroke sequence to clear the screen is: [GRAPH] [MORE] [F2 [DRAW]] [MORE] [MORE] [F1 [CLDRW]] [EXIT] [EXIT].

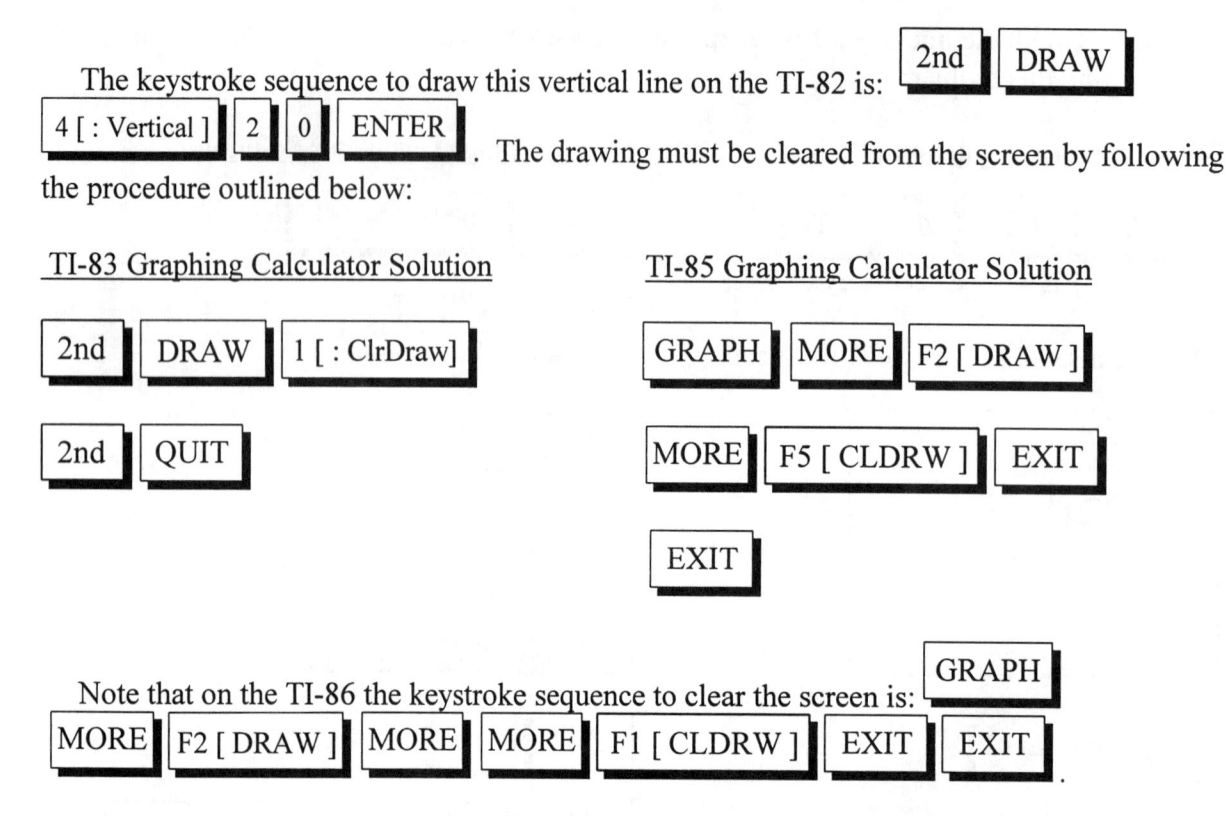

Linear inequalities in two unknowns can be graphed on the TI calculators by first graphing the equation of the line and then shading the proper side of the line. The **Shade** feature on the TI calculators has the following syntax: Shade(algebraic expression on the *less than* side of the inequality, algebraic expression on the *greater than* side of the inequality).

Problem 4: Graph the linear inequality $y \leq 2x - 5$ on the standard viewing window. Remember to clear all graphs before starting this problem.

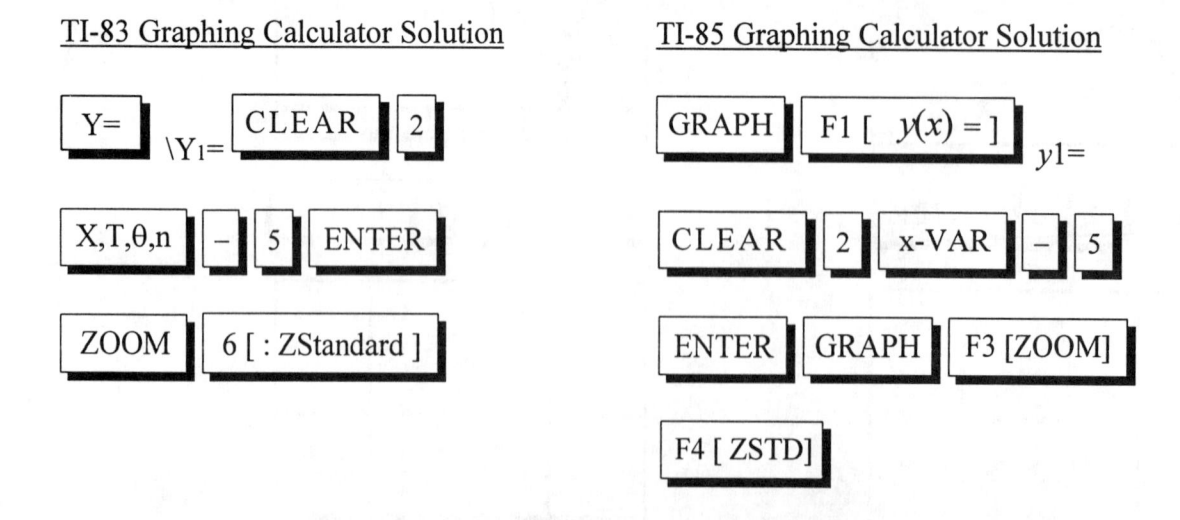

TI-83 Graphing Calculator Solution

[Y=] \Y₁= [CLEAR] [2]

[X,T,θ,n] [−] [5] [ENTER]

[ZOOM] [6 [: ZStandard]]

TI-85 Graphing Calculator Solution

[GRAPH] [F1 [y(x) =]] y1=

[CLEAR] [2] [x-VAR] [−] [5]

[ENTER] [GRAPH] [F3 [ZOOM]]

[F4 [ZSTD]]

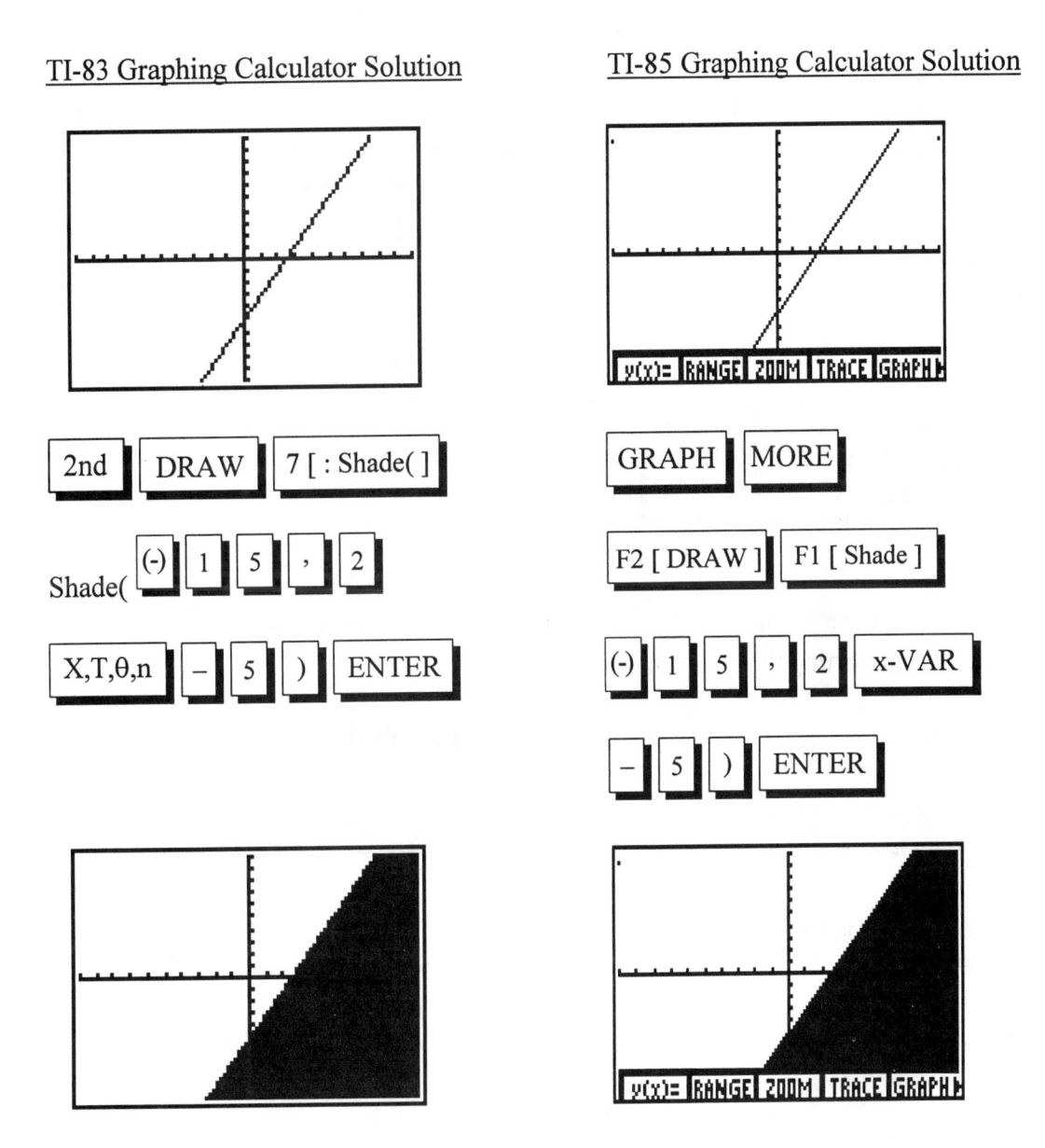

LABORATORY 4: INTRODUCTION TO GRAPHING: THE EQUATION OF A LINE

TI-83 Graphing Calculator Solution

TI-85 Graphing Calculator Solution

[2nd] [DRAW] [7 [: Shade( ]]

Shade([(-)] [1] [5] [,] [2]

[X,T,θ,n] [–] [5] [ )] [ENTER]

[GRAPH] [MORE]

[F2 [DRAW]] [F1 [Shade]]

[(-)] [1] [5] [,] [2] [x-VAR]

[–] [5] [ )] [ENTER]

Note the fact that the first argument in the **Shade** command is -15 in this problem. The calculator will shade everything above the line $y = -15$, but below the line $y = 2x - 5$. The number -15 was chosen since it is outside the current viewing window. The procedure to clear the graph from the screen is as follows:

[2nd] [DRAW] [1 [: ClrDraw]]

[GRAPH] [MORE] [F2 [DRAW]]

LABORATORY 4: INTRODUCTION TO GRAPHING: THE EQUATION OF A LINE

TI-83 Graphing Calculator Solution

|Y=| \Y₁= |CLEAR|

|2nd| |QUIT|

TI-85 Graphing Calculator Solution

|MORE| |F5 [CLDRW]|

|GRAPH| |F1 [y(x) =]| y1=

|CLEAR| |EXIT| |EXIT|

Recall that the keystroke sequence to clear the shading from the graph on the TI-86 is:

|GRAPH| |MORE| |F2 [DRAW]| |MORE| |MORE| |F1 [CLDRW]|.

Problem 5: Graph the linear inequality $y \geq 2 - x$ on the standard viewing window. Remember to clear all graphs before starting this problem.

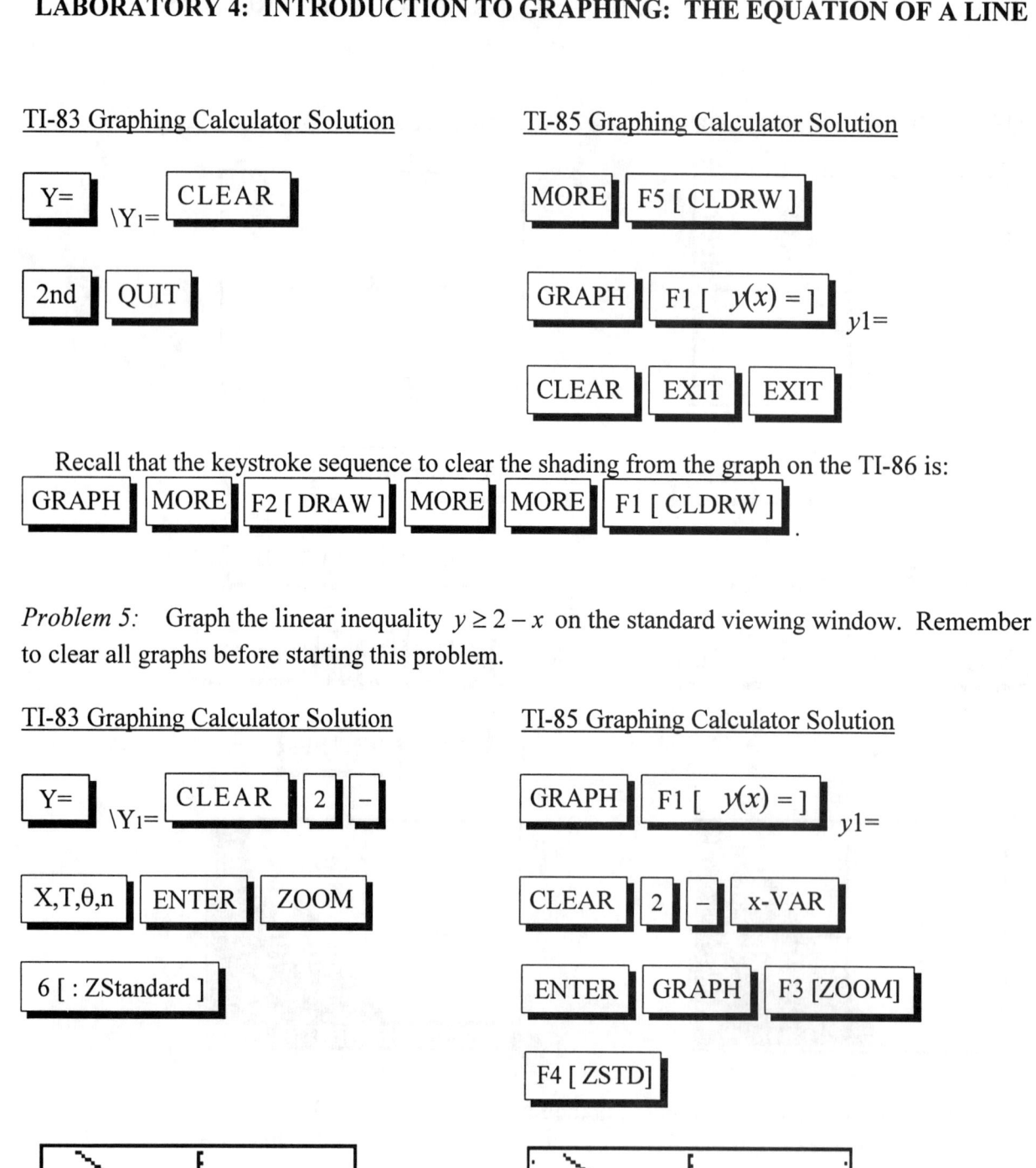

TI-83 Graphing Calculator Solution

|Y=| \Y₁= |CLEAR| |2| |−|

|X,T,θ,n| |ENTER| |ZOOM|

|6 [: ZStandard]|

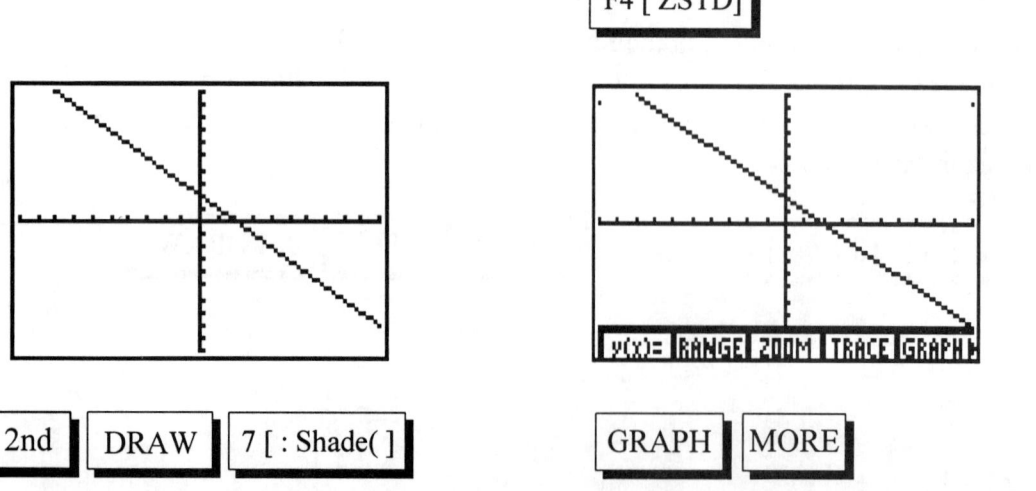

TI-85 Graphing Calculator Solution

|GRAPH| |F1 [y(x) =]| y1=

|CLEAR| |2| |−| |x-VAR|

|ENTER| |GRAPH| |F3 [ZOOM]|

|F4 [ZSTD]|

|2nd| |DRAW| |7 [: Shade( ]|

|GRAPH| |MORE|

44

LABORATORY 4: INTRODUCTION TO GRAPHING: THE EQUATION OF A LINE

TI-83 Graphing Calculator Solution

Shade([2] [−] [X,T,θ,n] [,] [1] [5] [)] [ENTER]

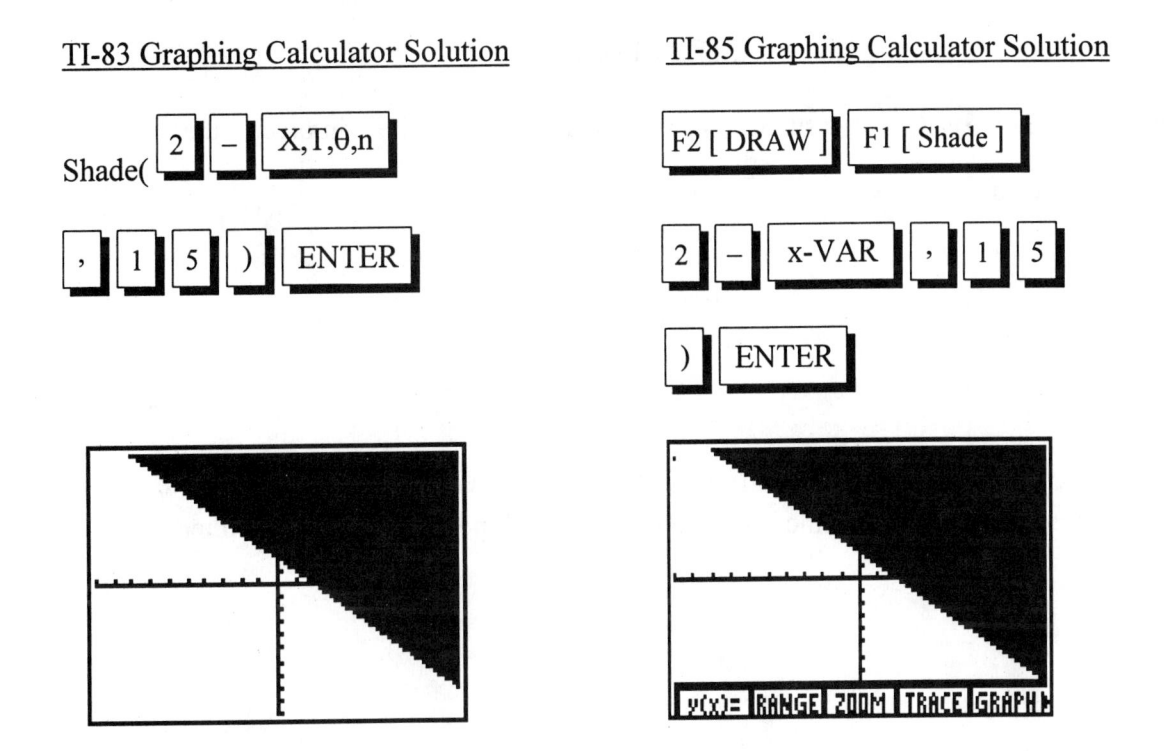

TI-85 Graphing Calculator Solution

[F2 [DRAW]] [F1 [Shade]]
[2] [−] [x-VAR] [,] [1] [5] [)] [ENTER]

Note the fact that the second argument in the **Shade** command is 15 in this problem. The calculator will shade everything above the line $y = 2 - x$, but below the line $y = 15$. The number 15 was chosen since it is outside the current viewing window.

LABORATORY 4: INTRODUCTION TO GRAPHING: THE EQUATION OF A LINE

EXERCISES

1. Graph the line $3x - 5y = 15$ on a suitable viewing window. Find the *x*-intercept, the *y*-intercept, and evaluate at $x = -3$, $x = \dfrac{2}{3}$, and $x = 35$.

2. Graph the lines $y = 5x - 4$ and $2x + 10y = -3$ on the same set of axes. Find the slope, *x*-intercept, and *y*-intercept of each line. Are the lines parallel, perpendicular, or neither?

3. Graph the vertical lines $x = -3$, $x = 4$, and $x = \dfrac{5}{7}$ on the same set of axes.

4. Graph the linear inequality $y \leq 9x - 5$ on a suitable viewing window. Is the origin in the solution set of this inequality?

5. Graph the linear inequality $2x + 9y \geq 18$ on a suitable viewing window.

LABORATORY 5: DATA ANALYSIS AND STATISTICAL APPROACH TO THE EQUATION OF A LINE

Purpose:

The purpose of this laboratory is to analyze data via statistical methods in order to determine the equation of a line. Given either a point and a slope, or two points, we will derive the equation of the line determined by the data. A brief discussion of linear regression by the least squares method will be included.

Analytic Approach:

In the real world we are often interested in testing hypotheses and predicting outcomes based on a known data set. For example, we might be interested in studying the relationship between performance, x, on a placement exam taken at the beginning of the semester and grade point average, y, at the conclusion of the semester. If such a relationship can be determined, it will be a useful tool in decision-making processes. One of the simplest relationships between two such variables is a linear equation of the form $y = ax + b$. Given a set of data points we can construct the *scatter plot* of the ordered pairs, (x,y), and estimate the equation of the line that *best fits* the data. The points need not lie on the line. It is necessary and sufficient to construct the line for which the sum of the squares of the distances of the data points from the line is minimized. The method by which this line is constructed via estimates for a and b is called *the method of least squares*.

Consider a set of data consisting of only two points, (x_1, y_1) and (x_2, y_2). The regression line for this data is simply the line that passes through these two points. The slope of the line could be calculated algebraically. The resulting slope could then be substituted for a into the equation $y = ax + b$, and one of the given points could be substituted for (x,y). We could then solve for the only unknown left, b. Although this is not a particularly difficult task to perform, it can become tedious. It is important to recognize that the regression analysis features of the TI calculators can be employed to focus our attention where it belongs, on the analysis of the data, rather than on the mathematical manipulation of numbers and equations in order to produce a solution. Understanding concept and procedure are just as important as being able to produce the correct solution. A firm grasp of the concept takes us beyond *number crunching* and makes us aware of the importance of recognizing patterns and relationships useful in *predicting* and *correcting* estimated data.

The statistics menu on the TI calculators can be found by pressing the STAT key. The features of this menu include various functions which are useful in editing lists, making calculations, and performing tests. We will be concerned only with those features that are necessary for graphing a line through two known points. However, the power of the **STAT** features on the TI calculators will be quite evident in the examples that follow. It is the intent of this laboratory to encourage you to explore these features further since they will be useful in many real world applications.

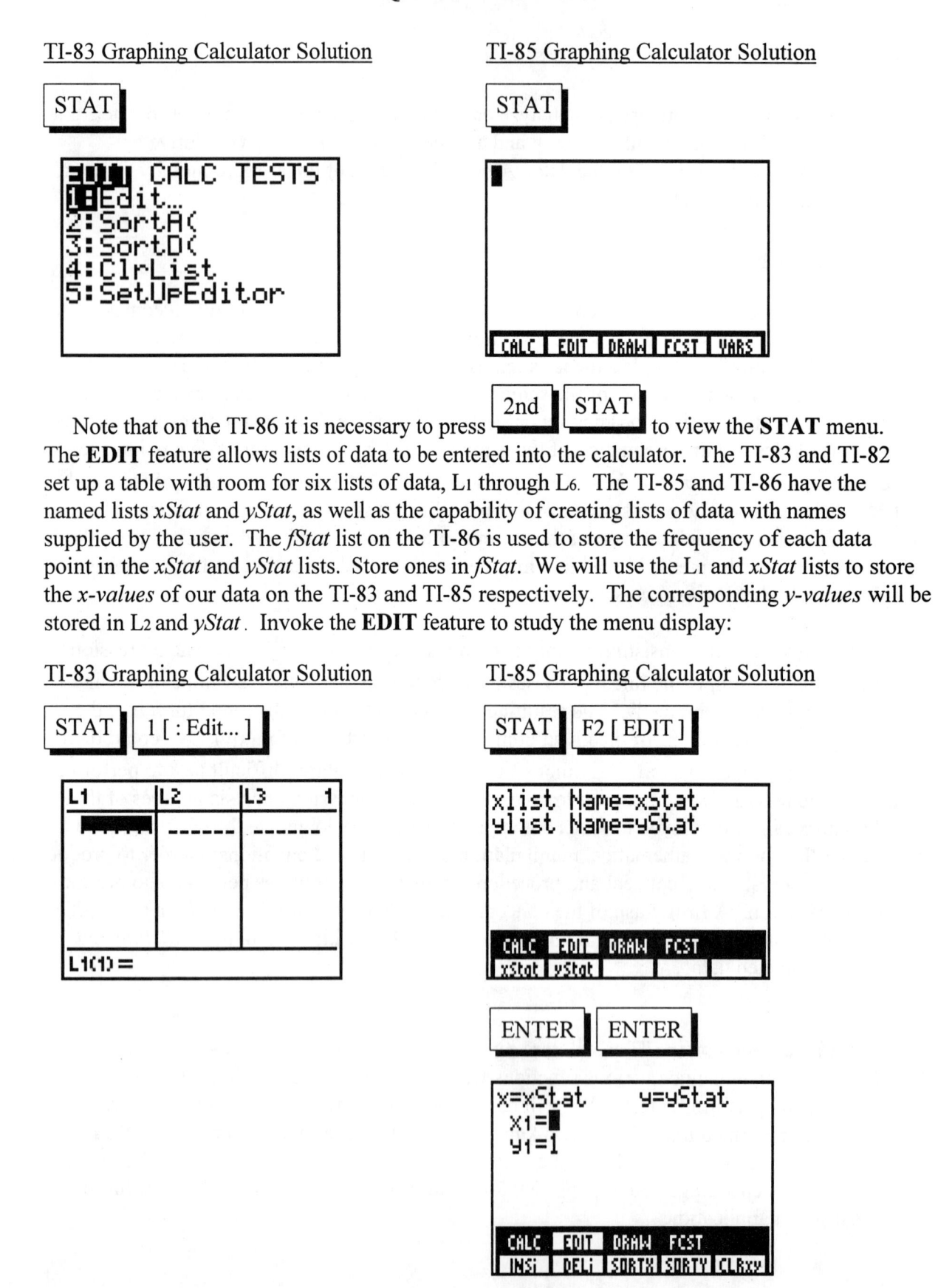

LABORATORY 5: DATA ANALYSIS AND STATISTICAL APPROACH TO THE EQUATION OF A LINE

TI-83 Graphing Calculator Solution

[STAT]

```
EDIT CALC TESTS
1:Edit…
2:SortA(
3:SortD(
4:ClrList
5:SetUpEditor
```

TI-85 Graphing Calculator Solution

[STAT]

```
|
CALC | EDIT | DRAW | FCST | VARS
```

Note that on the TI-86 it is necessary to press [2nd] [STAT] to view the **STAT** menu. The **EDIT** feature allows lists of data to be entered into the calculator. The TI-83 and TI-82 set up a table with room for six lists of data, L_1 through L_6. The TI-85 and TI-86 have the named lists *xStat* and *yStat*, as well as the capability of creating lists of data with names supplied by the user. The *fStat* list on the TI-86 is used to store the frequency of each data point in the *xStat* and *yStat* lists. Store ones in *fStat*. We will use the L_1 and *xStat* lists to store the *x-values* of our data on the TI-83 and TI-85 respectively. The corresponding *y-values* will be stored in L_2 and *yStat*. Invoke the **EDIT** feature to study the menu display:

TI-83 Graphing Calculator Solution

[STAT] [1 [: Edit…]]

```
L1    L2    L3    1
----  ----  ----

L1(1)=
```

TI-85 Graphing Calculator Solution

[STAT] [F2 [EDIT]]

```
xlist Name=xStat
ylist Name=yStat




CALC | EDIT | DRAW | FCST
xStat | yStat
```

[ENTER] [ENTER]

```
x=xStat    y=yStat
 x1=■
 y1=1



CALC | EDIT | DRAW | FCST
INSi | DELi | SORTx | SORTy | CLRxy
```

48

LABORATORY 5: DATA ANALYSIS AND STATISTICAL APPROACH TO THE EQUATION OF A LINE

To display the **EDIT** menu on the TI-86 press: [2nd] [STAT] [F2 [EDIT]].
The data points can now be entered into the lists displayed on the calculator screen by using the arrow keys and entering each number in the appropriate location in the list. Note that the TI-86 calculator uses the table entry format described in the TI-83 calculator solution below. Enter the data points into the *xStat* and *yStat* lists and fix the frequency to one in the *fStat* list on the TI-86. Once the data is entered, a calculation of the regression equation can be made through the **CALC** option on the **STAT** menu.

TI-83 Graphing Calculator Solution

[STAT] [▶]

```
EDIT CALC TESTS
1: 1-Var Stats
2: 2-Var Stats
3: Med-Med
4: LinReg(ax+b)
5: QuadReg
6: CubicReg
7↓QuartReg
```

TI-85 Graphing Calculator Solution

[STAT] [F1 [CALC]] [ENTER]

[ENTER]

```
x=xStat      y=yStat




CALC  EDIT  DRAW  FCST
1-VAR LINR  LNR  EXPR PWRR
```

The keystroke sequence on the TI-86 would be: [2nd] [STAT] [F1 [CALC]]. It is important to take a moment to explain the form in which the TI calculators store the linear regression equation. Notice that the menus contain several different types of regression equations. We will only be interested in the *linear* regression equation. Item number 4 on the TI-83 menu displayed above, calculates the linear regression equation in the form $y = ax + b$.

Pressing [4 [: LinReg(ax+b)]] [ENTER] will execute this function and display the regression equation. (This function is item number 5 on the corresponding TI-82 menu.) The linear regression equation may be calculated in the alternate form $y = a + bx$ by scrolling to item number 8 on the TI-83 menu and pressing [8 [: LinReg(a+bx)]] [ENTER]. (This corresponds to item number 9 on the TI-82 menu.) The TI-85 calculator, however, calculates the linear regression equation in the form $y = a + bx$, by pressing [F2 [LINR]]. The **LinR** function is located on the **F3** key of the corresponding **STAT** submenu on the TI-86. The screen displays the values of a and b. The TI-85 does not remind you of the form in which the equation is calculated, while the TI-82, TI-83, and TI-86 do.

LABORATORY 5: DATA ANALYSIS AND STATISTICAL APPROACH TO THE EQUATION OF A LINE

The regression equation is stored in a variable called **RegEQ** on the TI-83 and TI-82 calculators and in a variable called **RegEq** on the TI-85 and TI-86 calculators. The variable can be used to enter the equation to the graphing menu in order to graph the regression line. A scatter plot of the data points can also be drawn.

A. Finding the Equation of a Line Given Two Points on the Line

Problem 1: Plot the points (-8,-2) and (6,8) on the standard viewing window. Calculate and draw the equation of the line that passes through these two points. Remember to clear all graphs and drawings before starting this problem.

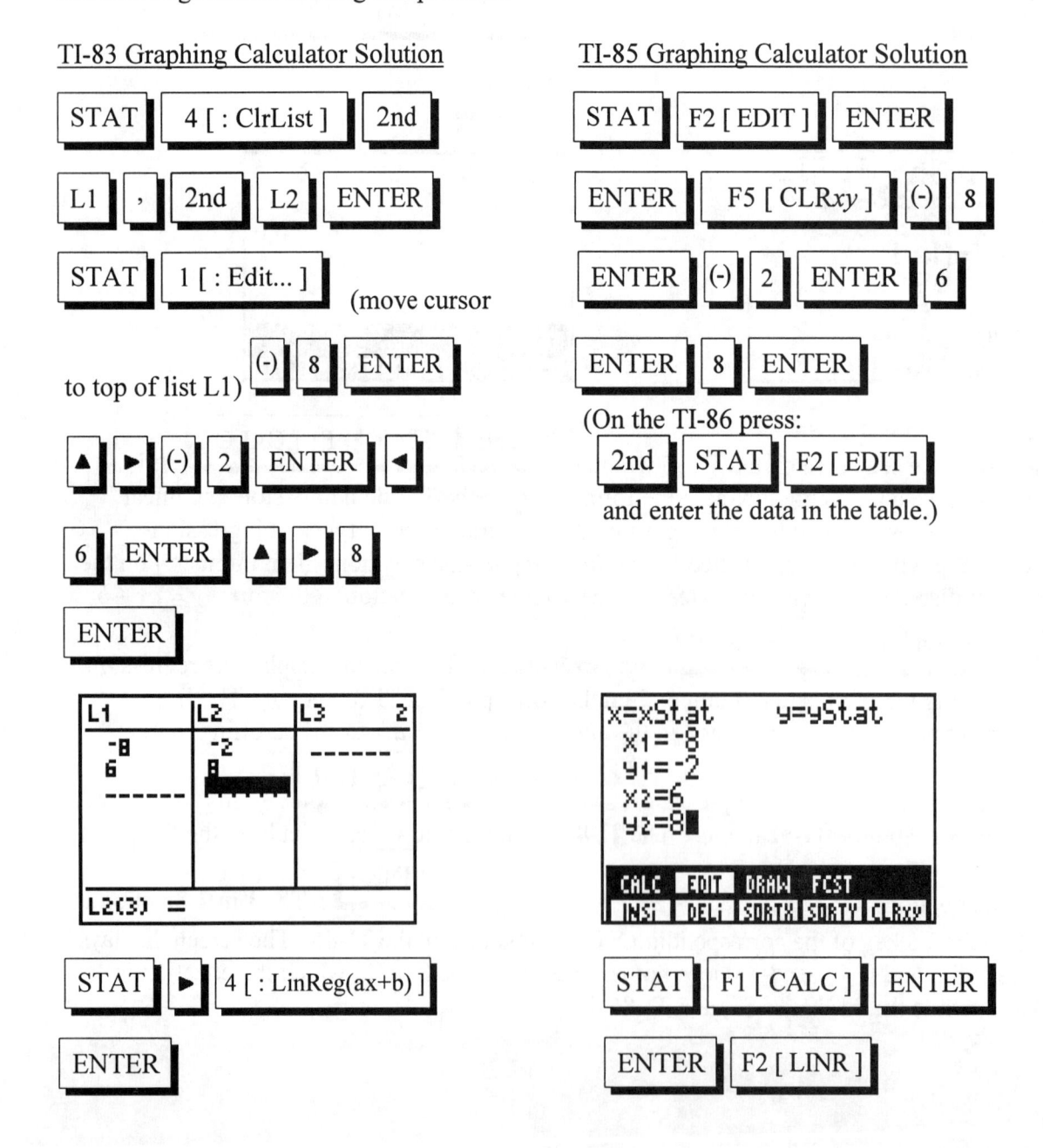

TI-83 Graphing Calculator Solution

| STAT | 4 [: ClrList] | 2nd |

| L1 | , | 2nd | L2 | ENTER |

| STAT | 1 [: Edit...]   (move cursor to top of list L1) | (-) | 8 | ENTER |

| ▲ | ▶ | (-) | 2 | ENTER | ◀ |

| 6 | ENTER | ▲ | ▶ | 8 |

| ENTER |

| STAT | ▶ | 4 [: LinReg(ax+b)] |

| ENTER |

TI-85 Graphing Calculator Solution

| STAT | F2 [EDIT] | ENTER |

| ENTER | F5 [CLRxy] | (-) | 8 |

| ENTER | (-) | 2 | ENTER | 6 |

| ENTER | 8 | ENTER |

(On the TI-86 press:

| 2nd | STAT | F2 [EDIT] |

and enter the data in the table.)

| STAT | F1 [CALC] | ENTER |

| ENTER | F2 [LINR] |

50

LABORATORY 5: DATA ANALYSIS AND STATISTICAL APPROACH TO THE EQUATION OF A LINE

TI-83 Graphing Calculator Solution

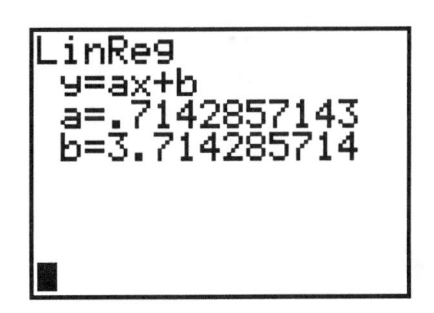

TI-85 Graphing Calculator Solution

```
LinReg
 y=ax+b
 a=.7142857143
 b=3.714285714
```

```
LinR
 a=3.71428571429
 b=.714285714286
 corr=1
 n=2
```

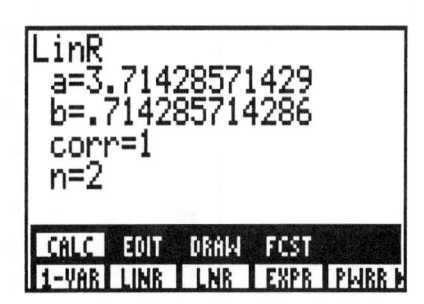

Recall that the linear regression feature described above is item number 5 on the corresponding TI-82 menu. On the TI-86, **EXIT** the **EDIT** mode. The **LinR** feature is located on the **F3** key of the corresponding **STAT** submenu. We see that a and b are rational numbers. They can be converted to fractional form by following the procedure outlined below. Note that on the TI-82, a is item number one, b is item number two, and the regression equation is item number seven on the given menu. Therefore, the appropriate adjustments will need to be made in the keystroke sequence. Furthermore, recall that the **STAT** feature is a [2nd] option on the TI-86.

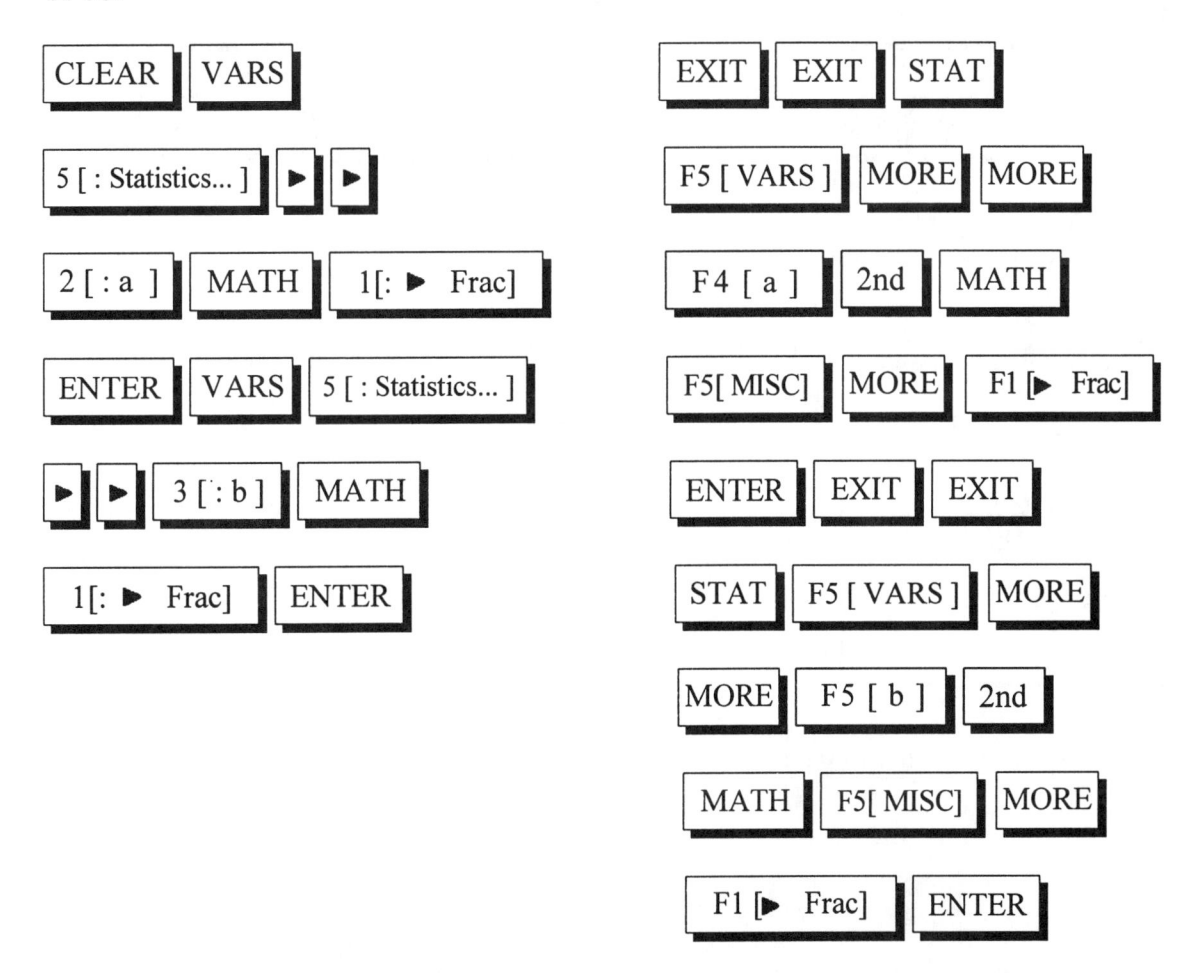

[CLEAR] [VARS]	[EXIT] [EXIT] [STAT]
[5 [: Statistics...]] [▶] [▶]	[F5 [VARS]] [MORE] [MORE]
[2 [: a]] [MATH] [1[: ▶ Frac]]	[F4 [a]] [2nd] [MATH]
[ENTER] [VARS] [5 [: Statistics...]]	[F5[MISC]] [MORE] [F1 [▶ Frac]]
[▶] [▶] [3 [: b]] [MATH]	[ENTER] [EXIT] [EXIT]
[1[: ▶ Frac]] [ENTER]	[STAT] [F5 [VARS]] [MORE]
	[MORE] [F5 [b]] [2nd]
	[MATH] [F5[MISC]] [MORE]
	[F1 [▶ Frac]] [ENTER]

LABORATORY 5: DATA ANALYSIS AND STATISTICAL APPROACH TO THE EQUATION OF A LINE

TI-83 Graphing Calculator Solution

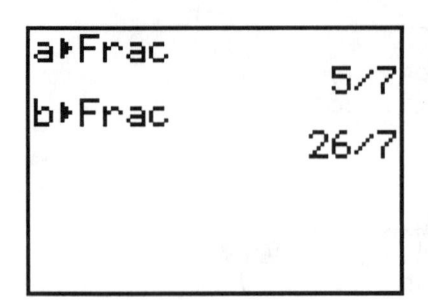

TI-85 Graphing Calculator Solution

Thus, the equation of the regression line can be written as $y = \frac{5}{7}x + \frac{26}{7}$. We will now draw the scatter plot and the equation of the regression line on the standard axes.

TI-83 Graphing Calculator Solution

Y= \Y₁= CLEAR 2nd

STAT PLOT 1 [:Plot1...]

ENTER ▼ ENTER ▼

2nd L1 ENTER 2nd

L2 ENTER ENTER

TI-85 Graphing Calculator Solution

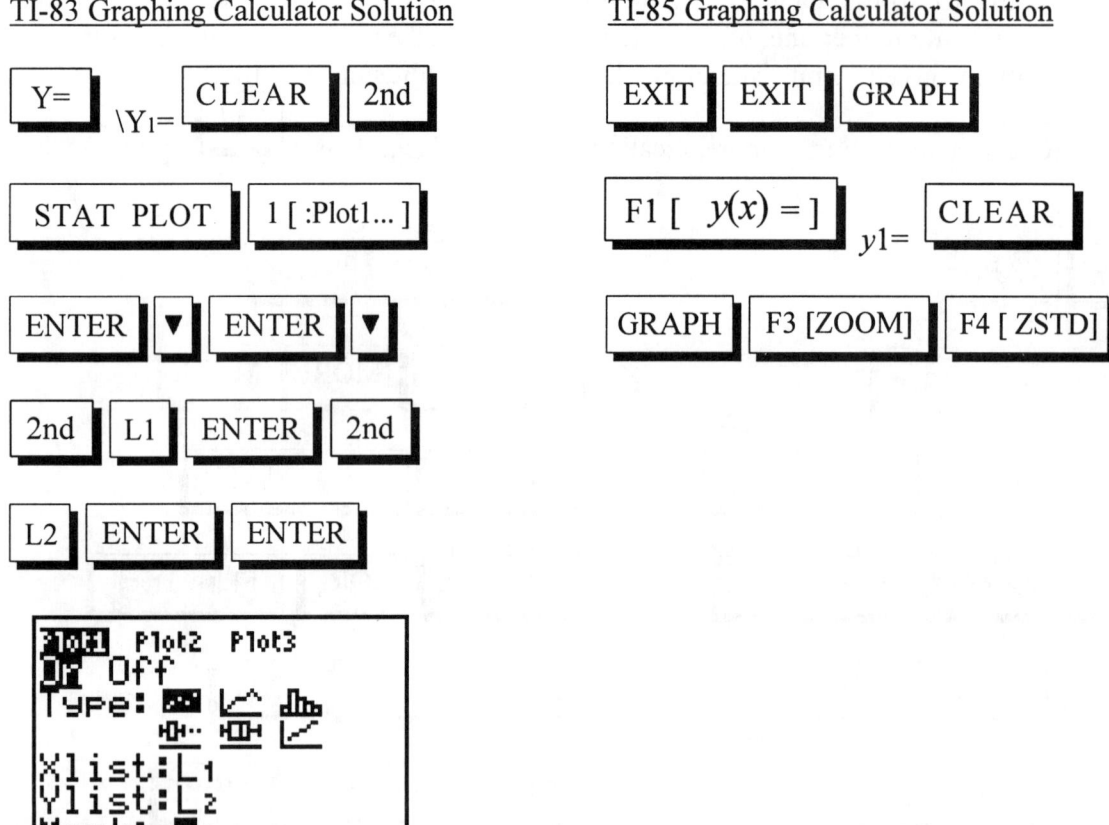

EXIT EXIT GRAPH

F1 [y(x) =] y1= CLEAR

GRAPH F3 [ZOOM] F4 [ZSTD]

Note that on the TI-82 it will be necessary to highlight **L1** as the Xlist and **L2** as the Ylist.

LABORATORY 5: DATA ANALYSIS AND STATISTICAL APPROACH TO THE EQUATION OF A LINE

TI-83 Graphing Calculator Solution

| ZOOM | 6 [: ZStandard] |

| Y= | \Y₁= | VARS |

| 5 [: Statistics...] | ▶ | ▶ |

| 1 [: RegEQ] | ENTER |

| GRAPH |

TI-85 Graphing Calculator Solution

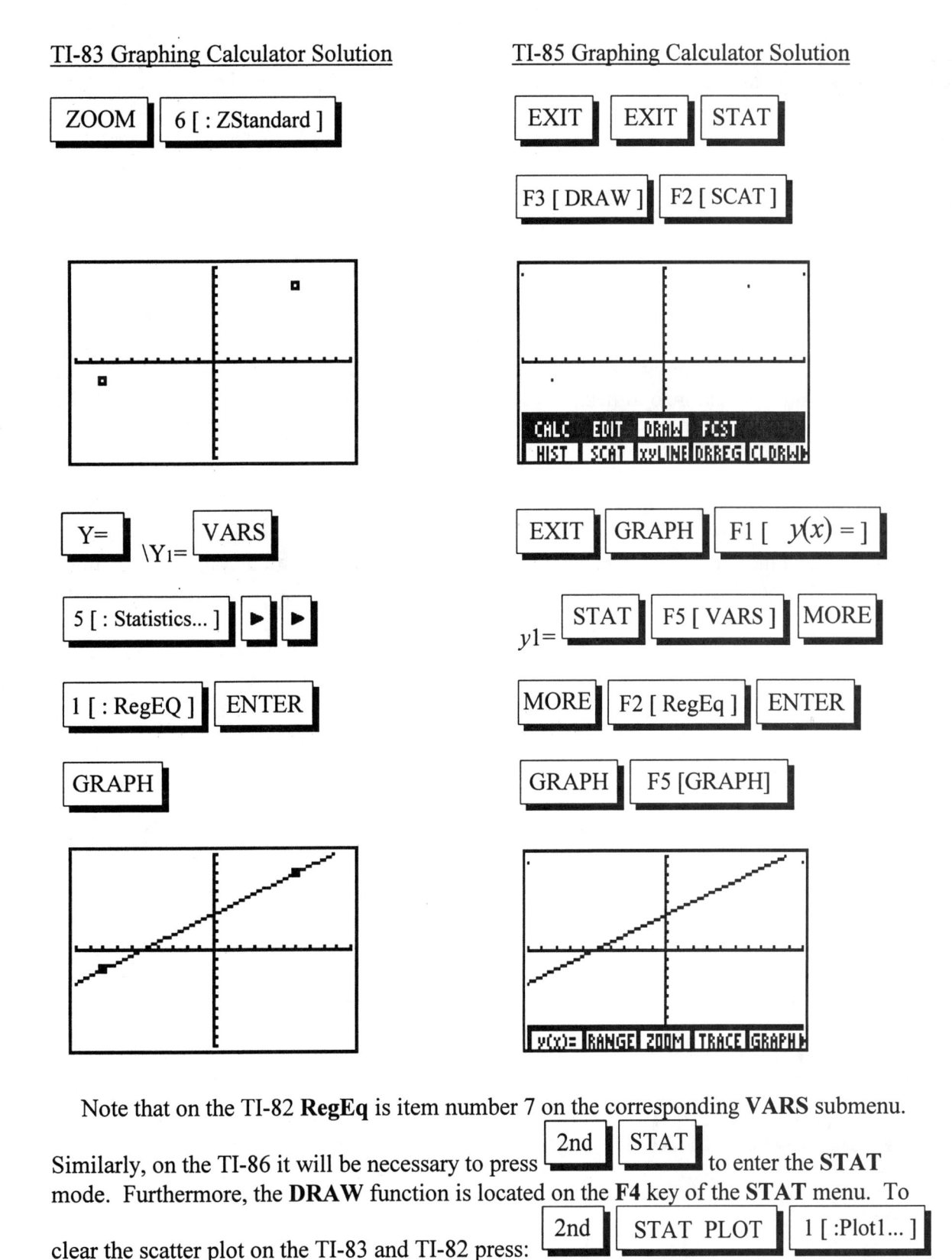

| EXIT | EXIT | STAT |

| F3 [DRAW] | F2 [SCAT] |

| EXIT | GRAPH | F1 [y(x) =] |

| y1= | STAT | F5 [VARS] | MORE |

| MORE | F2 [RegEq] | ENTER |

| GRAPH | F5 [GRAPH] |

Note that on the TI-82 **RegEq** is item number 7 on the corresponding **VARS** submenu. Similarly, on the TI-86 it will be necessary to press [2nd] [STAT] to enter the **STAT** mode. Furthermore, the **DRAW** function is located on the **F4** key of the **STAT** menu. To clear the scatter plot on the TI-83 and TI-82 press: [2nd] [STAT PLOT] [1 [:Plot1...]]

LABORATORY 5: DATA ANALYSIS AND STATISTICAL APPROACH TO THE EQUATION OF A LINE

▶ ENTER . On the TI-85 the corresponding keystroke sequence would be: STAT F3 [DRAW] F5 [CLDRW] . Recall that on the TI-86 the **DRAW** function is located on the **F4** key of the **STAT** menu and the **CLDRW** function is located on the **F2** key of the second **DRAW** submenu.

B. Finding the Equation of a Line Given a Point on the Line and the Slope

Suppose that instead of being asked to determine the equation of the line through two given points, we are asked to determine the equation of a line through a given point, with a particular slope. We can use the given point and the slope to determine a second point on the line and then follow the same procedure outlined above. In general, if we are given a point, (x_1, y_1) and a slope $m = \dfrac{\Delta y}{\Delta x}$, we can construct a second point $(x_2, y_2) = (x_1 + \Delta x, y_1 + \Delta y)$.

Problem 2: Find and graph the equation of the line that passes through the point (-2, -3) with slope equal to -3. Find the value of y when $x = 1$. Remember to clear all graphs and drawings before starting this problem.

First we construct the point $(x_2, y_2) = (-2 + 1, -3 + (-3)) = (-1, -6)$.

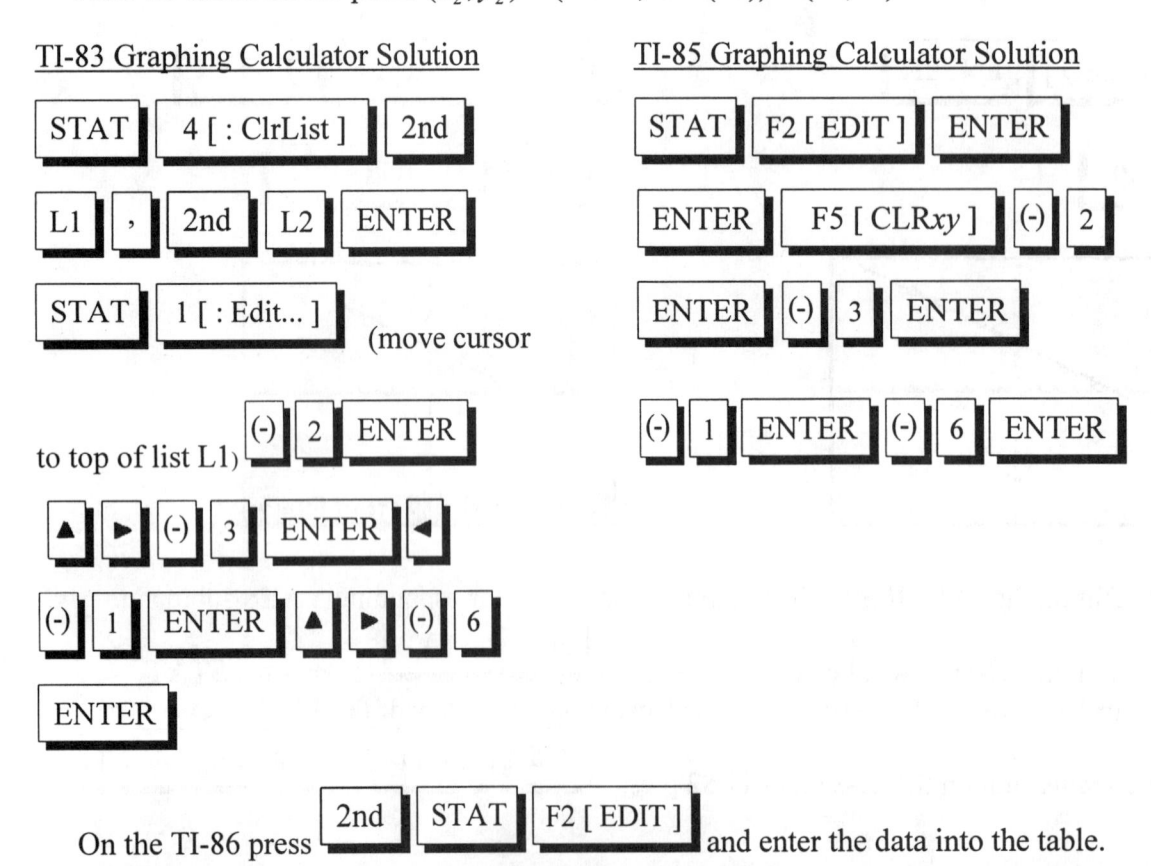

TI-83 Graphing Calculator Solution

STAT 4 [: ClrList] 2nd

L1 , 2nd L2 ENTER

STAT 1 [: Edit...]
(move cursor to top of list L1) (-) 2 ENTER

▲ ▶ (-) 3 ENTER ◀

(-) 1 ENTER ▲ ▶ (-) 6

ENTER

TI-85 Graphing Calculator Solution

STAT F2 [EDIT] ENTER

ENTER F5 [CLRxy] (-) 2

ENTER (-) 3 ENTER

(-) 1 ENTER (-) 6 ENTER

On the TI-86 press 2nd STAT F2 [EDIT] and enter the data into the table.

LABORATORY 5: DATA ANALYSIS AND STATISTICAL APPROACH TO THE EQUATION OF A LINE

TI-83 Graphing Calculator Solution

STAT ▶ 4 [: LinReg(ax+b)]

ENTER

TI-85 Graphing Calculator Solution

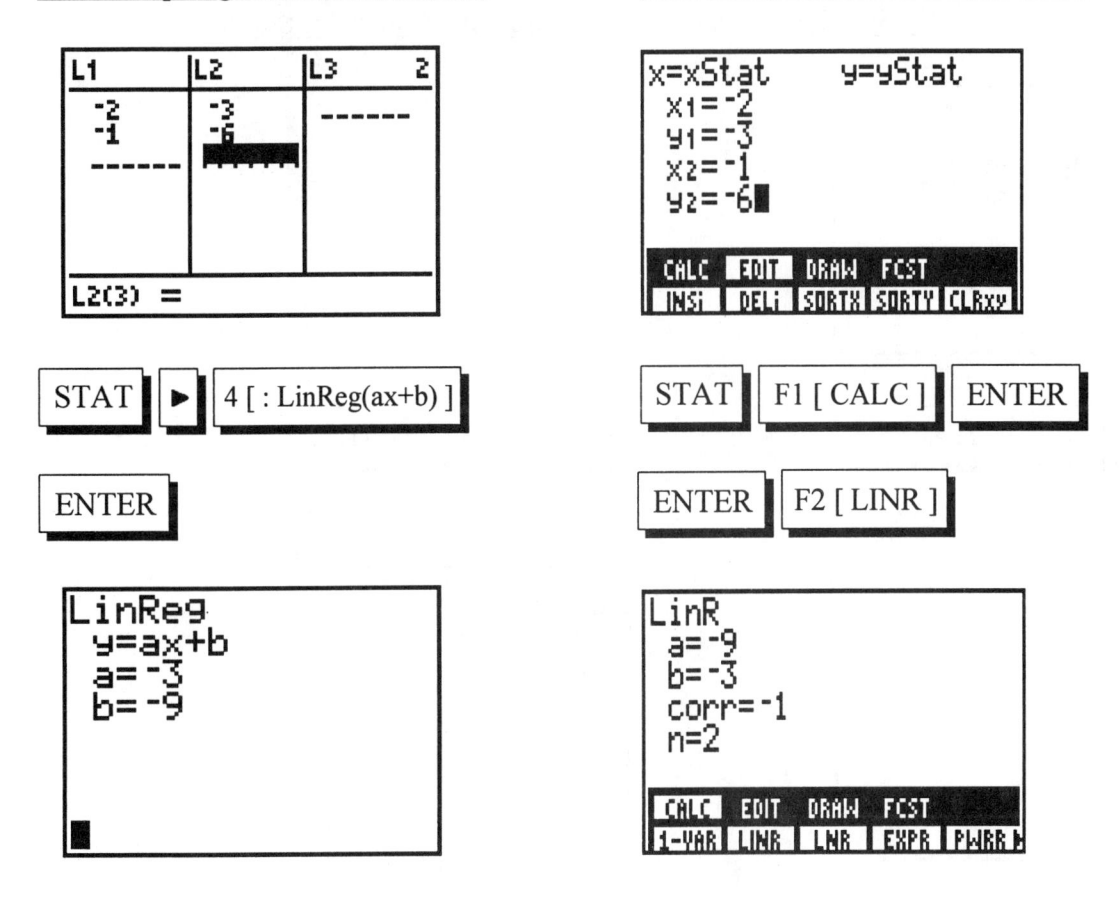

STAT F1 [CALC] ENTER

ENTER F2 [LINR]

Recall that the linear regression feature described above is item number 5 on the corresponding TI-82 menu. On the TI-86, the **LinR** feature is located on the **F3** key of the corresponding **STAT** submenu. Thus, the equation of the regression line can be written as $y1 = -3x - 9$. We will now draw the scatter plot and the equation of the regression line on the standard axes.

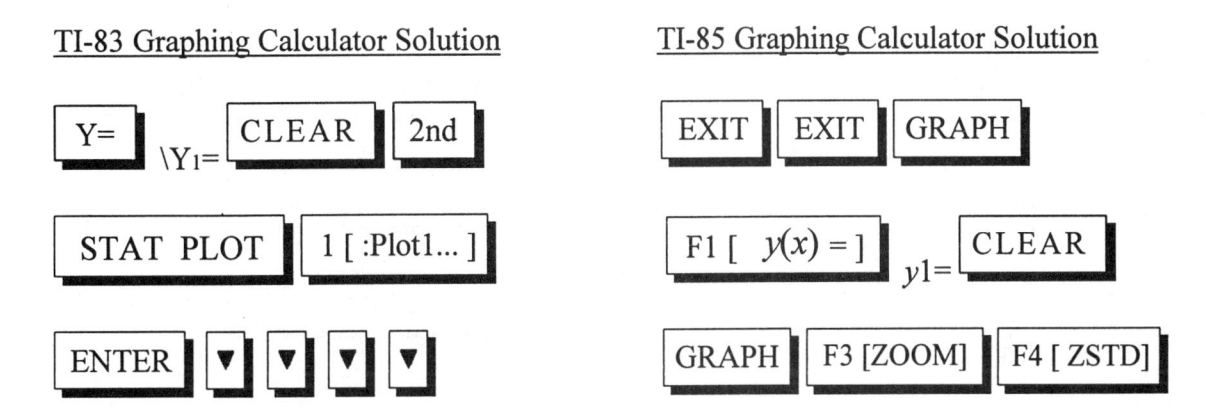

TI-83 Graphing Calculator Solution

Y= \Y₁= CLEAR 2nd

STAT PLOT 1 [:Plot1...]

ENTER ▼ ▼ ▼ ▼

TI-85 Graphing Calculator Solution

EXIT EXIT GRAPH

F1 [$y(x)$ =] y1= CLEAR

GRAPH F3 [ZOOM] F4 [ZSTD]

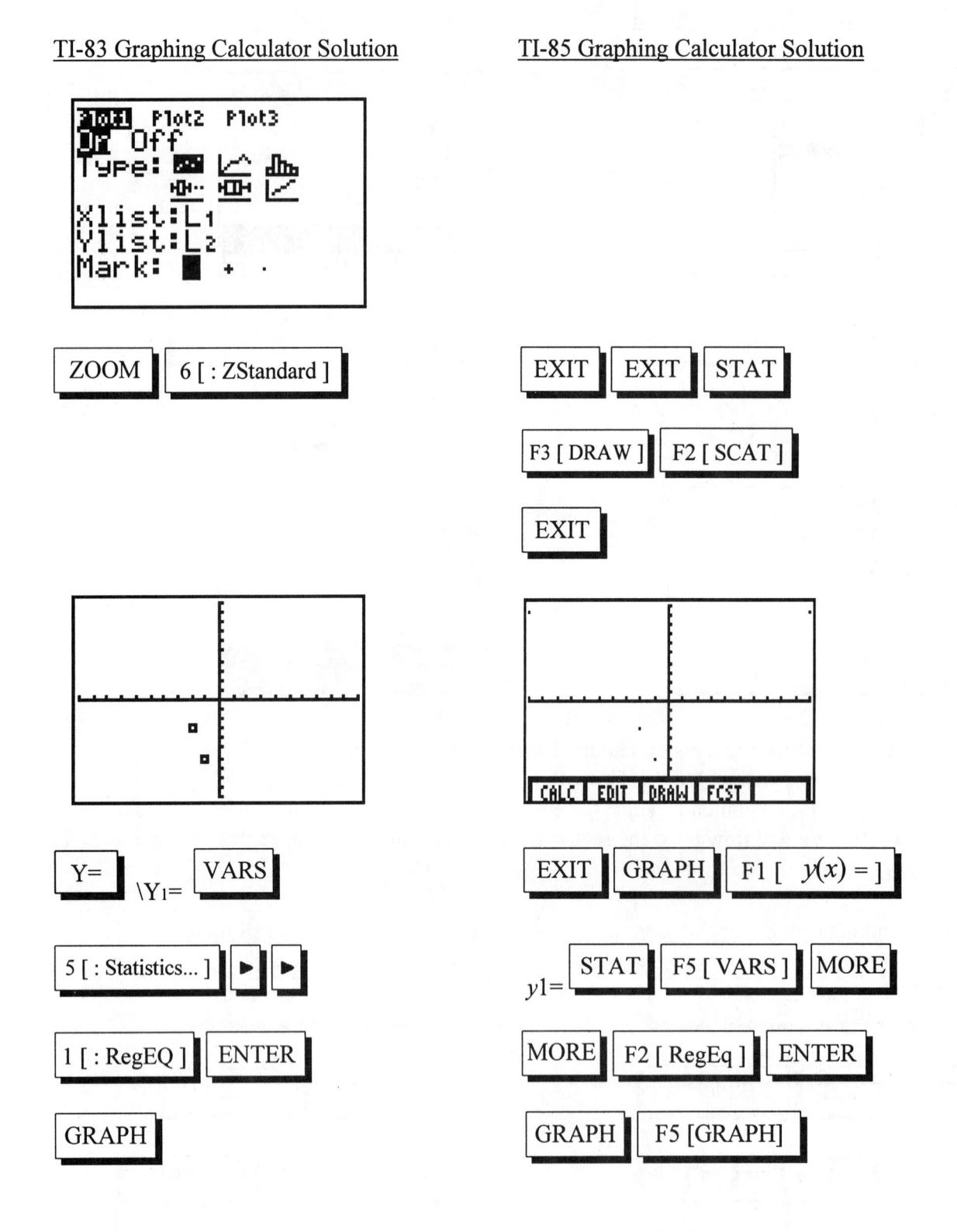

LABORATORY 5: DATA ANALYSIS AND STATISTICAL APPROACH TO THE EQUATION OF A LINE

TI-83 Graphing Calculator Solution

```
Plot1  Plot2  Plot3
On Off
Type:
Xlist: L1
Ylist: L2
Mark:     +  ·
```

[ZOOM] [6 [: ZStandard]]

[Y=] \Y₁= [VARS]

[5 [: Statistics...]] [▶] [▶]

[1 [: RegEQ]] [ENTER]

[GRAPH]

TI-85 Graphing Calculator Solution

[EXIT] [EXIT] [STAT]

[F3 [DRAW]] [F2 [SCAT]]

[EXIT]

[EXIT] [GRAPH] [F1 [$y(x)$ =]]

$y1=$ [STAT] [F5 [VARS]] [MORE]

[MORE] [F2 [RegEq]] [ENTER]

[GRAPH] [F5 [GRAPH]]

LABORATORY 5: DATA ANALYSIS AND STATISTICAL APPROACH TO THE EQUATION OF A LINE

TI-83 Graphing Calculator Solution

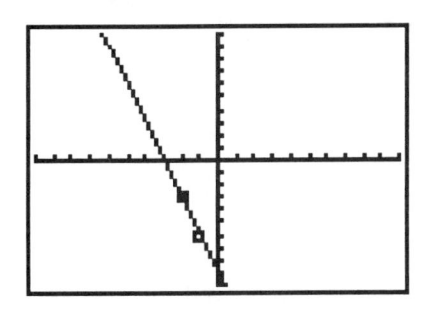

TI-85 Graphing Calculator Solution

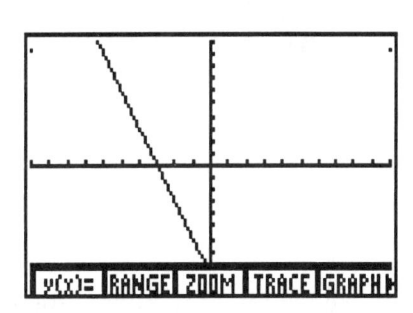

Recall that on the TI-82 the regression equation is item number seven on the given menu. Therefore, the appropriate adjustment will need to be made in the keystroke sequence. On the TI-86, the **DRAW** function is located on the **F4** key of the **STAT** menu. We will now find the value of y when $x = 1$.

TI-83 Graphing Calculator Solution

[2nd] [CALC] [1 [: value]]

X= [1] [ENTER]

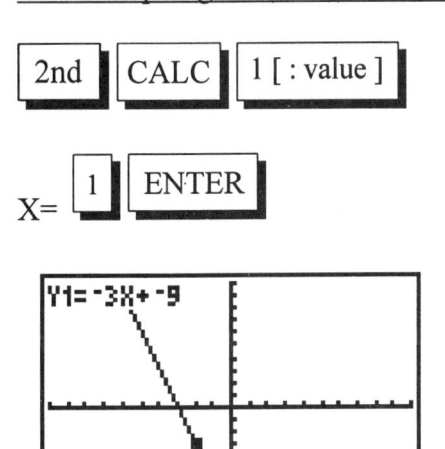

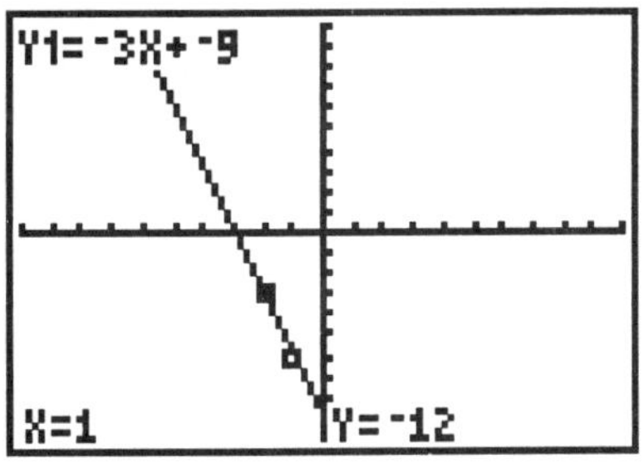

TI-85 Graphing Calculator Solution

[GRAPH] [MORE] [MORE]

[F1 [EVAL]] Eval $x =$ [1] [ENTER]

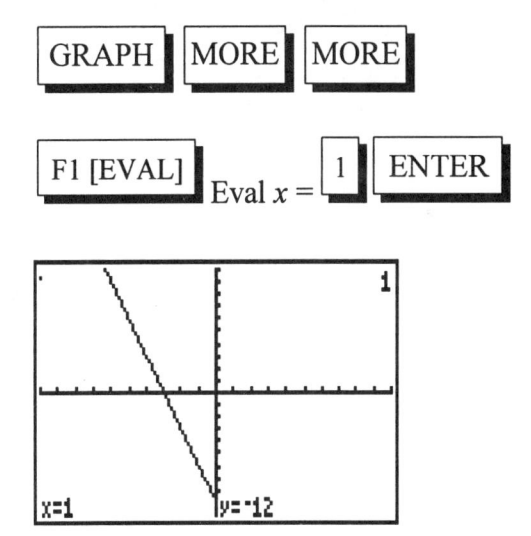

Therefore, the value of y when $x = 1$ is -12. This can also be accomplished on the TI-85 and TI-86 by using the **forecast** option on the **STAT** menu. The keystroke sequence on the TI-85 is: [STAT] [F4 [FCST]] $x=$ [1] [ENTER] [F5 [SOLVE]]. The **forecast** option is located on the **F1** key of the second submenu of the **STAT** menu on the TI-86. The **forecast** option is particularly useful for determining the value of x for a particular value of y.

When you are finished drawing scatter plots on the TI-83 and TI-82 you will have to turn the plots off by pressing: [2nd] [STAT PLOT] [1 [:Plot1...]] [▶] [ENTER].

57

LABORATORY 5: DATA ANALYSIS AND STATISTICAL APPROACH TO THE EQUATION OF A LINE

EXERCISES

1. Find and graph the equation of the line through the points (-2,5) and (8,1).

2. Find and graph the equation of the line through the point (2,4) with slope equal to -1.

3. Find and graph the equation of the line parallel to the line $y = -2x - \frac{5}{3}$ and passing through the point (3,2).

4. Find and graph the equation of the line perpendicular to the line $2x + 5y = 7$ and passing through the point (-1,-4).

5. Find and graph the equation of the line passing through the origin with slope equal to 2. Forecast the value of y when $x = 2$.

LABORATORY 6: SYSTEMS OF LINEAR EQUATIONS

Purpose:

The purpose of this laboratory is to study systems of linear equations as well as features on the TI calculators which are useful in determining their solution. Graphing and numerical approaches will be explored for systems of linear equations in two unknowns. The solution of systems of linear equations in three or more unknowns will be approached via numerical algorithms.

Analytic Approach:

Systems of linear equations are useful in analyzing many real world applications. For example, in business we know that there is an important relationship between cost, profit, and revenue. Profit is defined as revenue minus cost ($P = R - C$). The break-even point is the point at which profit is equal to zero. Equivalently, the break-even point occurs at the point at which revenue is equal to cost. If the revenue function and the cost function can be described as linear equations in two unknowns we may find the solution of this system of equations either graphically or algebraically. A solution to this system of equations would correspond to a point at which the equations of the lines describing the revenue function and the cost function intersect.

Similarly, systems of linear equations in three or more unknowns are often necessary to describe a particular scenario. For example, consider a movie theater that sells three types of tickets: children's, adults', and seniors'. The manager of the theater may find it useful to analyze the pricing of these tickets with respect to attendance. This type of analysis would involve a system of equations in three unknowns.

The examples which follow will provide useful insight into the solution of systems of linear equations. Algebraic solutions of such systems of equations can often be quite tedious when done by hand. The TI calculators offer several solution options which allow the focus in a problem to be placed on the analysis of the system and the relationship between the variables rather than on the manipulation of the specific numbers involved. This redirection of the focus is an important advantage of the calculator approach outlined below.

A. Graphing Approach to the Solution of Systems of Linear Equations in Two Unknowns

Consider the graph of a system of two linear equations in two unknowns. Each of the linear equations could be graphed on the same set of axes. A solution to this system would correspond to a point that the two lines have in common, or equivalently, a point of intersection.

There are three possible scenarios that we must consider. The scenario of particular interest is the one in which the two lines intersect in a single point. The other two scenarios involve special circumstances. The first of these scenarios is the one in which the two lines are parallel;

LABORATORY 6: SYSTEMS OF LINEAR EQUATIONS

in which case they do not intersect, and there are no solutions. This type of system of equations is referred to as *inconsistent*. The second of these scenarios occurs when the graphs of the two lines are the same. This occurs when the two equations are equivalent. In this case there are infinitely many solutions. Such a system of equations is referred to as *dependent*.

Problem 1: Graph the following system of equations on the standard axes and solve:

$$6x + 2y = 4$$
$$8x - 4y = 12$$

Remember to clear all graphs and to isolate the variable y in each line before starting the problem.

TI-83 Graphing Calculator Solution

$\boxed{Y=}$ $\backslash Y_1 =$ $\boxed{CLEAR}$ $\boxed{(}$ $\boxed{4}$ $\boxed{-}$

$\boxed{6}$ $\boxed{X,T,\theta,n}$ $\boxed{)}$ $\boxed{\div}$ $\boxed{2}$

$\boxed{ENTER}$ $\backslash Y_2=$ $\boxed{CLEAR}$ $\boxed{(}$ $\boxed{1}$

$\boxed{2}$ $\boxed{-}$ $\boxed{8}$ $\boxed{X,T,\theta,n}$ $\boxed{)}$ $\boxed{\div}$ $\boxed{(-)}$

$\boxed{4}$ $\boxed{ENTER}$ $\boxed{ZOOM}$

$\boxed{6 \ [\ : ZStandard \ ]}$

TI-85 Graphing Calculator Solution

$\boxed{GRAPH}$ $\boxed{F1\ [\ y(x) = \ ]}$ $y1=$

$\boxed{CLEAR}$ $\boxed{(}$ $\boxed{4}$ $\boxed{-}$ $\boxed{6}$

$\boxed{x\text{-VAR}}$ $\boxed{)}$ $\boxed{\div}$ $\boxed{2}$ $\boxed{ENTER}$

$y2=$ $\boxed{CLEAR}$ $\boxed{(}$ $\boxed{1}$ $\boxed{2}$ $\boxed{-}$ $\boxed{8}$

$\boxed{x\text{-VAR}}$ $\boxed{)}$ $\boxed{\div}$ $\boxed{(-)}$ $\boxed{4}$ $\boxed{ENTER}$

$\boxed{GRAPH}$ $\boxed{F3\ [ZOOM]}$ $\boxed{F4\ [\ ZSTD]}$

LABORATORY 6: SYSTEMS OF LINEAR EQUATIONS

The calculator will display the graph of both lines on the same axes. Now locate the point of intersection. Note that each curve must be selected by pressing ENTER.

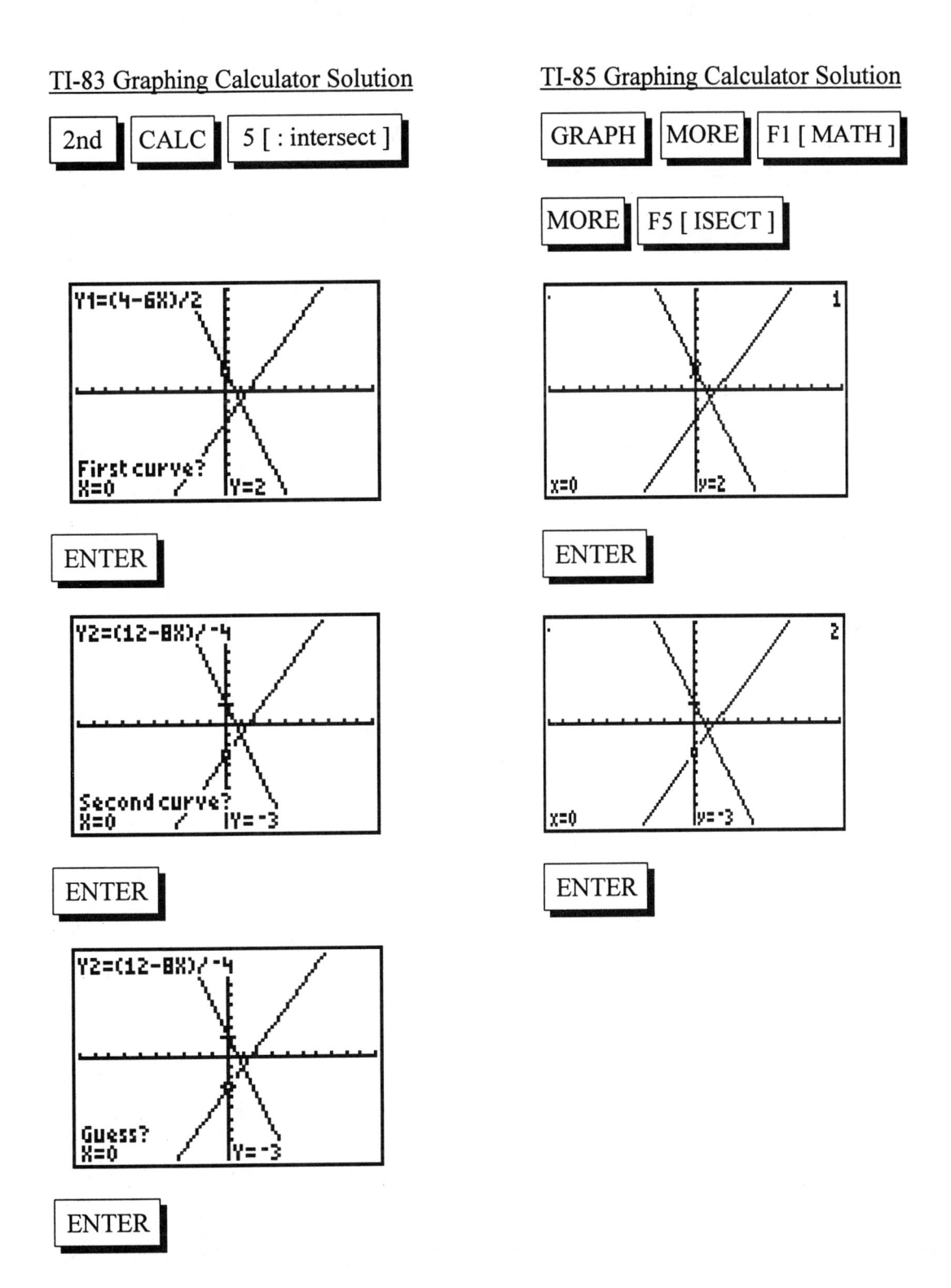

TI-83 Graphing Calculator Solution

2nd CALC 5 [: intersect]

ENTER

ENTER

ENTER

TI-85 Graphing Calculator Solution

GRAPH MORE F1 [MATH]

MORE F5 [ISECT]

ENTER

ENTER

61

LABORATORY 6: SYSTEMS OF LINEAR EQUATIONS

Note that on the TI-86 the **ISECT** function is located on the **F3** key on the corresponding submenu. It requires the first curve, second curve, and a guess to be entered.

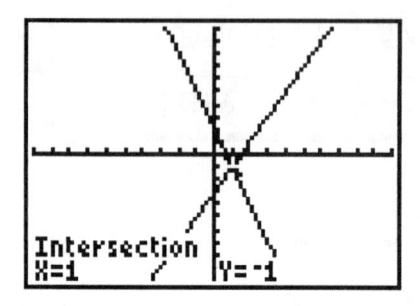

TI-83 Graphing Calculator Solution   TI-85 Graphing Calculator Solution

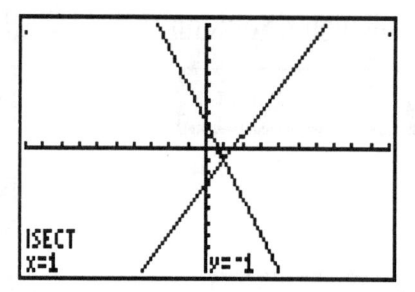

We see that the solution to this system of two linear equations in two unknowns is the unique point $(x, y) = (1, -1)$.

B. Numerical Approach to Systems of Linear Equations in Two or More Unknowns

Consider a system of n linear equations in n unknowns:

$$a_{11}x_1 + a_{12}x_2 + a_{13}x_3 + \ldots + a_{1n}x_n = b_1$$
$$a_{21}x_1 + a_{22}x_2 + a_{23}x_3 + \ldots + a_{2n}x_n = b_2$$
$$\ldots\ldots\ldots\ldots\ldots\ldots\ldots\ldots\ldots\ldots\ldots\ldots\ldots\ldots$$
$$a_{n1}x_1 + a_{n2}x_2 + a_{n3}x_3 + \ldots + a_{nn}x_n = b_n$$

Form the n x n coefficient matrix A and the n x 1 column vectors x and b. The system may be described by the matrix equation $Ax = b$. If the coefficient matrix A has a unique inverse, A^{-1}, (such that $AA^{-1} = A^{-1}A = I$, where I is the n x n identity matrix with ones on the diagonal, a_{kk}, for $k = 1, 2, \ldots, n$, and zeroes elsewhere), then the system has $x = A^{-1}b$ as a solution. A matrix that does not have such an inverse is said to be *singular*. The **MATRX** features of the TI calculators may be used to solve systems of n linear equations in n unknowns.

Problem 2: Solve the following system of linear equations:

$$2x + 4y = 5$$
$$6x + 12y = 7$$

TI-83 Graphing Calculator Solution   TI-85 Graphing Calculator Solution

| MATRX | ▶ | ▶ | | 2nd | MATRX | F2 [EDIT] |

LABORATORY 6: SYSTEMS OF LINEAR EQUATIONS

Use the **EDIT** option to enter a 2 x 2 matrix *A* and a 2 x 1 column vector *B*.

TI-83 Graphing Calculator Solution

[1 [:A]] [2] [ENTER] [2]

[ENTER]

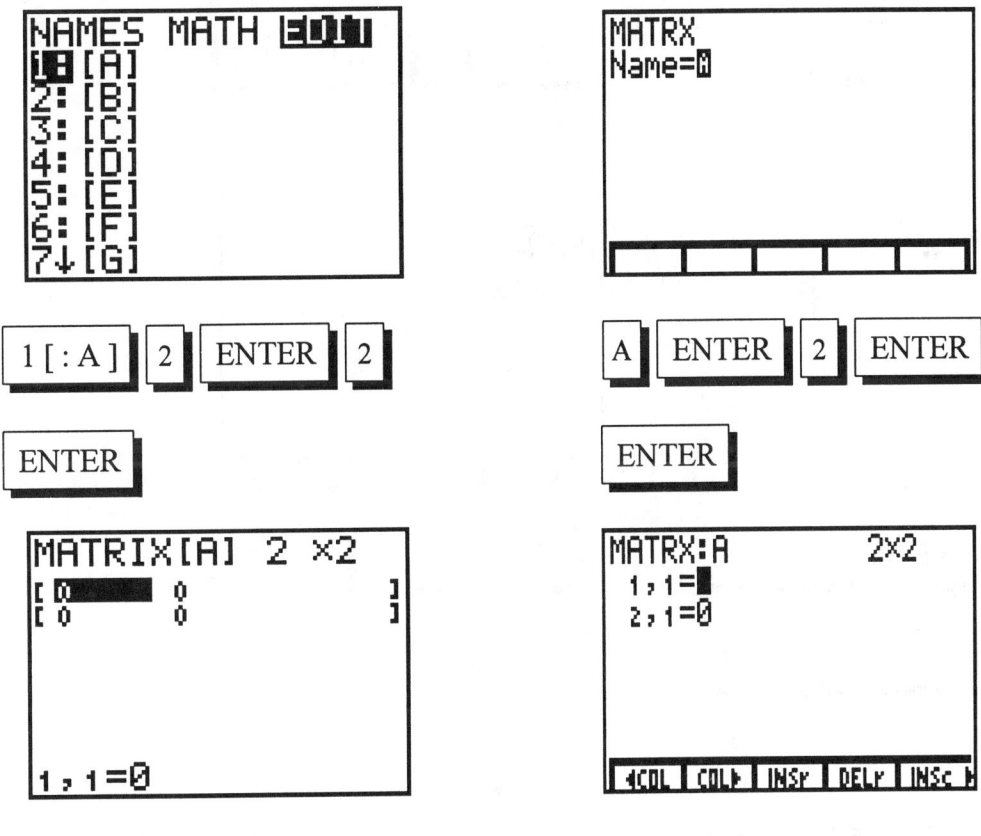

Enter numbers row by row in display.

[2] [ENTER] [4] [ENTER]

[6] [ENTER] [1] [2] [ENTER]

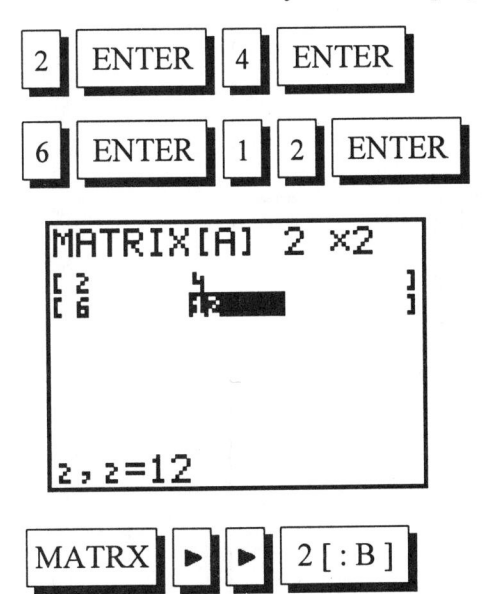

[MATRX] [▶] [▶] [2 [:B]]

TI-85 Graphing Calculator Solution

[A] [ENTER] [2] [ENTER] [2]

[ENTER]

Enter numbers row by row.

[2] [ENTER] [4] [ENTER]

[6] [ENTER] [1] [2] [ENTER]

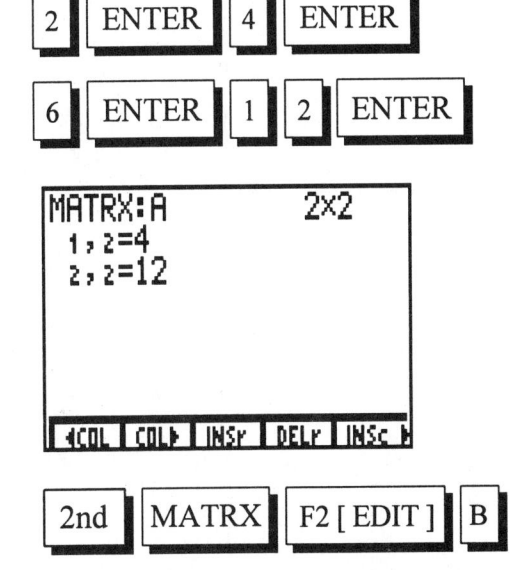

[2nd] [MATRX] [F2 [EDIT]] [B]

LABORATORY 6: SYSTEMS OF LINEAR EQUATIONS

TI-83 Graphing Calculator Solution

| 2 | ENTER | 1 | ENTER |

| 5 | ENTER | 7 | ENTER |

```
MATRIX[B] 2 ×1
[ 5       ]
[ ▓       ]



2,1=7
```

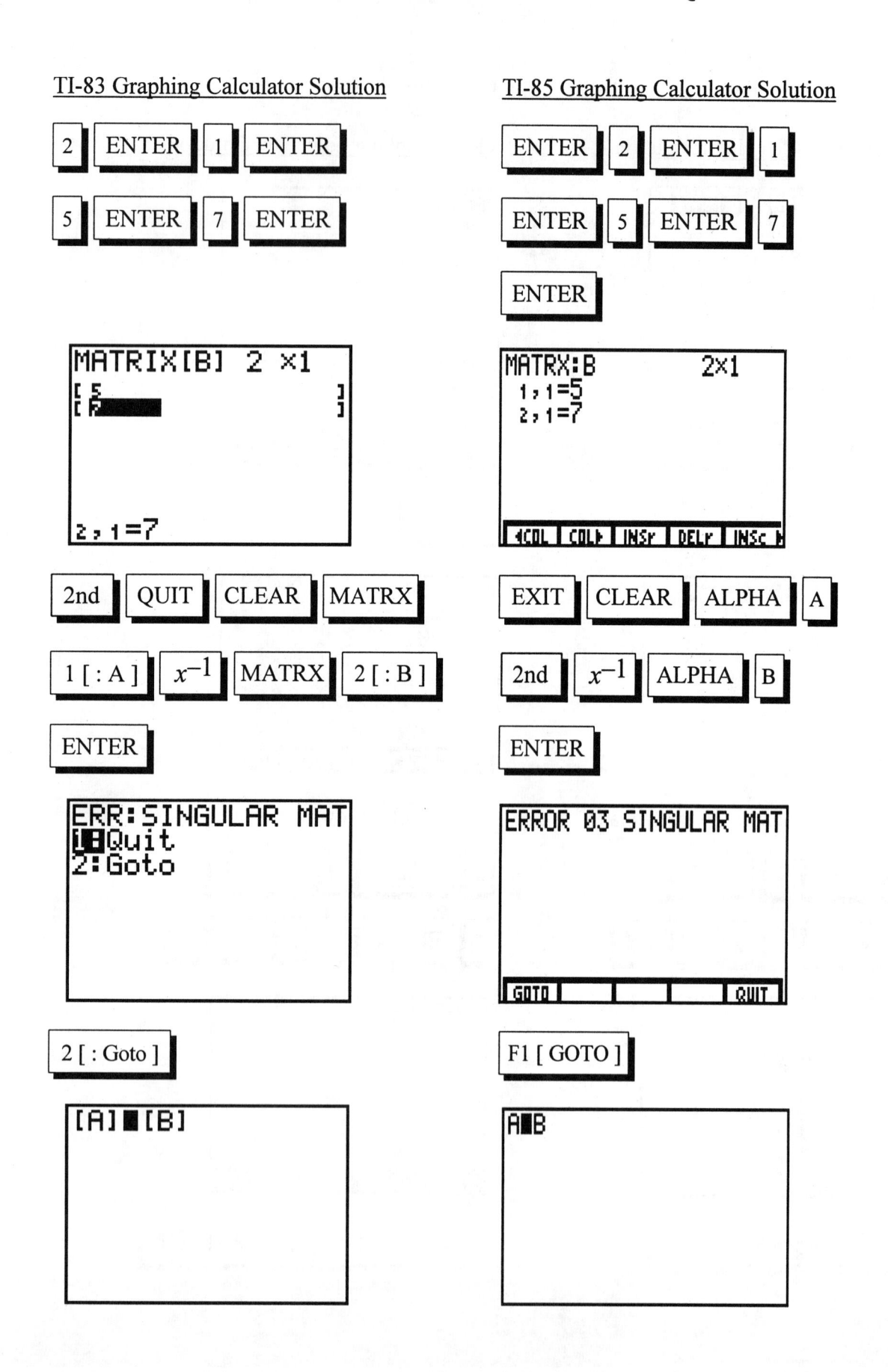

| 2nd | QUIT | CLEAR | MATRX |

| 1 [:A] | x^{-1} | MATRX | 2 [:B] |

| ENTER |

```
ERR:SINGULAR MAT
1:Quit
2:Goto
```

| 2 [:Goto] |

```
[A]▓[B]
```

TI-85 Graphing Calculator Solution

| ENTER | 2 | ENTER | 1 |

| ENTER | 5 | ENTER | 7 |

| ENTER |

```
MATRX:B           2×1
 1,1=5
 2,1=7



◄COL COL► INSr DELr INSc►
```

| EXIT | CLEAR | ALPHA | A |

| 2nd | x^{-1} | ALPHA | B |

| ENTER |

```
ERROR 03 SINGULAR MAT




GOTO              QUIT
```

| F1 [GOTO] |

```
A▓B
```

64

LABORATORY 6: SYSTEMS OF LINEAR EQUATIONS

The error message indicates that the matrix A is singular. Follow the procedure outlined in Problem 1 above to graph this system of equations. The lines in this system are parallel. Therefore, the system is inconsistent and there is no solution. What is the slope of each of the lines? Note that the error message on the TI-82 shows "Goto" as item number one and "Quit" as item number two on the menu and the appropriate keystroke adjustments must be made. Furthermore, matrix editing on the TI-86 follows the format for the TI-83 and TI-82 rather than the TI-85.

Problem 3: Solve the following system of linear equations:

$$2x + 3y = 2$$
$$4x + 6y = 4$$

TI-83 Graphing Calculator Solution

| MATRX | ▶ | ▶ |

| NAMES MATH EDIT |
| 1: [A] 2×2 |
| 2: [B] 2×1 |
| 3: [C] |
| 4: [D] |
| 5: [E] |
| 6: [F] |
| 7↓[G] |

| 1 [: A] | 2 | ENTER | 2 |

| ENTER |

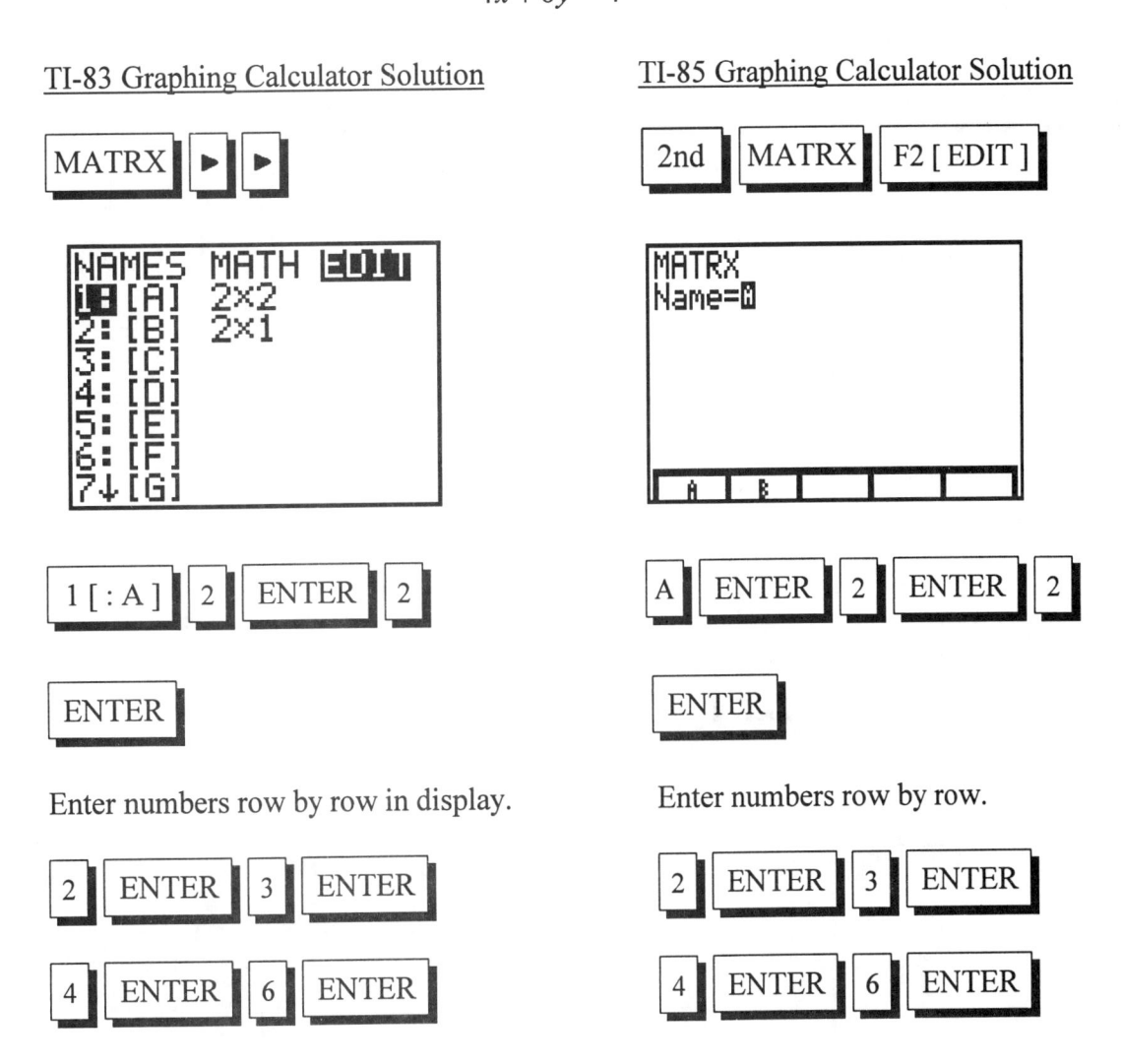

Enter numbers row by row in display.

| 2 | ENTER | 3 | ENTER |

| 4 | ENTER | 6 | ENTER |

TI-85 Graphing Calculator Solution

| 2nd | MATRX | F2 [EDIT] |

| MATRX |
| Name=█ |

| A | ENTER | 2 | ENTER | 2 |

| ENTER |

Enter numbers row by row.

| 2 | ENTER | 3 | ENTER |

| 4 | ENTER | 6 | ENTER |

LABORATORY 6: SYSTEMS OF LINEAR EQUATIONS

TI-83 Graphing Calculator Solution

```
MATRIX[A] 2 ×2
[ 2   3   ]
[ 4   6   ]


2,2=6
```

| MATRX | ▶ | ▶ | 2 [:B] |

| 2 | ENTER | 1 | ENTER |

| 2 | ENTER | 4 | ENTER |

```
MATRIX[B] 2 ×1
[ 2 ]
[ 4 ]


2,1=4
```

| 2nd | QUIT | CLEAR | MATRX |

| 1 [:A] | x^{-1} | MATRX | 2 [:B] |

| ENTER |

```
ERR:SINGULAR MAT
1:Quit
2:Goto
```

TI-85 Graphing Calculator Solution

```
MATRX:A        2×2
 1,2=3
 2,2=6




◀COL  COL▶  INS▶  DEL▶  INSc▶
```

| 2nd | MATRX | F2 [EDIT] | B |

| ENTER | 2 | ENTER | 1 |

| ENTER | 2 | ENTER | 4 |

| ENTER |

```
MATRX:B        2×1
 1,1=2
 2,1=4




◀COL  COL▶  INS▶  DEL▶  INSc▶
```

| EXIT | CLEAR | ALPHA | A |

| 2nd | x^{-1} | ALPHA | B |

| ENTER |

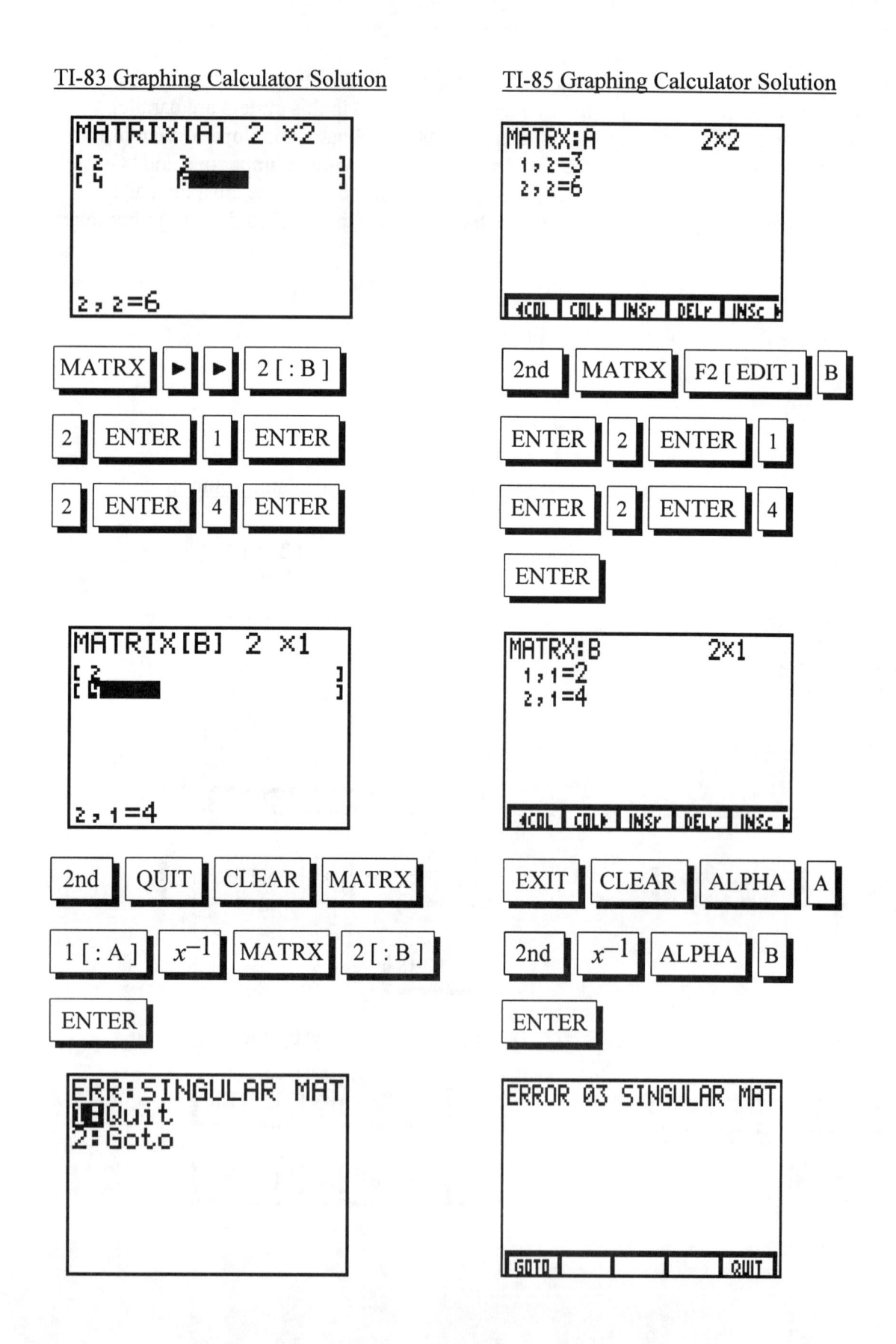

```
ERROR 03 SINGULAR MAT




GOTO                    QUIT
```

LABORATORY 6: SYSTEMS OF LINEAR EQUATIONS

TI-83 Graphing Calculator Solution

2 [: Goto]

[A]■[B]

TI-85 Graphing Calculator Solution

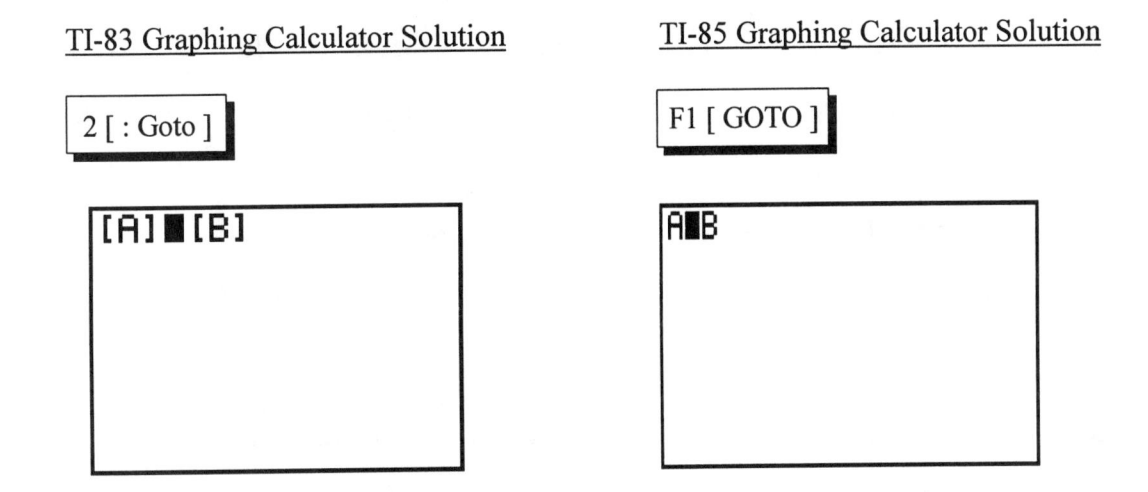

F1 [GOTO]

A■B

The error message indicates that the matrix A is singular. Recall that the error message on the TI-82 has the options appearing in the reverse order. Verify that the lines in this system are the same and that the system is therefore dependent. How many solutions to this system are there? List a few of the solutions.

Problem 4: Solve the following system of three linear equations in three unknowns:

$$2x + 2y + 4z = 0$$
$$4x - 2y - 2z = 2$$
$$2x + 4y + 6z = 2$$

TI-83 Graphing Calculator Solution

MATRX ▶ ▶

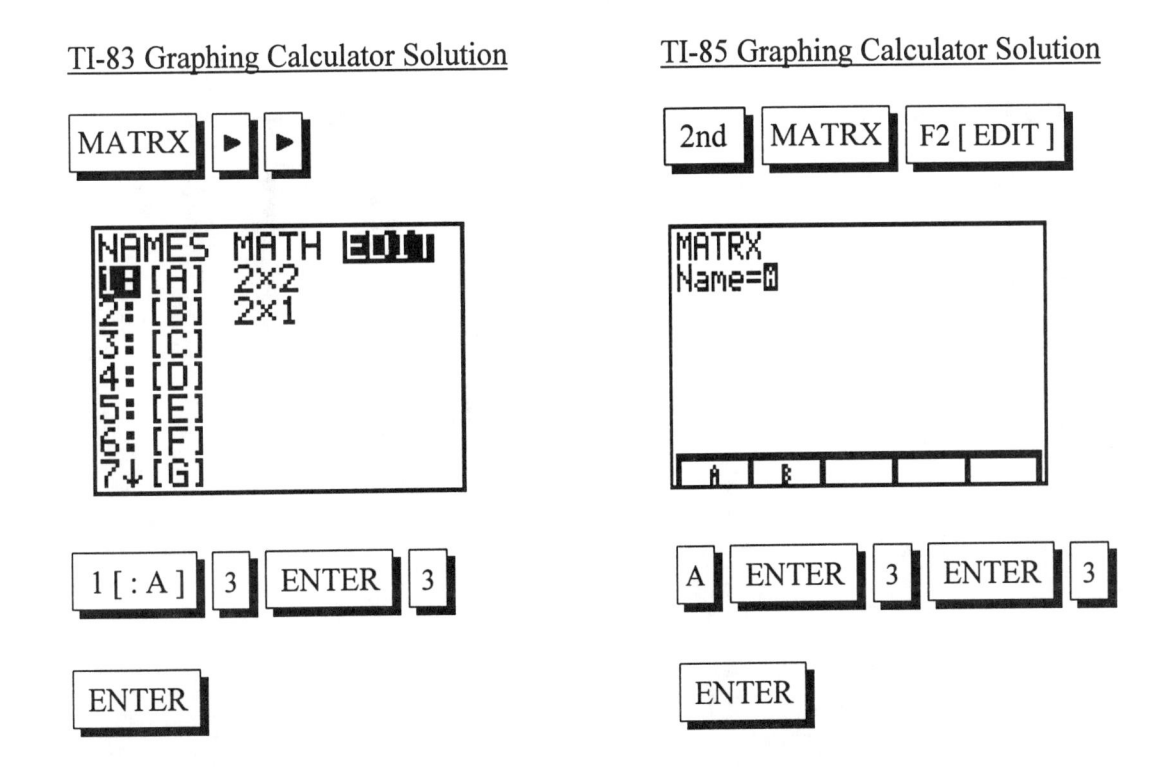

```
NAMES  MATH  EDIT
1: [A]   2×2
2: [B]   2×1
3: [C]
4: [D]
5: [E]
6: [F]
7↓[G]
```

1 [: A]   3   ENTER   3

ENTER

TI-85 Graphing Calculator Solution

2nd   MATRX   F2 [EDIT]

```
MATRX
Name=■
```

A   ENTER   3   ENTER   3

ENTER

67

LABORATORY 6: SYSTEMS OF LINEAR EQUATIONS

TI-83 Graphing Calculator Solution

Enter numbers row by row in display.

[2] [ENTER] [2] [ENTER] [4]

[ENTER] [4] [ENTER] [(-)] [2]

[ENTER] [(-)] [2] [ENTER] [2]

[ENTER] [4] [ENTER] [6]

[ENTER]

```
MATRIX[A]  3 x3
[ 2    2    4  ]
[ 4   -2   -2  ]
[ 2    4    6  ]

3,3=6
```

[MATRX] [►] [►] [2 [:B]]

[3] [ENTER] [1] [ENTER]

[0] [ENTER] [2] [ENTER]

[2] [ENTER]

TI-85 Graphing Calculator Solution

Enter numbers row by row.

[2] [ENTER] [2] [ENTER] [4]

[ENTER] [4] [ENTER] [(-)] [2]

[ENTER] [(-)] [2] [ENTER] [2]

[ENTER] [4] [ENTER] [6]

[ENTER]

```
MATRX:A        3x3
1,3=4
2,3=-2
3,3=6


◄COL  COL►  INSr  DELr  INSc►
```

[2nd] [MATRX] [F2 [EDIT]] [B]

[ENTER] [3] [ENTER] [1]

[ENTER] [0] [ENTER] [2]

[ENTER] [2] [ENTER]

LABORATORY 6: SYSTEMS OF LINEAR EQUATIONS

TI-83 Graphing Calculator Solution

TI-85 Graphing Calculator Solution

```
MATRIX[B] 3 x1
[ 0         ]
[ 2         ]
[ 2         ]


3,1=2
```

```
MATRX:B        3x1
 1,1=0
 2,1=2
 3,1=2



◂COL  COL▸  INSr  DELr  INSc ▸
```

[2nd] [QUIT] [CLEAR]

[MATRX] [1 [:A]] [x^{-1}]

[MATRX] [2 [:B]] [ENTER]

[EXIT] [CLEAR] [ALPHA] [A]

[2nd] [x^{-1}] [ALPHA] [B]

[ENTER]

```
[A]⁻¹[B]
          [[1  ]
           [3  ]
           [-2]]
■
```

```
A⁻¹B
          [[1  ]
           [3  ]
           [-2]]
```

We see that the solution to this system of three linear equations in three unknowns is $x = 1$, $y = 3$, and $z = -2$. The TI-85 and TI-86 calculators are equipped with a **SIMULT** feature which can be used to solve systems of n equations in n unknowns. This alternate solution on the TI-85 and TI-86 is outlined below. Enter the number of equations in the system first, followed by the system information, row by row. Be sure to enter the n variables in the given row of the A matrix as well as the entry in the corresponding row of the B column vector.

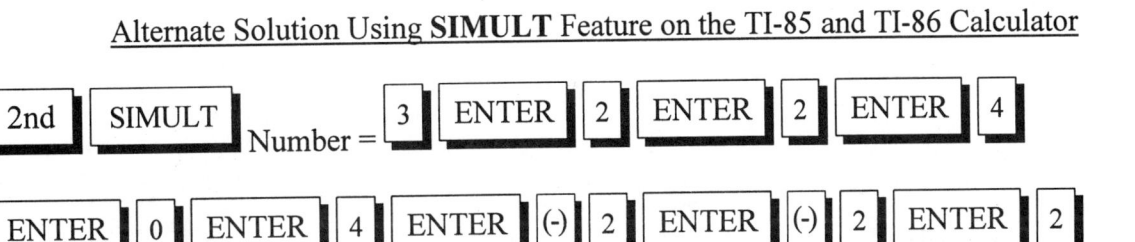

Alternate Solution Using SIMULT Feature on the TI-85 and TI-86 Calculator

[2nd] [SIMULT] Number = [3] [ENTER] [2] [ENTER] [2] [ENTER] [4]

[ENTER] [0] [ENTER] [4] [ENTER] [(-)] [2] [ENTER] [(-)] [2] [ENTER] [2]

LABORATORY 6: SYSTEMS OF LINEAR EQUATIONS

<u>Alternate Solution Using **SIMULT** Feature on the TI-85 and TI-86 Calculator</u>

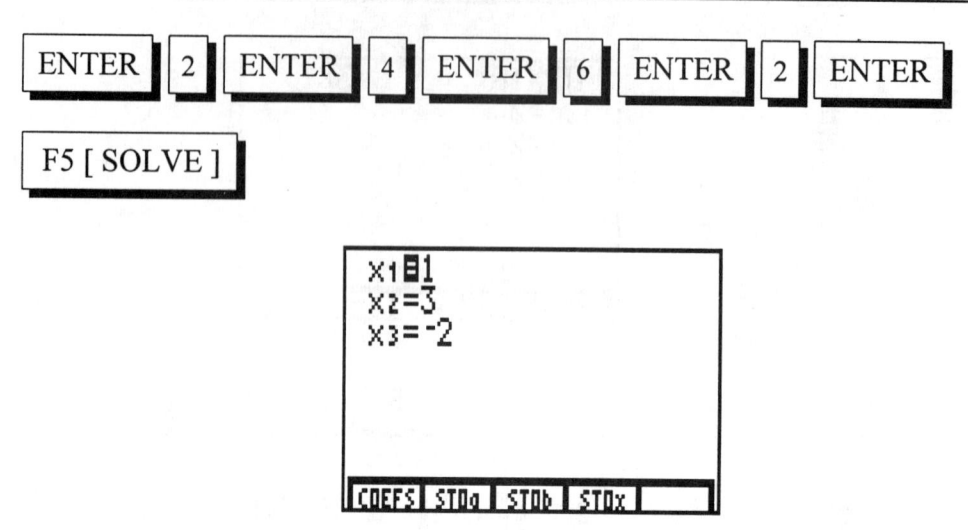

| ENTER | 2 | ENTER | 4 | ENTER | 6 | ENTER | 2 | ENTER |

| F5 [SOLVE] |

```
x1=1
x2=3
x3=-2


COEFS  STOa  STOb  STOx
```

Note that this is equivalent to the solution found via the **MATRX** procedure outlined above. Try Problems 1, 2, and 3 again using the **SIMULT** feature on your TI-85 or TI-86 calculator. Compare and contrast the use of each of these methods.

LABORATORY 6: SYSTEMS OF LINEAR EQUATIONS

EXERCISES

1. Solve the following system of two linear equations in two unknowns by graphing:

$$3x - 4y = 12$$
$$3x + 4y = 36$$

2. Solve the system of equations in Exercise 1 above via **MATRX** methods. Compare your answers.

3. Solve the following system of two linear equations in two unknowns via any method studied in this laboratory:

$$\tfrac{5}{2}x - y = 4$$
$$\tfrac{3}{4}x + \tfrac{1}{4}y = \tfrac{7}{4}$$

4. Solve the following system of three linear equations in three unknowns:

$$3x + 3y + 6z = 0$$
$$4x - 2y - 2z = 2$$
$$3x + 6y + 9z = 3$$

5. Solve the following system of three linear equations in three unknowns:

$$6x + 3y + 9z = 6$$
$$2x - 2y + 4z = -8$$
$$3x + 9y - 3z = 3$$

LABORATORY 7: ADVANCED GRAPHING TECHNIQUES - THE QUADRATIC EQUATION

Purpose:

The purpose of this laboratory is to study the quadratic equation as well as the graph of a parabola. Advanced graphing techniques will be discussed as they pertain to the analysis of the graph of a parabola. Further applications involving polynomials in general will be discussed in Laboratory 9.

Analytic Approach:

A *quadratic equation* is an equation which can be written in the form $ax^2 + bx + c = 0$, where a, b, and c are real numbers and $a \neq 0$. There are many methods which can be used to solve quadratic equations, including factoring and setting each factor equal to zero, completing the square, using the quadratic formula $x = \dfrac{-b \pm \sqrt{b^2 - 4ac}}{2a}$, and locating the x-intercepts on the graph of $y = ax^2 + bx + c$. In this laboratory we will be exploring the graph of $y = ax^2 + bx + c$, where a, b, and c are real numbers and $a \neq 0$. The graph of this equation is the graph of a *parabola*. It can be shown that if $a > 0$, then the graph of the parabola turns up and the *vertex* or turning point is a *relative minimum*. Similarly, if $a < 0$, then the graph of the parabola turns down and the vertex is a *relative maximum*. Furthermore, it can be shown that the x-coordinate of the vertex is $x = \dfrac{-b}{2a}$, while the y-coordinate of the vertex is $y = \dfrac{4ac - b^2}{4a}$. The y-intercept occurs at $y = c$.

A quadratic equation with real coefficients (where the coefficient of x^2 is not equal to zero) has at most two real roots. In fact, we will see from the examples below that such an equation has two real roots if the graph has its vertex above the x-axis and turns down or if it has its vertex below the x-axis and turns up. It will have no real roots if its vertex is below the x-axis and the graph turns down or if it has its vertex above the x-axis and its graph turns up. Finally, the equation will have exactly one real root if its vertex is on the x-axis.

Problem 1: Graph $y = x^2 - 4x + 4$ on the standard axes. Find the x-intercepts, y-intercept, and vertex. Determine whether the vertex is a relative maximum or a relative minimum. Remember to clear all graphs before starting this problem.

TI-83 Graphing Calculator Solution

TI-85 Graphing Calculator Solution

| Y= | \Y1= | CLEAR | X,T,θ,n |

| GRAPH | F1 [y(x) =] | y1=

LABORATORY 7: ADVANCED GRAPHING TECHNIQUES - THE QUADRATIC EQUATION

TI-83 Graphing Calculator Solution

| x^2 | – | 4 | X,T,θ,n | + | 4 |

| ENTER | ZOOM |

| 6 [: ZStandard] |

Find the y-intercept by evaluating $x = 0$.

| 2nd | CALC | 1 [: value] |

X= | 0 | ENTER |

TI-85 Graphing Calculator Solution

| CLEAR | x-VAR | x^2 | – | 4 |

| x-VAR | + | 4 | ENTER |

| GRAPH | F3 [ZOOM] | F4 [ZSTD] |

Find the y-intercept via **YICPT**.

| GRAPH | MORE | F1 [MATH] |

| MORE | F4 [YICPT] | ENTER |

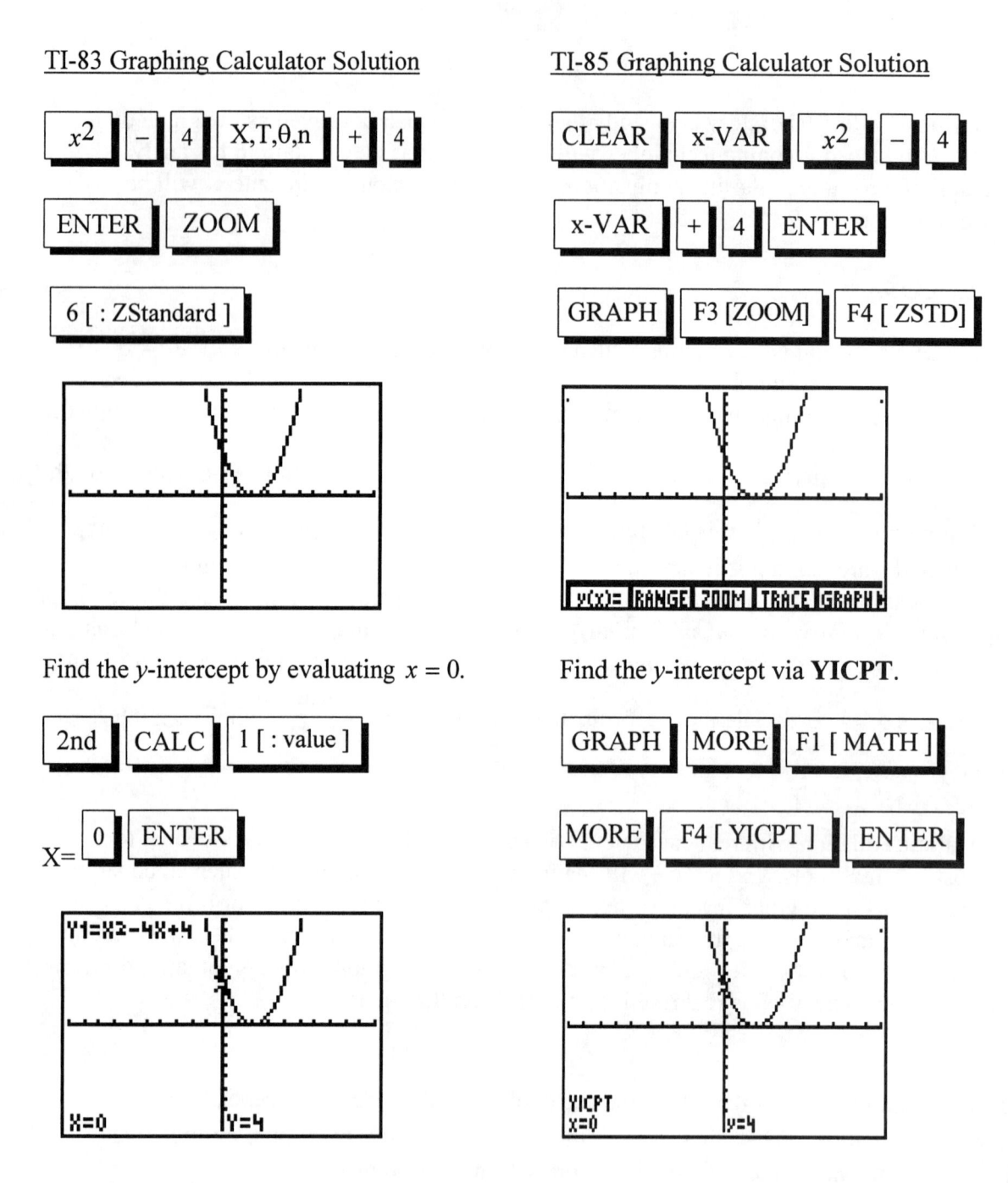

The **YICPT** function can be found on the **F2** key of the corresponding TI-86 menu. The y-intercept is located at the point (0,4). Note that this is consistent with the fact that the y-intercept occurs at $y = c$ and in this case $c = 4$. The root appears to be at the vertex of this parabola. Verify this by using the equations given above to find the location of the vertex, then use the **zero** or **ROOT** feature to find the x-intercept.

LABORATORY 7: ADVANCED GRAPHING TECHNIQUES - THE QUADRATIC EQUATION

TI-83 Graphing Calculator Solution

| 2nd | | CALC | | 2 [: zero] |

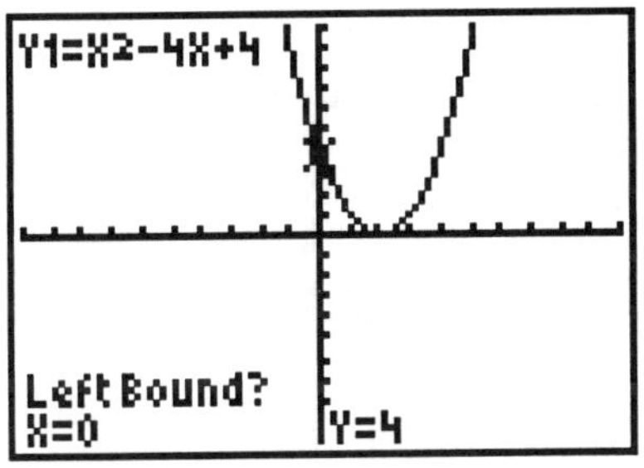

(The cursor is already to the left of the root.)

| ENTER |

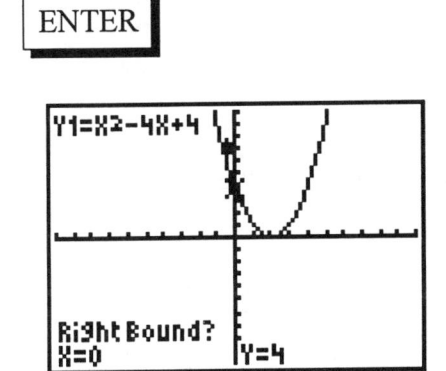

(Move the cursor to the right of the root.)

| ENTER |

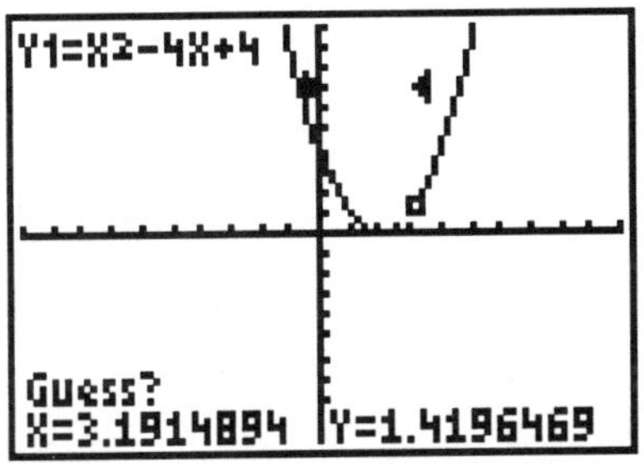

TI-85 Graphing Calculator Solution

| GRAPH | | MORE | | F1 [MATH] |

| F3 [ROOT] |

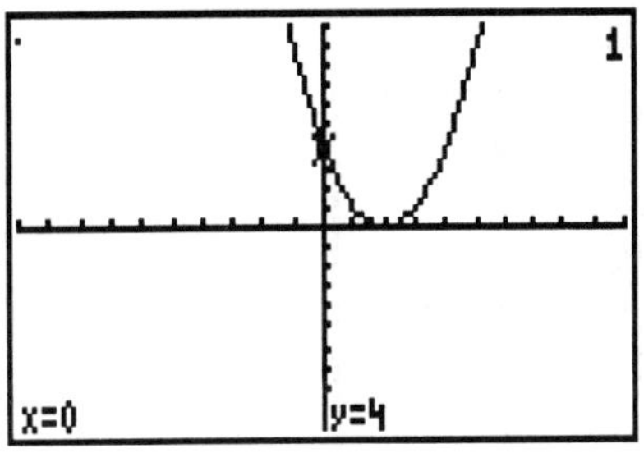

(There is a root in this window.)

LABORATORY 7: ADVANCED GRAPHING TECHNIQUES - THE QUADRATIC EQUATION

The **ROOT** function is located on the **F1** key of the corresponding TI-86 menu. Furthermore, recall that the **ROOT** function on the TI-86 requires entry of a left bound, right bound, and a guess as on the TI-83 and TI-82.

TI-83 Graphing Calculator Solution   TI-85 Graphing Calculator Solution

ENTER   ENTER

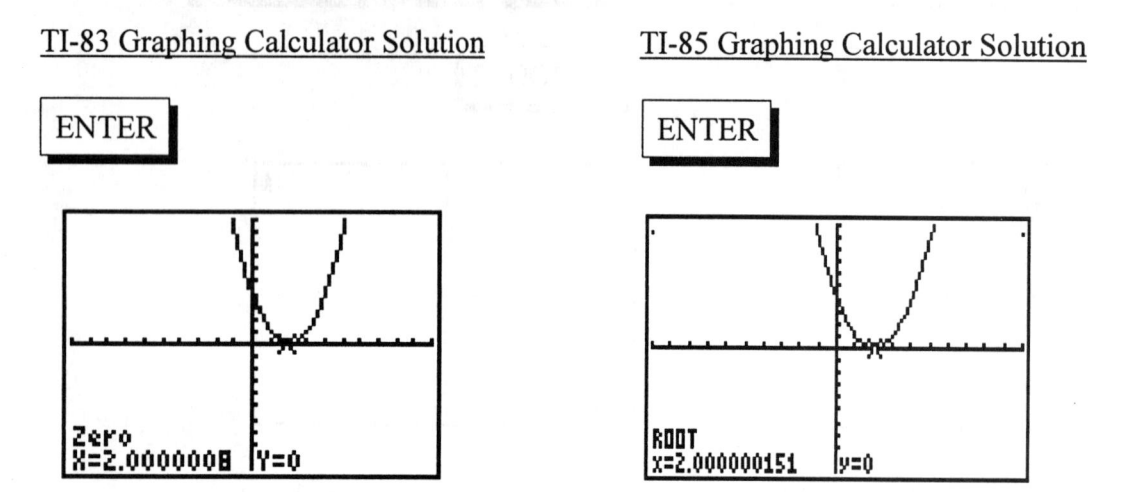

The root to this equation is located at (2,0). The vertex of this parabola is a relative minimum. We can use the features of the TI calculators to locate the relative minimum. This will also serve to verify the fact that the vertex of this parabola is the only root of the equation.

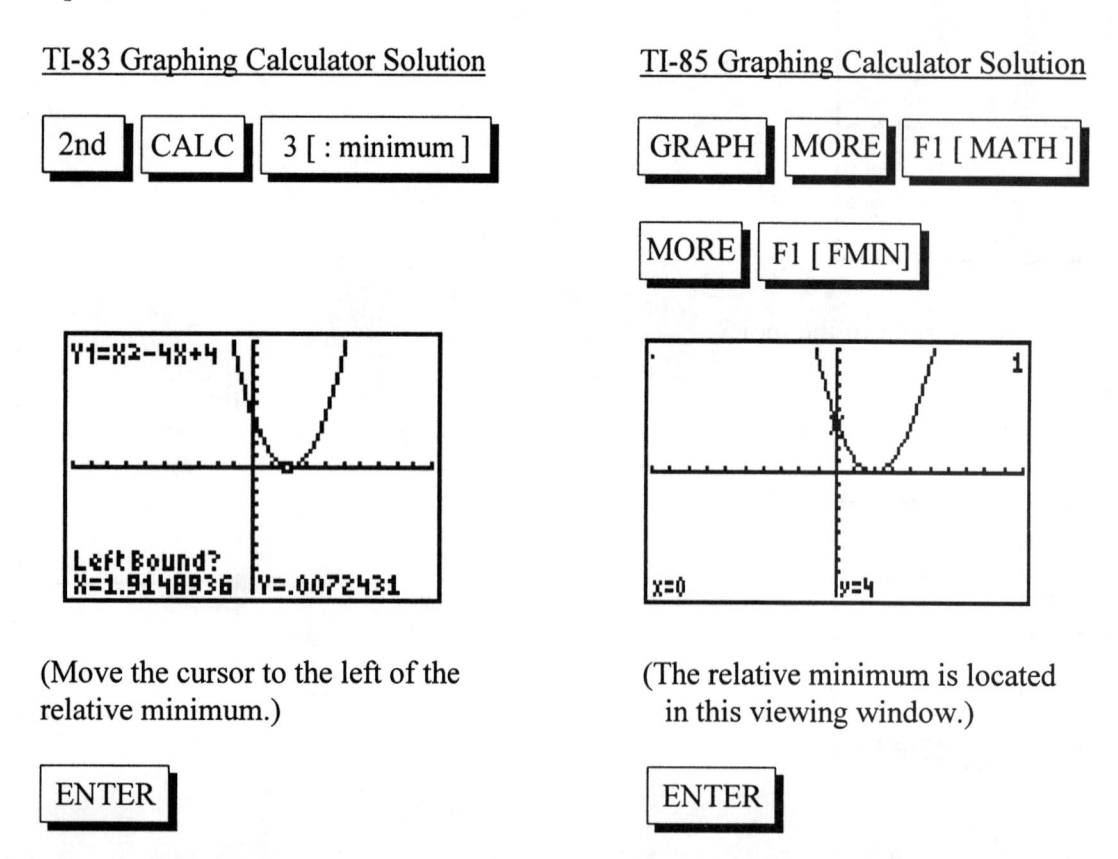

TI-83 Graphing Calculator Solution   TI-85 Graphing Calculator Solution

2nd   CALC   3 [: minimum]   GRAPH   MORE   F1 [MATH]

                                             MORE   F1 [FMIN]

(Move the cursor to the left of the   (The relative minimum is located
relative minimum.)   in this viewing window.)

ENTER   ENTER

LABORATORY 7: ADVANCED GRAPHING TECHNIQUES - THE QUADRATIC EQUATION

TI-83 Graphing Calculator Solution

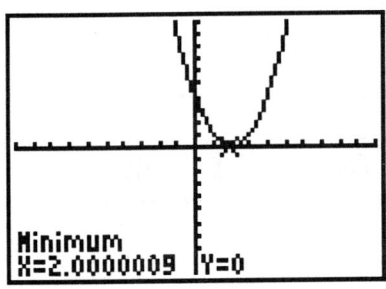

TI-85 Graphing Calculator Solution

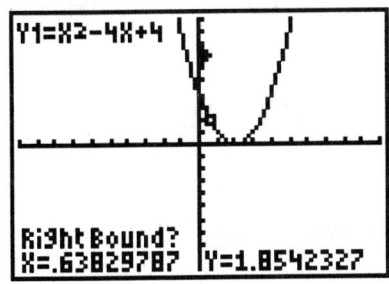

(Move the cursor to the right of the relative minimum.)

ENTER

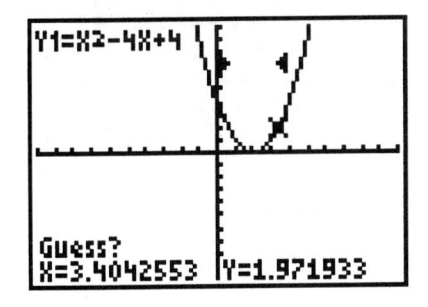

ENTER

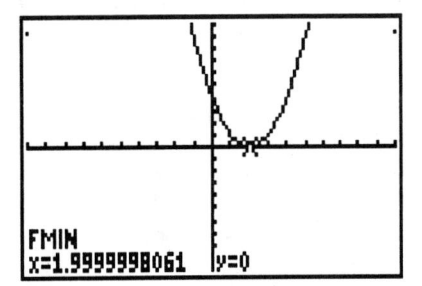

The **FMIN** function is located on the **F4** key of the corresponding TI-86 menu. Furthermore, the TI-86 requires entry of a left bound, right bound, and a guess as on the TI-83 and TI-82. The relative minimum occurs at the point (2,0), which is also the only root of this equation.

LABORATORY 7: ADVANCED GRAPHING TECHNIQUES - THE QUADRATIC EQUATION

Problem 2: Graph $y = x^2 - 4x - 4$ on the standard axes. Find the *x*-intercepts, *y*-intercept, and vertex. Determine whether the vertex is a relative maximum or a relative minimum. Remember to clear all graphs before starting this problem

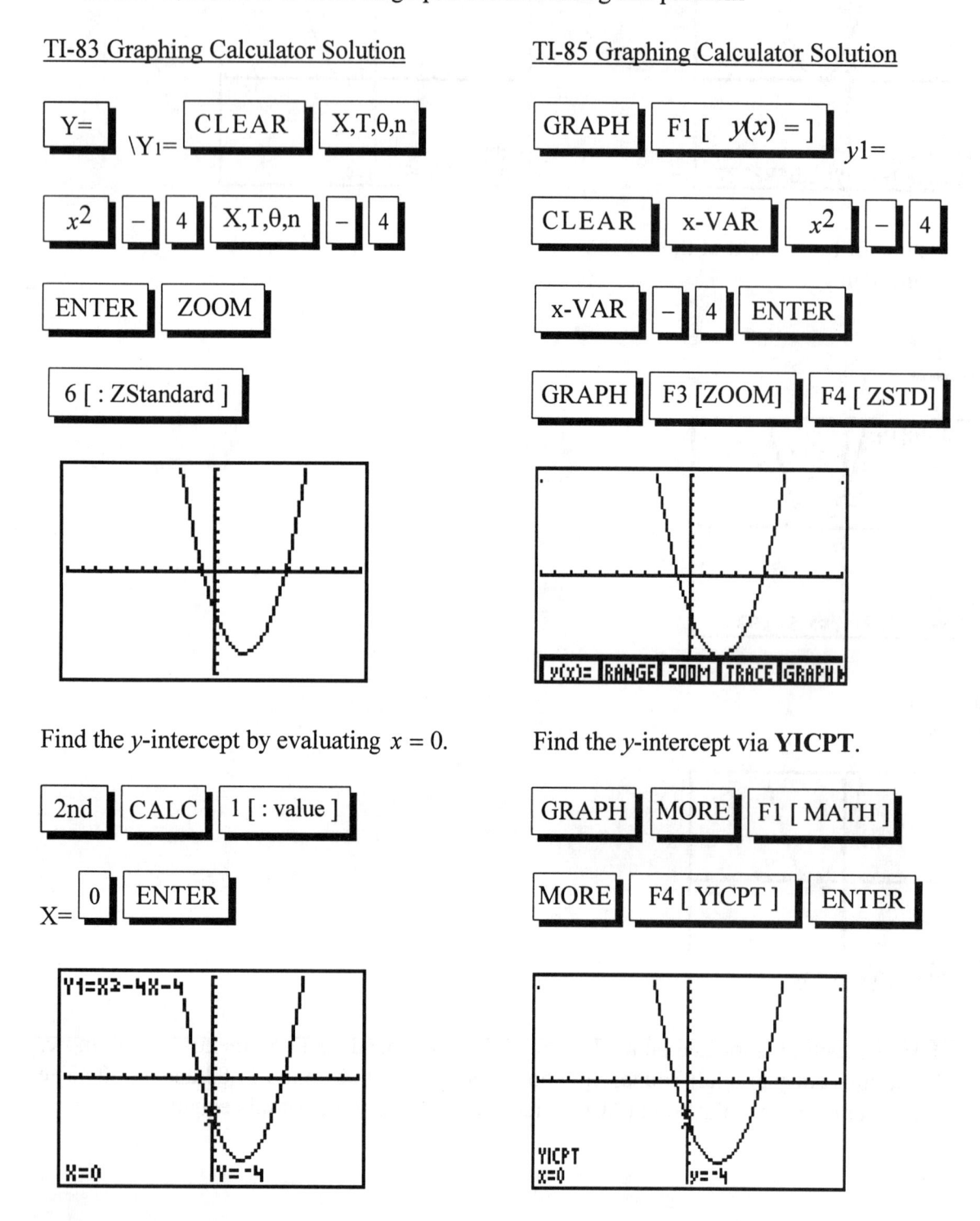

TI-83 Graphing Calculator Solution

| Y= | \Y₁= | CLEAR | X,T,θ,n |

| x^2 | − | 4 | X,T,θ,n | − | 4 |

| ENTER | ZOOM |

| 6 [: ZStandard] |

Find the *y*-intercept by evaluating $x = 0$.

| 2nd | CALC | 1 [: value] |

X= | 0 | ENTER |

TI-85 Graphing Calculator Solution

| GRAPH | F1 [$y(x) =$] | y1= |

| CLEAR | x-VAR | x^2 | − | 4 |

| x-VAR | − | 4 | ENTER |

| GRAPH | F3 [ZOOM] | F4 [ZSTD] |

Find the *y*-intercept via **YICPT**.

| GRAPH | MORE | F1 [MATH] |

| MORE | F4 [YICPT] | ENTER |

LABORATORY 7: ADVANCED GRAPHING TECHNIQUES - THE QUADRATIC EQUATION

Recall that the **YICPT** function is located on the **F2** key of the corresponding TI-86 menu. The y-intercept is located at the point (0,-4). Note that this is consistent with the fact that the y-intercept occurs at $y = c$ and in this case $c = -4$. There are two roots to this equation, one negative and one positive. Find the negative root first.

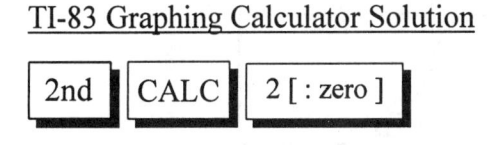

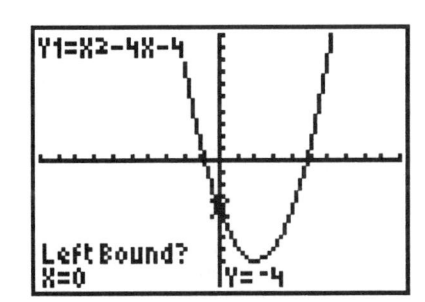

TI-83 Graphing Calculator Solution

| 2nd | CALC | 2 [: zero] |

(Move the cursor to the left of the negative root.)

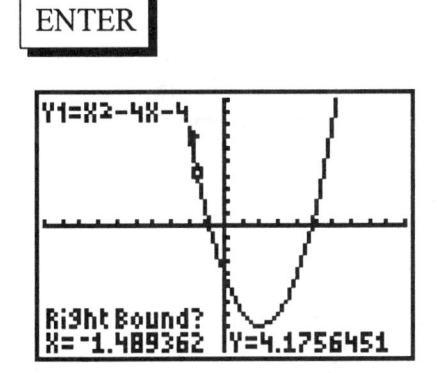

| ENTER |

(Move the cursor to the right of the negative root, but to the left of the positive root.)

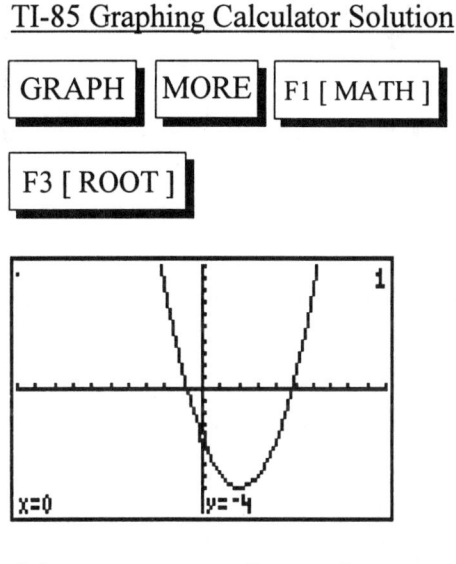

| ENTER |

TI-85 Graphing Calculator Solution

| GRAPH | MORE | F1 [MATH] |

| F3 [ROOT] |

(Move the cursor close to the negative root.)

| ENTER |

The **ROOT** function is located on the **F1** key of the corresponding TI-86 menu. Furthermore, recall that the **ROOT** function requires entry of a left bound, right bound, and a guess as on the TI-83 and TI-82.

79

LABORATORY 7: ADVANCED GRAPHING TECHNIQUES - THE QUADRATIC EQUATION

TI-83 Graphing Calculator Solution   TI-85 Graphing Calculator Solution

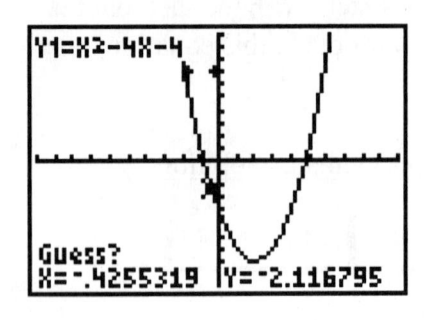

| ENTER |

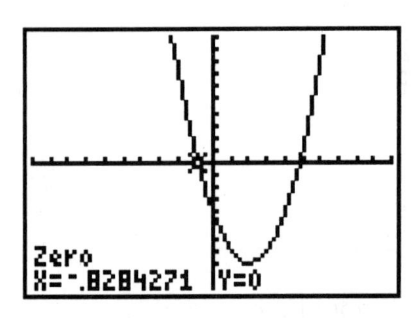

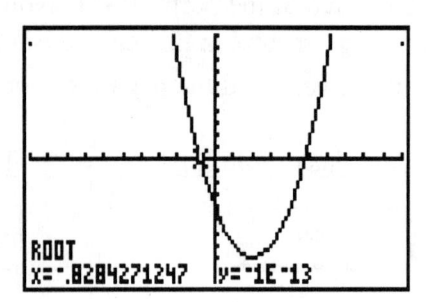

Note that the *x*-value of the negative root is approximately equal to -.8284271. Now find the positive root.

| 2nd | CALC | 2 [: zero] | | GRAPH | MORE | F1 [MATH] |

| F3 [ROOT] |

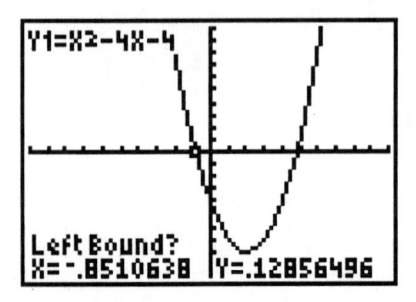

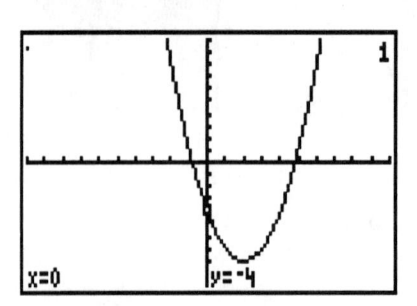

(Move the cursor to the left of the positive root, but to the right of the negative root.)

(Move the cursor closer to the positive root.)

| ENTER | | ENTER |

LABORATORY 7: ADVANCED GRAPHING TECHNIQUES - THE QUADRATIC EQUATION

TI-83 Graphing Calculator Solution   TI-85 Graphing Calculator Solution

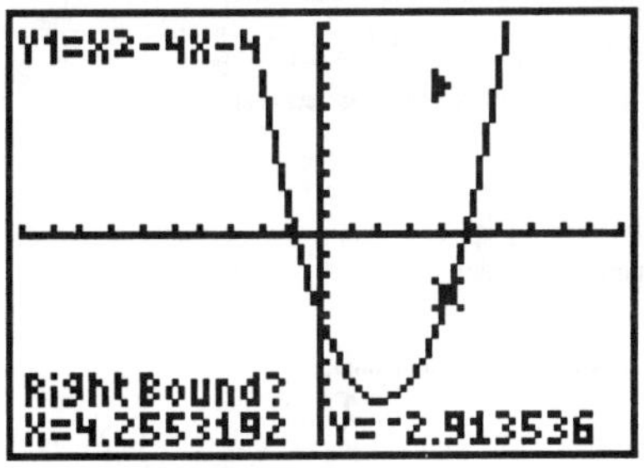

(Move the cursor to the right of the positive root.)

ENTER

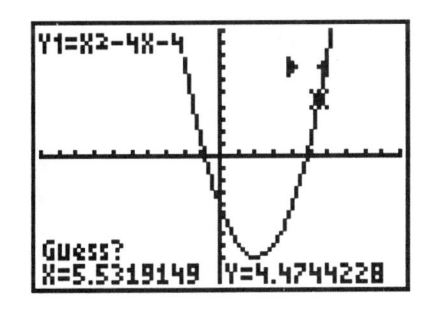

ENTER

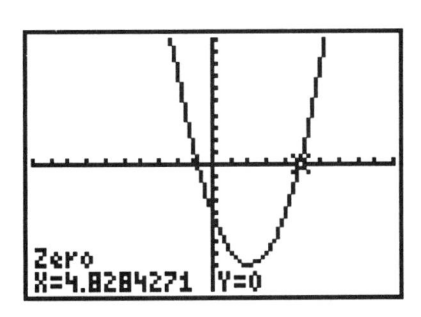

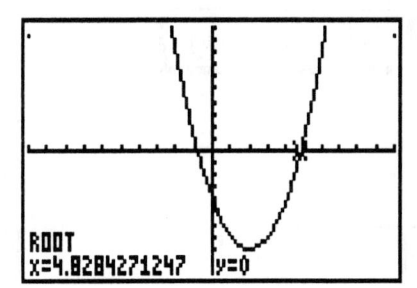

The **ROOT** function is located on the **F1** key of the corresponding TI-86 menu. Furthermore, recall that the **ROOT** function requires entry of a left bound, right bound, and a guess as on the TI-83 and TI-82. Note that the x-value of the positive root is approximately equal to 4.8284271. Now find the relative minimum.

LABORATORY 7: ADVANCED GRAPHING TECHNIQUES - THE QUADRATIC EQUATION

TI-83 Graphing Calculator Solution

| 2nd | CALC | 3 [: minimum] |

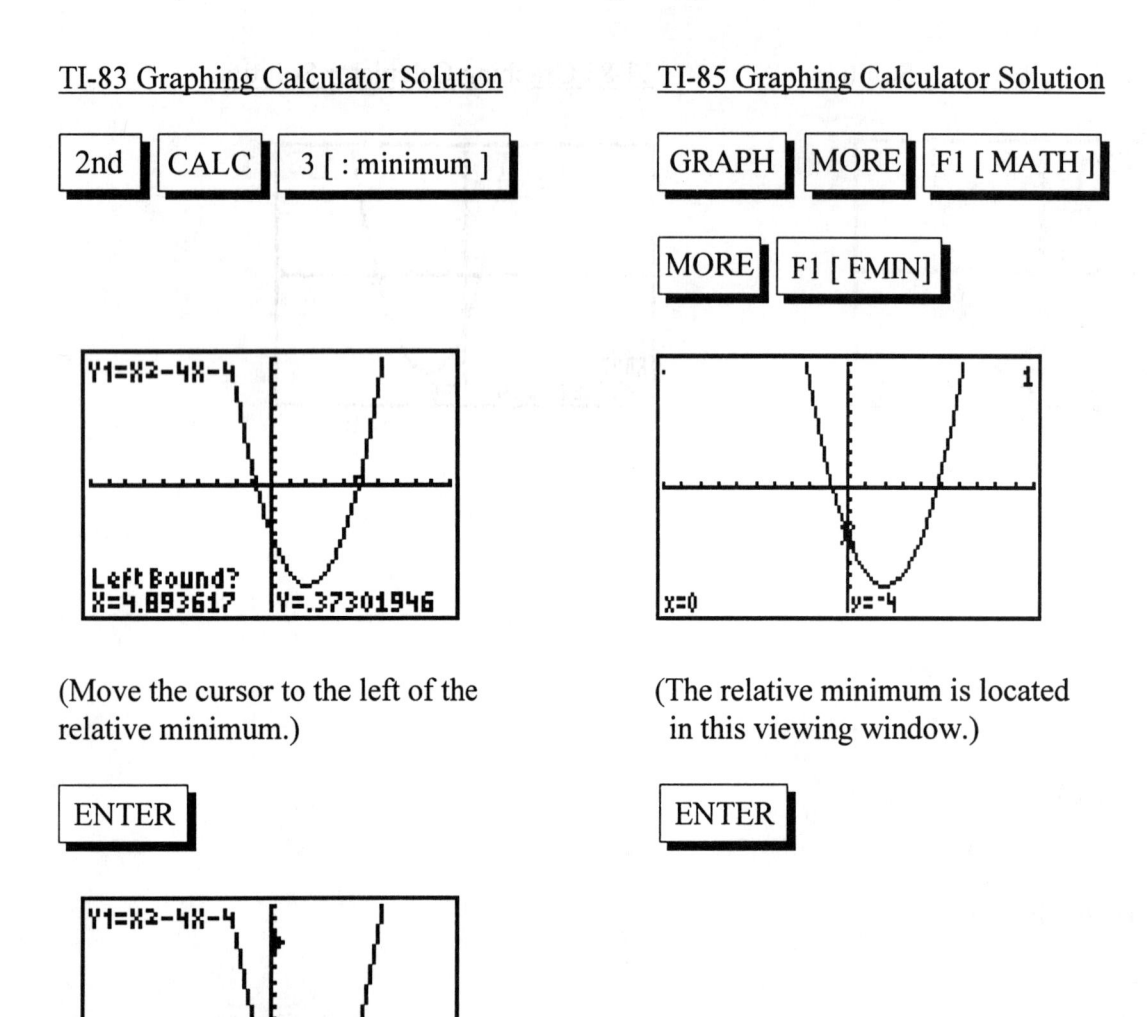

(Move the cursor to the left of the relative minimum.)

| ENTER |

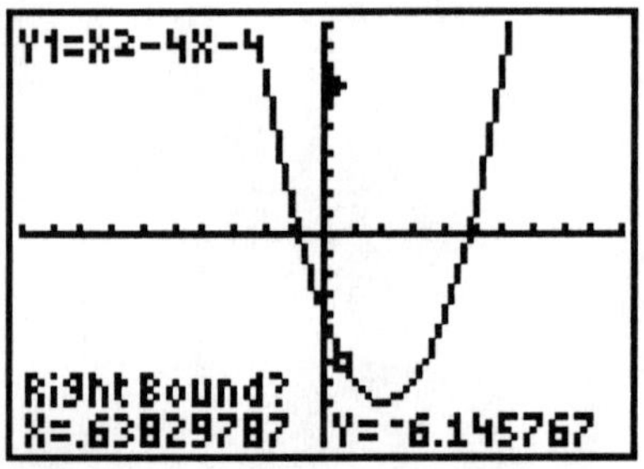

(Move the cursor to the right of the relative minimum.)

| ENTER |

TI-85 Graphing Calculator Solution

| GRAPH | MORE | F1 [MATH] |

| MORE | F1 [FMIN] |

(The relative minimum is located in this viewing window.)

| ENTER |

Recall that the **FMIN** function is located on the **F4** key of the corresponding TI-86 menu. Furthermore, the TI-86 requires entry of a left bound, right bound, and a guess as on the TI-83 and TI-82.

LABORATORY 7: ADVANCED GRAPHING TECHNIQUES - THE QUADRATIC EQUATION

TI-83 Graphing Calculator Solution   TI-85 Graphing Calculator Solution

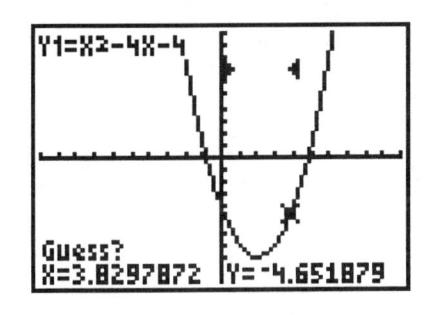

ENTER

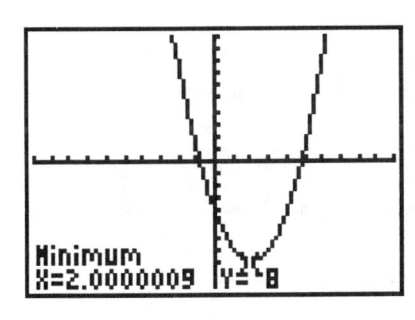

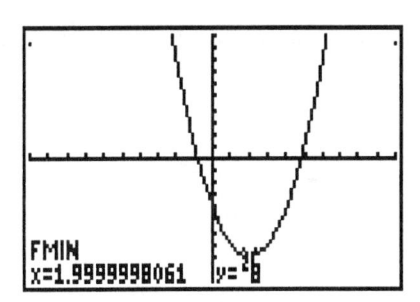

The relative minimum occurs at the point (2,-8). Verify using the formulas for the coordinates of the vertex outlined above.

Problem 3: Graph $y = -x^2 - 4x - 5$ on the standard axes. Find the *x*-intercepts, *y*-intercept, and vertex. Determine whether the vertex is a relative maximum or a relative minimum. Remember to clear all graphs before starting this problem

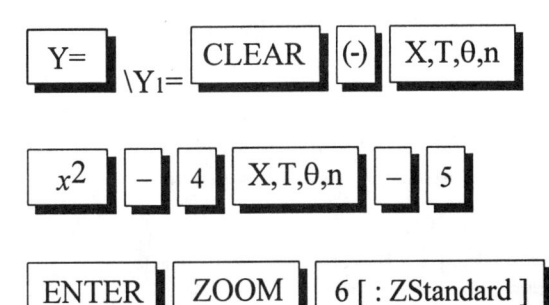

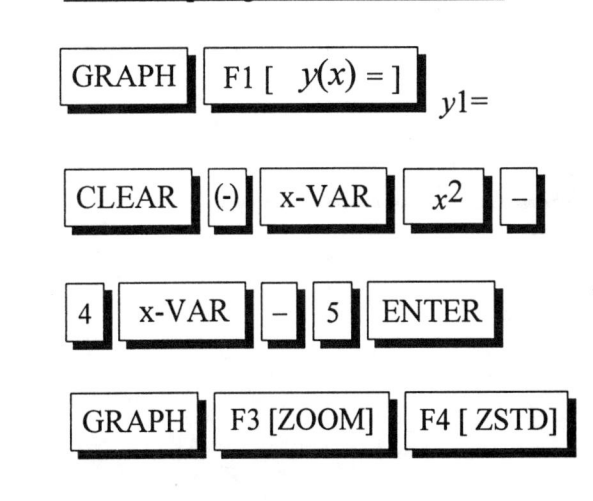

TI-83 Graphing Calculator Solution   TI-85 Graphing Calculator Solution

| Y= | \Y₁= | CLEAR | (-) | X,T,θ,n |

| x² | − | 4 | X,T,θ,n | − | 5 |

| ENTER | ZOOM | 6 [: ZStandard] |

| GRAPH | F1 [y(x) =] | y1= |

| CLEAR | (-) | x-VAR | x² | − |

| 4 | x-VAR | − | 5 | ENTER |

| GRAPH | F3 [ZOOM] | F4 [ZSTD] |

LABORATORY 7: ADVANCED GRAPHING TECHNIQUES - THE QUADRATIC EQUATION

TI-83 Graphing Calculator Solution

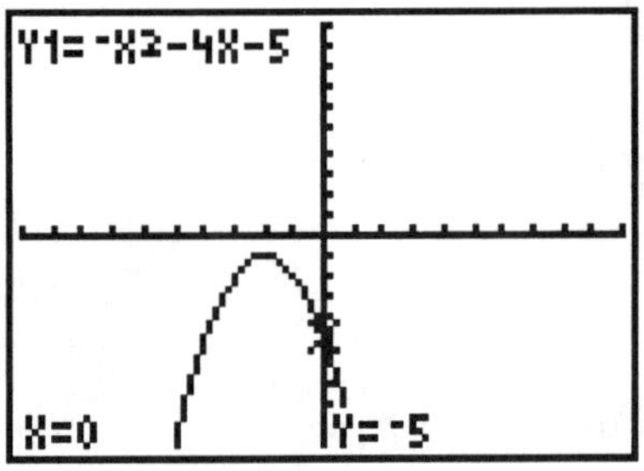

TI-85 Graphing Calculator Solution

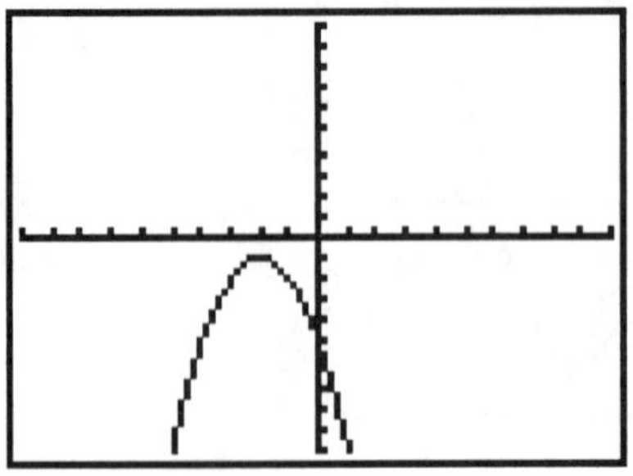

Find the *y*-intercept by evaluating $x = 0$.

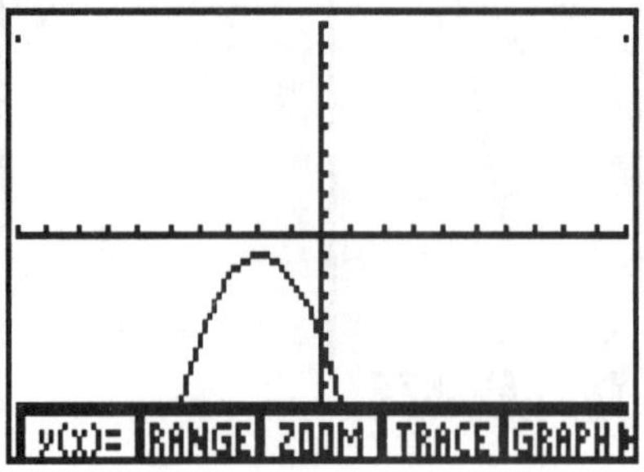

Find the *y*-intercept via **YICPT**.

| 2nd | CALC | 1 [: value] |

X= | 0 | ENTER |

| GRAPH | MORE | F1 [MATH] |

| MORE | F4 [YICPT] | ENTER |

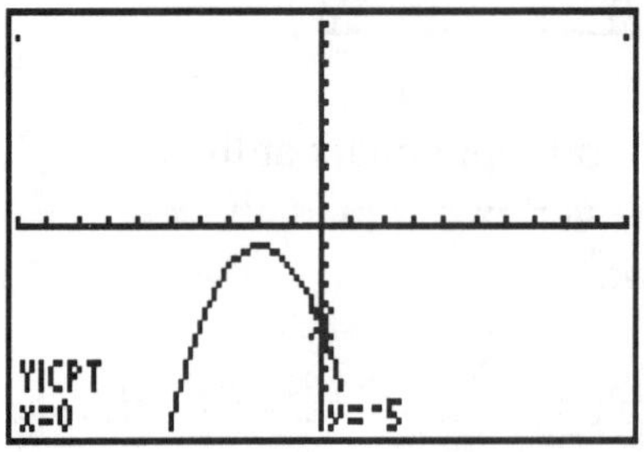

Recall that the **YICPT** function is located on the **F2** key of the corresponding TI-86 menu. The *y*-intercept is located at the point (0,-5). Note that this is consistent with the fact that the *y*-intercept occurs at $y = c$ and in this case $c = -5$. The graph of this parabola does not cross the *x*-axis. Therefore, there are no real roots, only complex roots. Find the relative maximum of this graph.

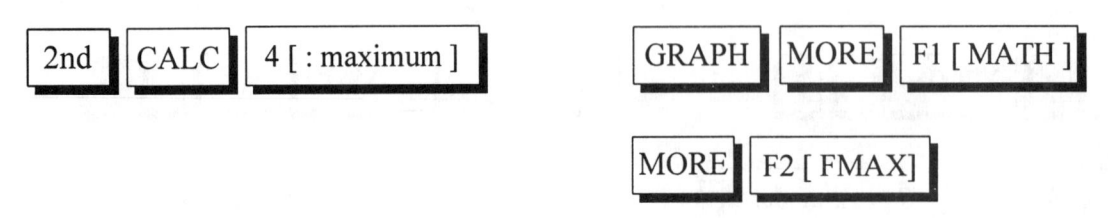

| 2nd | CALC | 4 [: maximum] |

| GRAPH | MORE | F1 [MATH] |

| MORE | F2 [FMAX] |

LABORATORY 7: ADVANCED GRAPHING TECHNIQUES - THE QUADRATIC EQUATION

TI-83 Graphing Calculator Solution

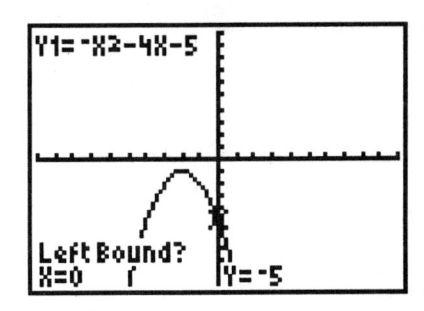

(Move the cursor to the left of the relative maximum.)

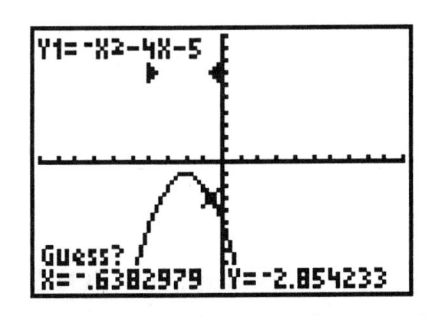

ENTER

TI-85 Graphing Calculator Solution

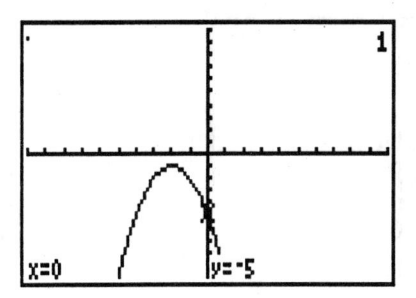

(The relative maximum is located in this viewing window.)

ENTER

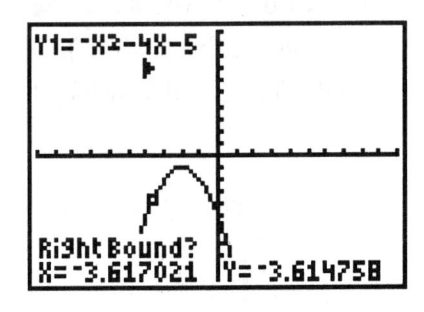

(Move the cursor to the right of the relative maximum.)

ENTER

ENTER

The **FMAX** function is located on the **F5** key of the corresponding TI-86 menu. The TI-86 requires entry of a left bound, right bound, and a guess as on the TI-83 and TI-82.

LABORATORY 7: ADVANCED GRAPHING TECHNIQUES - THE QUADRATIC EQUATION

TI-83 Graphing Calculator Solution   TI-85 Graphing Calculator Solution

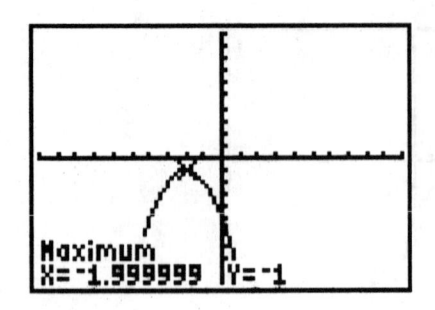

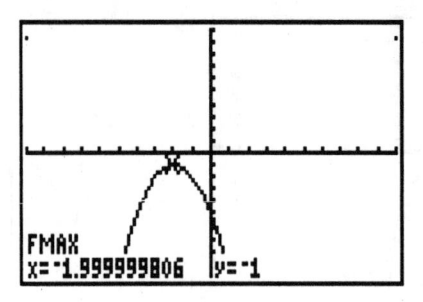

The relative maximum occurs at the point (-2,-1). Verify using the formulas for the coordinates of the vertex outlined above.

The TI-85 and TI-86 calculators are equipped with a **POLY** function which finds the roots (real and complex) of a polynomial of degree n. The **POLY** function will request the order or degree of the polynomial as well as the coefficients of the polynomial. Complex roots are displayed inside parentheses in the format (x, y) where the root $z = x + yi$ (with $i = \sqrt{-1}$). The complex roots of the polynomial in Problem 3 above can be found on the TI-85 and TI-86 by the procedure outlined below:

Solution to Finding the Complex Roots in Problem 3 on the TI-85 and TI-86

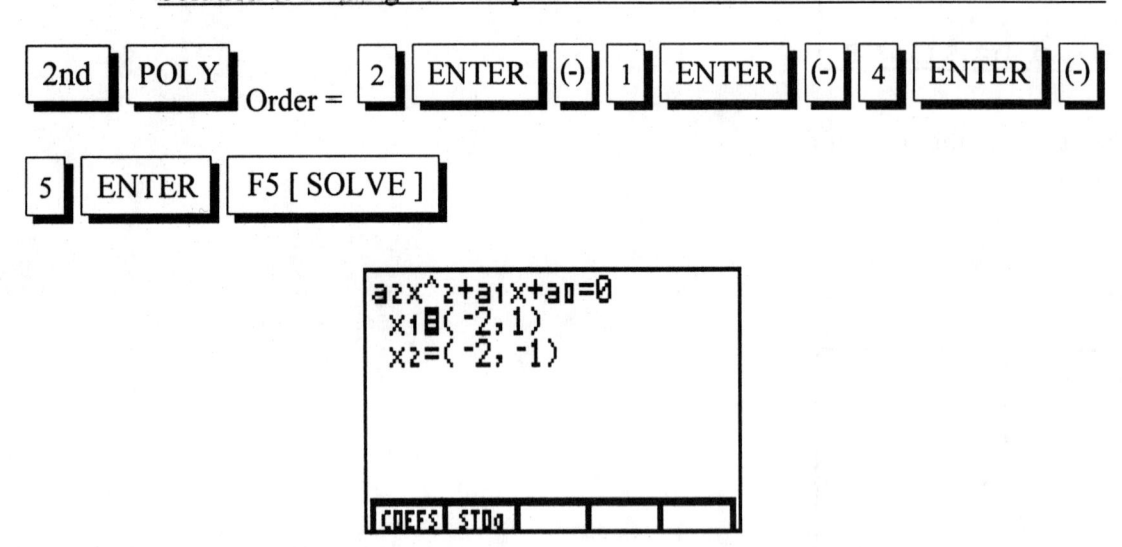

| 2nd | POLY | Order = | 2 | ENTER | (-) | 1 | ENTER | (-) | 4 | ENTER | (-) |

| 5 | ENTER | F5 [SOLVE] |

The two complex roots are $-2 + i$ and its complex conjugate $-2 - i$. Try using the **POLY** function to find the roots of the equations in Problem 1 and Problem 2 above.

LABORATORY 7: ADVANCED GRAPHING TECHNIQUES - THE QUADRATIC EQUATION

EXERCISES

1. Graph $y = 4x^2 - 6x - 2$ on the standard axes. Find the x-intercepts, y-intercept, and vertex. Determine whether the vertex is a relative maximum or a relative minimum.

2. Graph $y = -x^2 - 5x + 3$ on the standard axes. Find the x-intercepts, y-intercept, and vertex. Determine whether the vertex is a relative maximum or a relative minimum.

3. Subtract ten from the polynomial in Exercise 2 above and graph the resulting polynomial. Discuss the difference in the two graphs and find the x-intercepts, y-intercept, and relative maximum or relative minimum if appropriate.

4. Graph $y = 2x^2 + 6x - 40$ on an appropriate set of axes. Adjust the minimum value on the y-axis so that the vertex can be seen in the viewing window. Find the x-intercepts, y-intercept, and relative maximum or relative minimum.

5. Graph $y = x^2 - 10x + 25$ on the standard viewing window. Find the x-intercepts, y-intercept, and relative maximum or relative minimum.

LABORATORY 8: FUNCTIONS

Purpose:

The purpose of this laboratory is to look at the characteristics of functions using the TI-82/83 and TI-85/86.

Analytic Approach:

A good algebraic approach to functions and relations should be discussed simultaneously with the following graphical approach. In many instances, algebraic manipulations must be performed before any graphing techniques can be utilized.

Before beginning a discussion of functions, we should define a *relation*. A *relation* is a set of ordered pairs (x, y). For example, the set of ordered pairs $\{(1, 3), (2, 5), (4, 7), (2, -1)\}$ is a relation because for every x value there is a corresponding y value. The value of y is related to the value of x, usually by an equation or an inequality. $y = \sqrt{x}$ is a relation when $x \geq 0$.

Domain and Range of a Function

A **function** is a special type of relation. A *function* is a rule of correspondence between two variables, x and y, such that for every value of the variable, x, there exists *one and only one* value of the variable, y. The x variable is referred to as the **independent variable**, and the set of all x values is called the *domain* of the function. The y value is referred to as the **dependent variable**, and the set of all y values is called the *range* of the function. The *domain* of the relation $\{(1, 3), (2, 5), (4, 7), (2, -1)\}$ consists of the x values 1, 2, 4; while the *range* is composed of the y values -1, 3, 5, and 7. The set of ordered pairs $\{(1, 3), (2, 5), (4, 7), (2, -1)\}$, *is not a function* since the two ordered pairs $(2, 5)$ and $(2, -1)$ have the same x value but different y values. The equation $y = \sqrt{x}$ *is a function* because there is one, and only one, value for y for each value of $x \geq 0$. The equation $x = y^2$ is *not a function* of y, since there exists two different y values for a given x value. The equation $x = y^2$ is a relation.

A function is expressed by letting $y = f(x)$, where f is a *function* of x, y is the dependent variable, x is the independent variable, and $f(x)$ is the value of the function at x.

Note that the use of the word "*range*" of a function is the not the same as setting the **RANGE** on the **TI-85** or the **WINDOW** on the **TI-83, 82** and **TI-86**. Setting the **RANGE** on the TI-85 or the **WINDOW** on the TI-83, 82 and TI-86 means setting the minimum and maximum values of x and y which are to be graphed as well as the scale for the x and y axes.

A vertical line drawn through the graph of a relation is a test of whether a relation is also a function. A vertical line drawn through the graph must intersect the graph at one and only one point if the relation is a function.

Set the viewing rectangle to the standard setting for each of the problems below.

LABORATORY 8: FUNCTIONS

Problem 1: Is $y = x$ a function and if so, what is its domain and range?

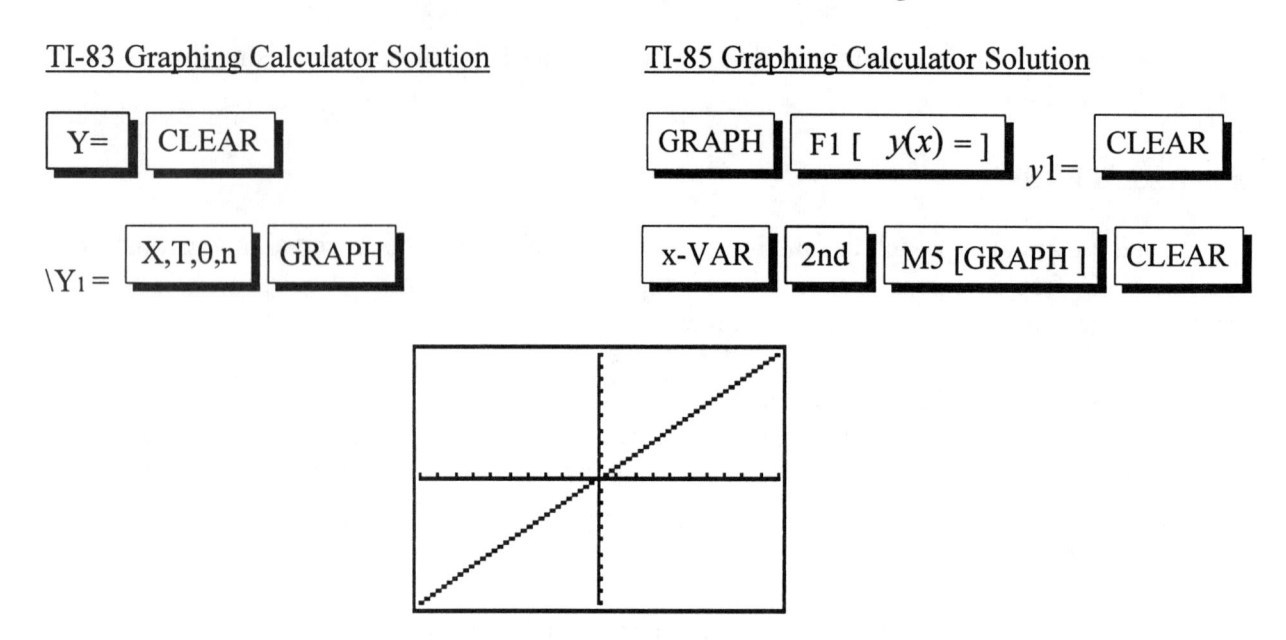

TI-83 Graphing Calculator Solution   TI-85 Graphing Calculator Solution

| Y= | CLEAR | | GRAPH | F1 [$y(x) =$] | $y1=$ | CLEAR |

\Y₁= | X,T,θ,n | GRAPH | | x-VAR | 2nd | M5 [GRAPH] | CLEAR |

To perform the vertical line test, press the following keys:

| 2nd | DRAW | 4 [: Vertical] | | MORE | F2 [DRAW] | F3 [VERT] |

Press the ▶ or the ◀ key to move the vertical line to the right or to the left along the graph. Continue moving the vertical line to the right or to the left by holding down either the right or left arrow key. The vertical line will continue to move in either direction as far as the range settings will allow it. Each time the vertical line intersects the graph, it does so at one and only one point on the line. The vertical line test proves that $y = x$ is a function.

Use | TRACE | to determine the domain.

From the calculator screens, we can see that the domain of the function is all real numbers and the range of the function $y = x$ is also all real numbers.

Problem 2: Is $y^2 = x$ a function? If it is a function, determine its domain and range.

Before we can graph the equation, the equation must be solved for y. Take the square root of both sides of the equation obtaining $y = \pm\sqrt{x}$. There are now two graphs to be drawn:

(a) $y = \sqrt{x}$ and (b) $y = -\sqrt{x}$.

LABORATORY 8: FUNCTIONS

Part (a)

TI-83 Graphing Calculator Solution

[Y=] [CLEAR] \Y₁= [2nd]

[√] [X,T,θ,n] [ )] [GRAPH]

TI-85 Graphing Calculator Solution

[GRAPH] [F1 [y(x) =]] y1= [CLEAR]

[2nd] [√] [x-VAR] [ENTER] [2nd]

[M5 [GRAPH]]

Part (b)

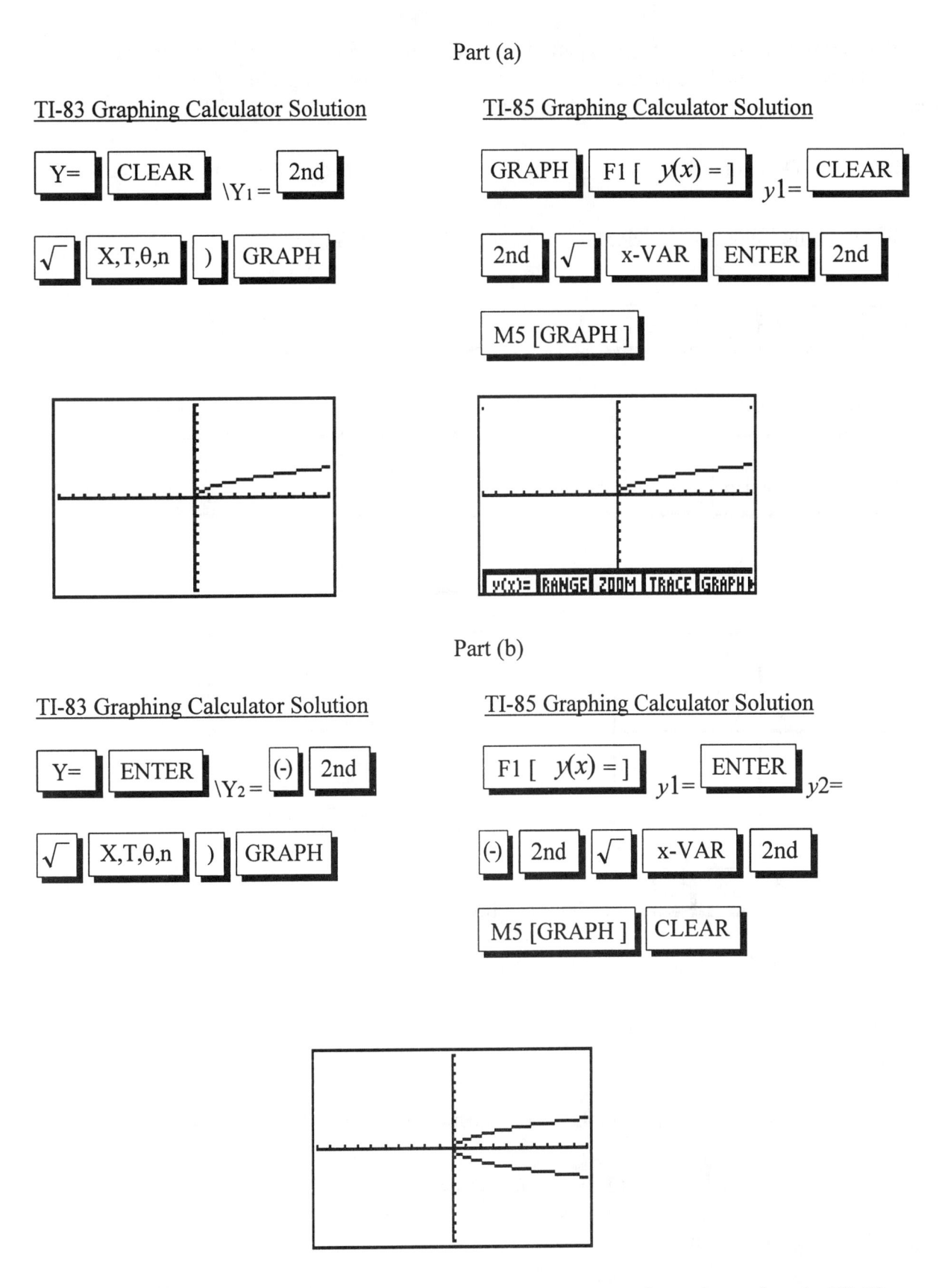

TI-83 Graphing Calculator Solution

[Y=] [ENTER] \Y₂= [(-)] [2nd]

[√] [X,T,θ,n] [ )] [GRAPH]

TI-85 Graphing Calculator Solution

[F1 [y(x) =]] y1= [ENTER] y2=

[(-)] [2nd] [√] [x-VAR] [2nd]

[M5 [GRAPH]] [CLEAR]

REMEMBER to *insert* a left parenthesis after the square root key when using the **TI-82**.

LABORATORY 8: FUNCTIONS

Perform a vertical line test. Each part of the equation when solved for *y* forms a function, but when the two parts of the equation $y^2 = x$ are looked at together, the graph clearly *is not a function*. $y^2 = x$ is a relation because there exist a value of *y* for every value of *x*.

Problem 3: Determine if $y = |x^2 - 4|$ is a function, and if so, what is its domain and range?

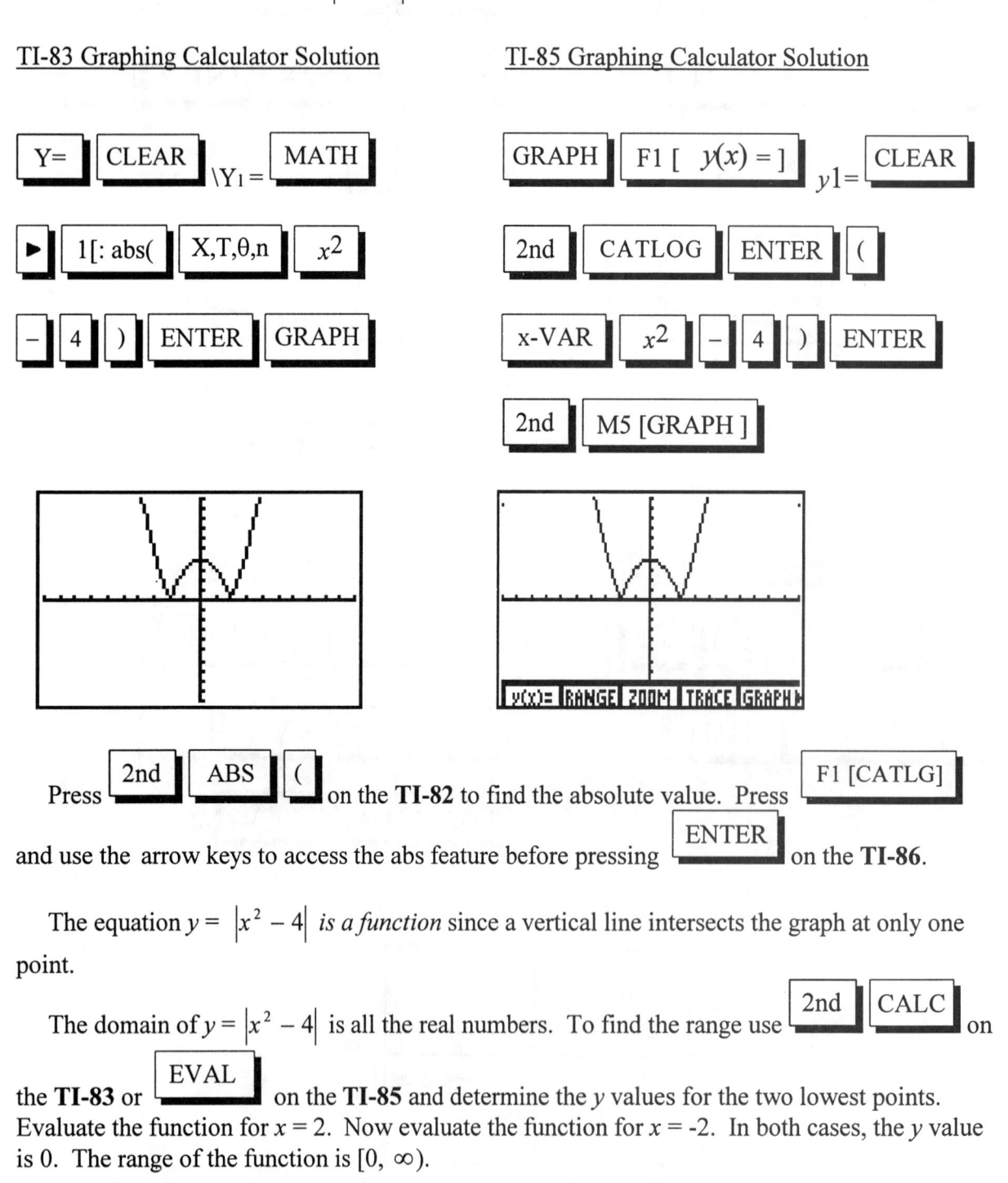

TI-83 Graphing Calculator Solution   TI-85 Graphing Calculator Solution

| Y= | CLEAR | \Y₁= | MATH | | GRAPH | F1 [y(x) =] | y1= | CLEAR |

| ▶ | 1[: abs(| X,T,θ,n | x² | | 2nd | CATLOG | ENTER | (|

| - | 4 |) | ENTER | GRAPH | | x-VAR | x² | - | 4 |) | ENTER |

                                            | 2nd | M5 [GRAPH] |

Press | 2nd | ABS | (| on the **TI-82** to find the absolute value. Press | F1 [CATLG] | and use the arrow keys to access the abs feature before pressing | ENTER | on the **TI-86**.

The equation $y = |x^2 - 4|$ *is a function* since a vertical line intersects the graph at only one point.

The domain of $y = |x^2 - 4|$ is all the real numbers. To find the range use | 2nd | CALC | on the **TI-83** or | EVAL | on the **TI-85** and determine the *y* values for the two lowest points. Evaluate the function for $x = 2$. Now evaluate the function for $x = -2$. In both cases, the *y* value is 0. The range of the function is $[0, \infty)$.

LABORATORY 8: FUNCTIONS

Increasing, Decreasing and Constant Functions

The more you know about the graph of a function, the more you actually know about the function itself. Once a graph has been drawn it is very easy to observe its characteristics. A graph may increase as *x* increases (*y* must also be increasing). The graph may remain the same as *x* increases (*y* must be constant). The graph may actually decrease as *x* increases which means that *y* is decreasing. If the graph is increasing, decreasing, or remaining constant, the function that the graph represents is also an increasing, decreasing, or constant function. A function can increase over a given interval, then decrease in a different interval, and finally remain constant over yet a different interval, *all* on the same graph.

Problem 4: Determine the interval on the graph where the function $y = 4 - x^2$ is increasing, decreasing, or remaining constant.

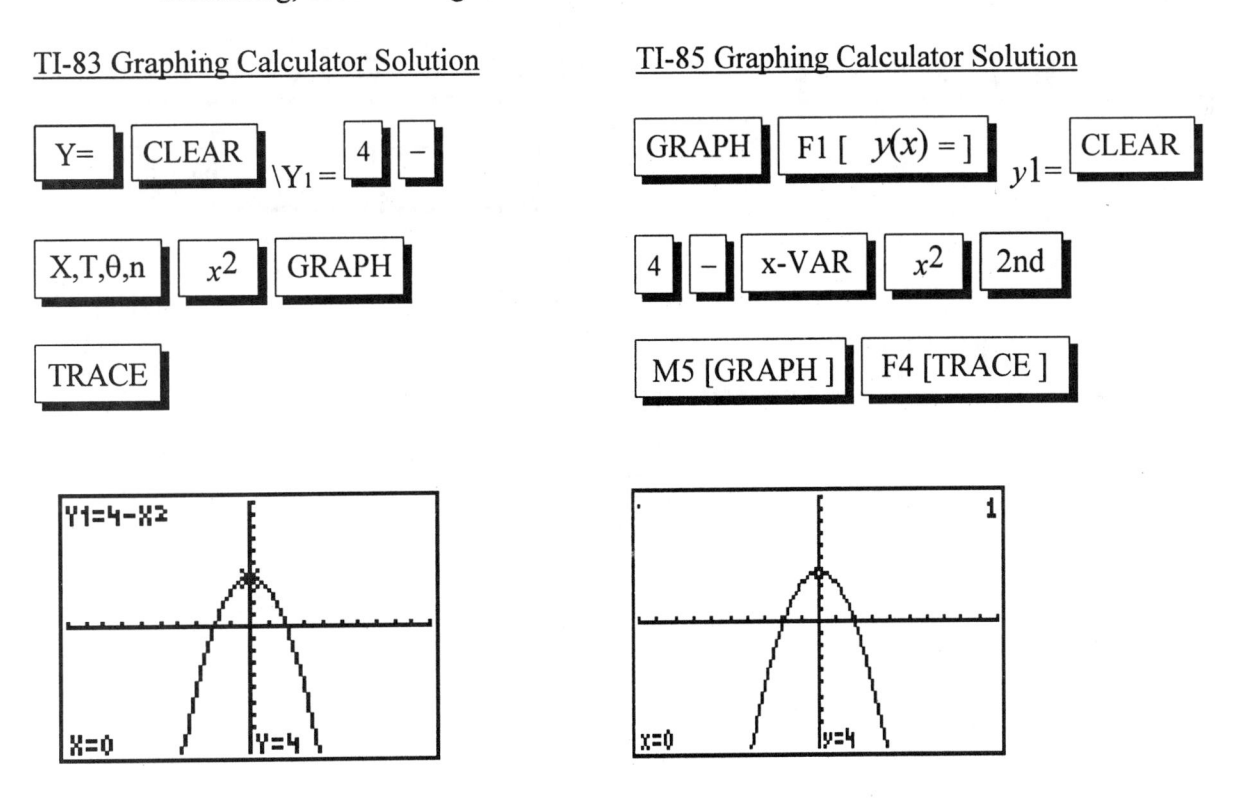

TI-83 Graphing Calculator Solution

| Y= | CLEAR | \Y₁ = | 4 | − |
| X,T,θ,n | x^2 | GRAPH |
| TRACE |

TI-85 Graphing Calculator Solution

GRAPH	F1 [y(x) =]	y1=	CLEAR
4	−	x-VAR	x^2	2nd
M5 [GRAPH]	F4 [TRACE]

Use TRACE with all the TI calculators to determine the interval where the function increases or decreases.

The description of the interval is always in terms of the *x* values (x_1, x_2). The function "turns" or "changes direction" at the point (0, 4). Therefore, the function increases over the interval (-∞, 0]. The function *changes direction* and decreases over the interval [0, ∞).

93

LABORATORY 8: FUNCTIONS

Problem 5: Determine the interval(s) where the function $f(x) = |x + 2| + |x - 2|$ increases, decreases, or remains constant.

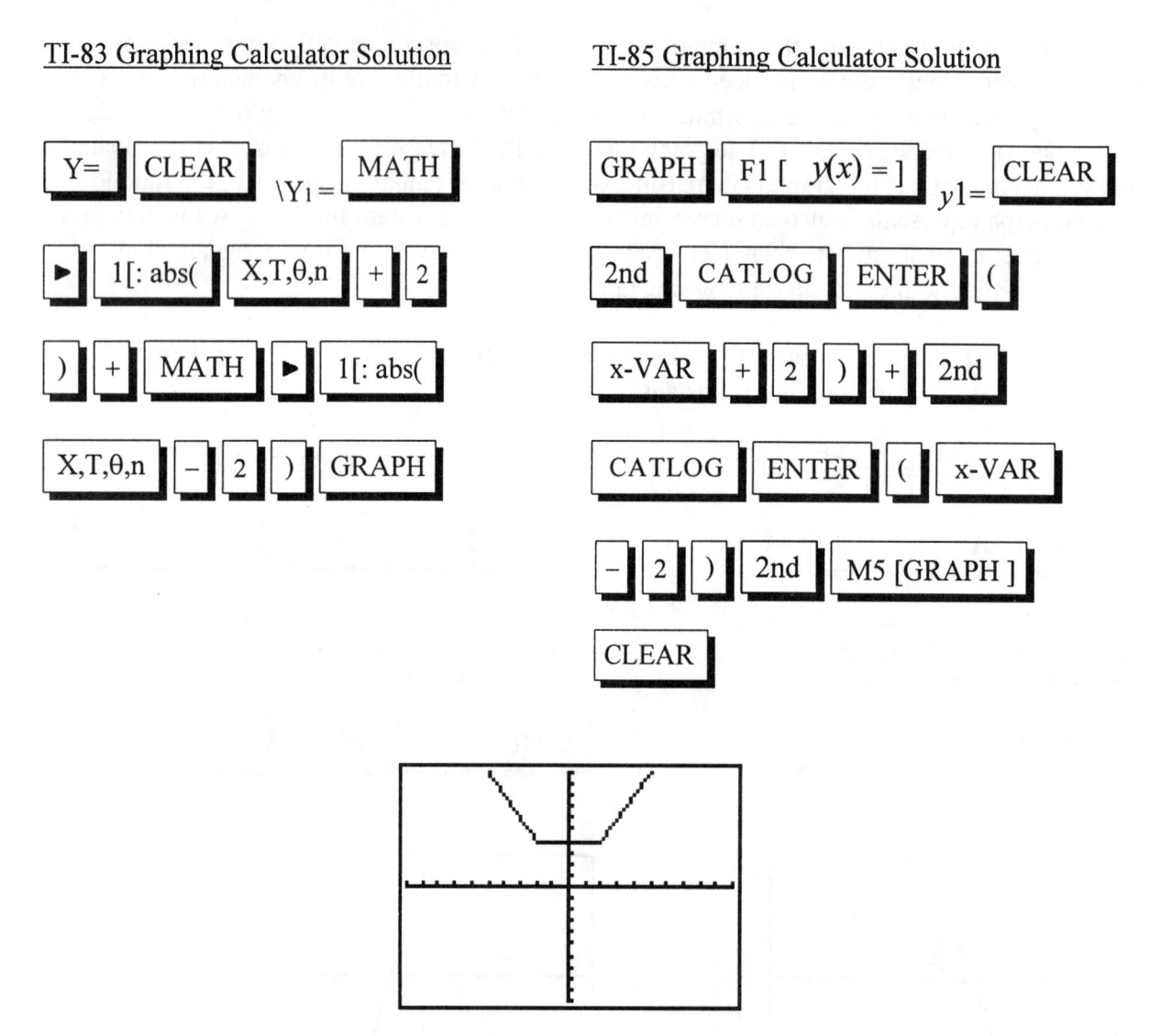

TI-83 Graphing Calculator Solution   TI-85 Graphing Calculator Solution

[Y=] [CLEAR] \Y₁= [MATH]   [GRAPH] [F1 [y(x) =]] y1= [CLEAR]

[▶] [1[: abs(] [X,T,θ,n] [+] [2]   [2nd] [CATLOG] [ENTER] [(]

[)] [+] [MATH] [▶] [1[: abs(]   [x-VAR] [+] [2] [)] [+] [2nd]

[X,T,θ,n] [−] [2] [)] [GRAPH]   [CATLOG] [ENTER] [(] [x-VAR]

[−] [2] [)] [2nd] [M5 [GRAPH]]

[CLEAR]

TI-82 and **TI-86** users should refer to *Problem 4* to make the proper keystroke adjustments for the absolute value.

Determine the interval where the graph shows the function to be increasing, decreasing or remaining constant. Use [2nd] [CALC] [1 [: value]] on the **TI-83/82** or [EVAL] on the **TI-85/86** to determine the point at which the function changes direction. *Remember* that you should always move from left to right along the *x*-axis.

The first *x* value that should be examined is *x* = -2. When *x* = -2, the direction of the graph changes.

94

LABORATORY 8: FUNCTIONS

TI-83 Graphing Calculator Solution

| 2nd | | CALC | | 1 [: value] |

| (-) | | 2 | | ENTER |

TI-85 Graphing Calculator Solution

| GRAPH | | MORE | | MORE | | F1 [EVAL] |

| (-) | | 2 | | ENTER |

Continue moving along the *x*-axis in a positive direction. Use | TRACE | to scroll from *x* = -2. Once the *y* direction changes, stop and use | EVAL | on the **TI-85/86** or | 2nd | | CALC | | 1 [: value] | on the **TI-83/82** to find the *y* value at the point where the graph started to change direction. The next value of *x* to be evaluated is 2.

The function is decreasing over the interval $(-\infty, -2]$, remains constant over the interval $[-2, 2]$, and is increasing over the interval $[2, \infty)$.

Graphical Determination of the Roots of a Function

On a graph the solutions of a function are called the *roots* or *zeros* of the function. Only the "real" roots can be found on the graph since only real numbers are used to name the *x* and *y* values. No matter what their name, *solutions, roots,* or *zeros,* the objective is still the same: solve the equation for *x*, when *y* = 0.

Problem 6: Given the function $f(x) = x^2 - 2$, find

    (a) the domain and range of the function
    (b) the real roots of $f(x)$. (Altogether, how many roots should there be?)
    (c) find $f(0)$, $f(3)$, $f\left(-\dfrac{1}{2}\right)$.

LABORATORY 8: FUNCTIONS

TI-83 Graphing Calculator Solution

| Y= | CLEAR | \Y₁= | X,T,θ,n |

| x^2 | − | 2 | GRAPH |

| TRACE |

TI-85 Graphing Calculator Solution

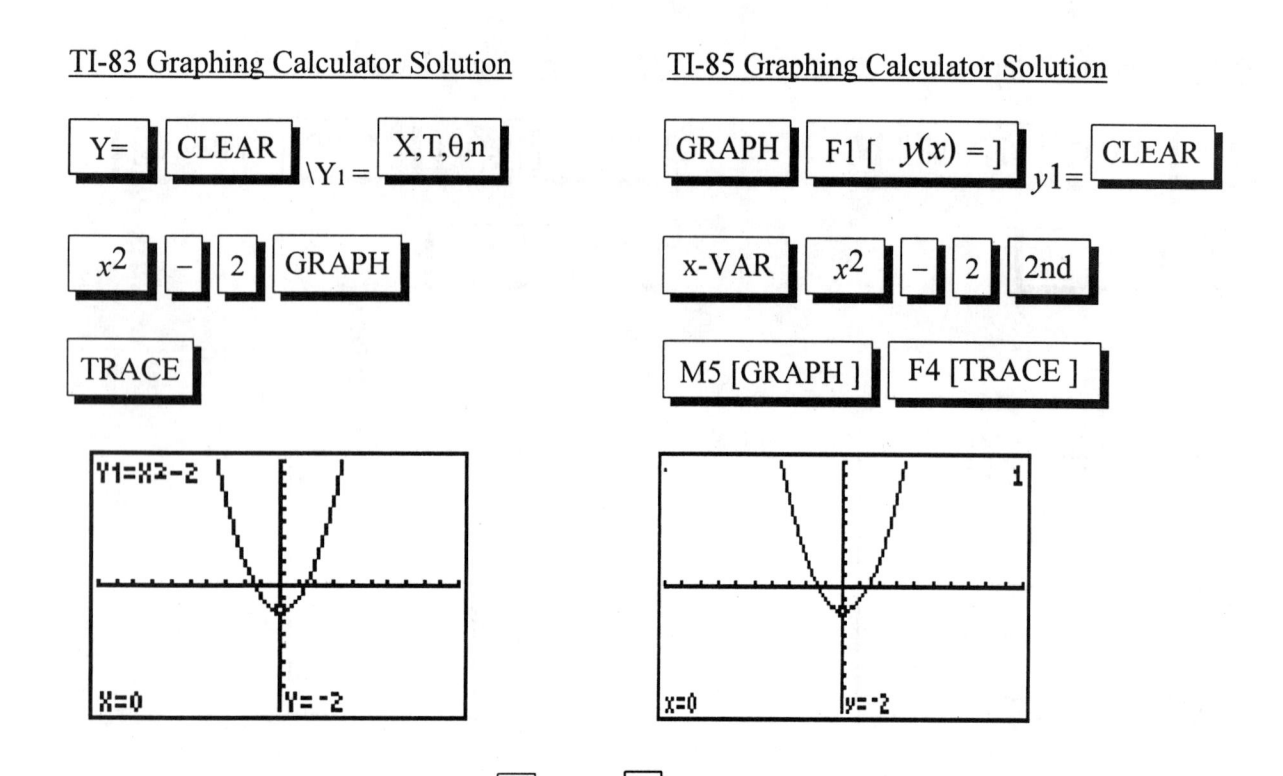

| GRAPH | F1 [y(x) =] | y1= | CLEAR |

| x-VAR | x^2 | − | 2 | 2nd |

| M5 [GRAPH] | F4 [TRACE] |

Scroll along the graph using the ▶ and ◀ arrow keys to determine the domain and range. The domain of the function is all real numbers, while the range is [-2, ∞).

Part (b)

From the screens above, we see that the graph of $f(x)$ crosses the x-axis twice. Each point corresponds to a real root of $f(x)$. In order to find each real root on the TI-83, TI-82, and TI-86, we must set the upper and lower limits of the interval in which the proposed root exists. This procedure provides the calculators with a "guess" to perform a numerical analysis to aid in the determination of the actual root.

We will first find the root to the right of the present position of the cursor.

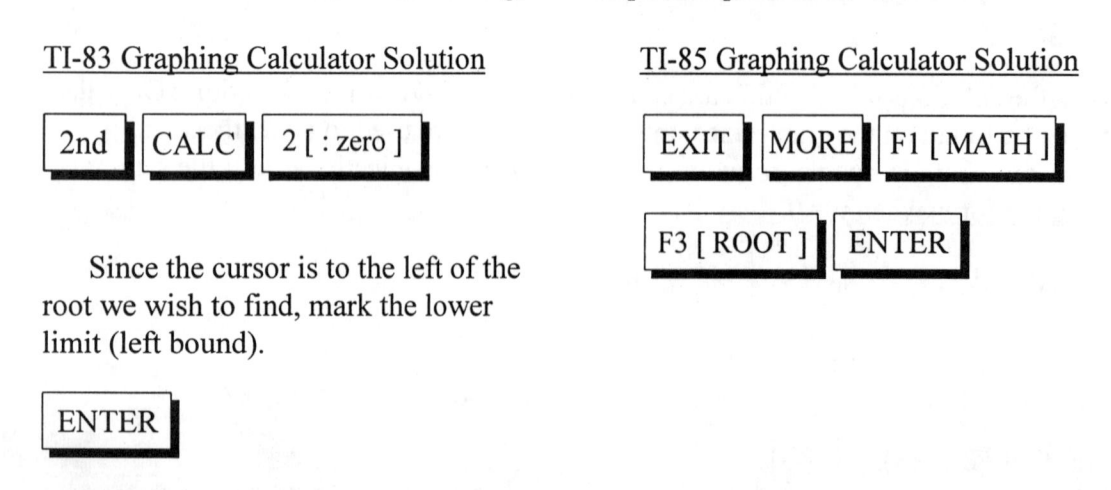

TI-83 Graphing Calculator Solution

| 2nd | CALC | 2 [: zero] |

Since the cursor is to the left of the root we wish to find, mark the lower limit (left bound).

| ENTER |

TI-85 Graphing Calculator Solution

| EXIT | MORE | F1 [MATH] |

| F3 [ROOT] | ENTER |

LABORATORY 8: FUNCTIONS

TI-83 Graphing Calculator Solution

Note that the lower limit is now marked on the viewing screen by a ▶.

We must now set the upper limit by pressing the ▶ key until the cursor is to the right of the zero.

| ENTER |

The upper limit is now marked on the viewing screen by a ◀.

TI-85 Graphing Calculator Solution

Users of the **TI-86** should remember to press **F1** to determine the root.

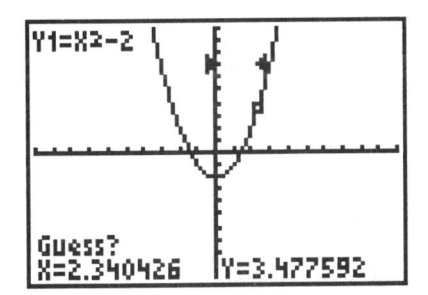

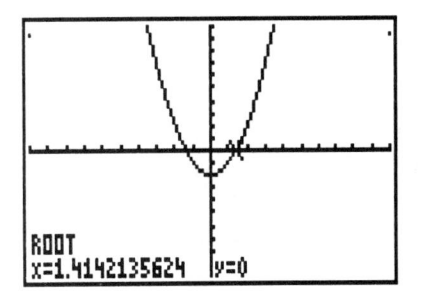

We can now find the zero of the function.

| ENTER |

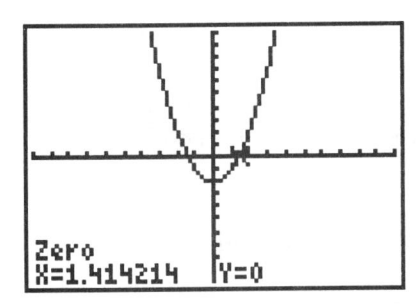

We see that the positive root of the function $f(x) = x^2 - 2$ is 1.414214.

The keystrokes needed to find the remaining negative root are now provided.

TI-83 Graphing Calculator Solution

| 2nd | CALC | 2 [: zero] |

TI-85 Graphing Calculator Solution

| EXIT | F3 [ROOT] | ◀ | ◀ | ENTER |

97

LABORATORY 8: FUNCTIONS

TI-83 Graphing Calculator Solution

Use the ◄ key move the cursor to the left of the other root.

Set the left bound by pressing

| ENTER |

Use the ► key to move the cursor to the right of the root. Set the right bound and find the root by pressing

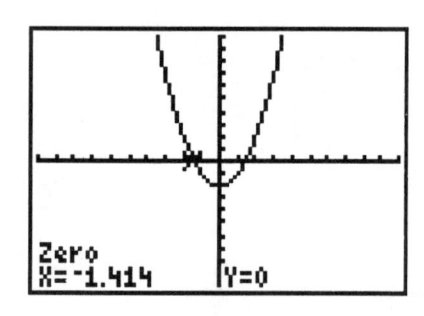

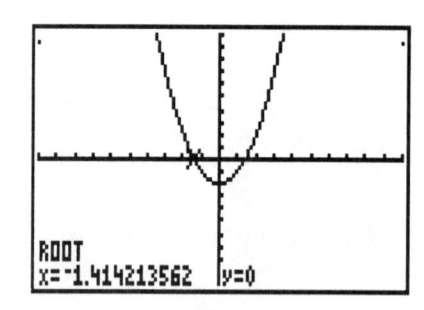

| ENTER | | ENTER |

TI-85 Graphing Calculator Solution

From these graphs, we conclude that the roots are $x = 1.414214$ and $x = -1.414214$.

Part (c)

For part (c), use the graph to evaluate the function when x has a value of 0, 3, and then $-1/2$. Press the following keys to find the three different x values.

TI-83 Graphing Calculator Solution

| 2nd | | CALC | | 1 [: value] |

| 0 | | ENTER |

TI-85 Graphing Calculator Solution

| GRAPH | | MORE | | MORE |

| F1 [EVAL] | | 0 | | ENTER |

98

LABORATORY 8: FUNCTIONS

TI-83 Graphing Calculator Screen

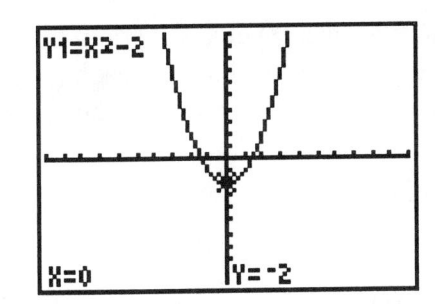

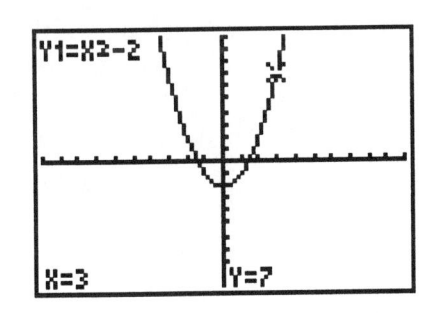

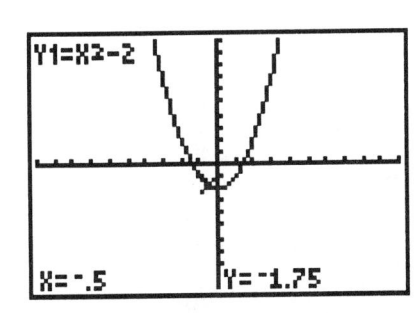

TI-85 Graphing Calculator Screen

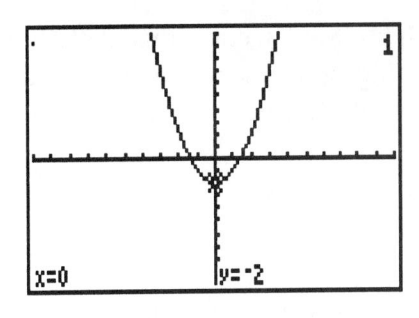

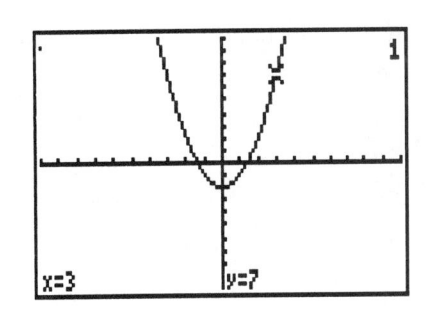

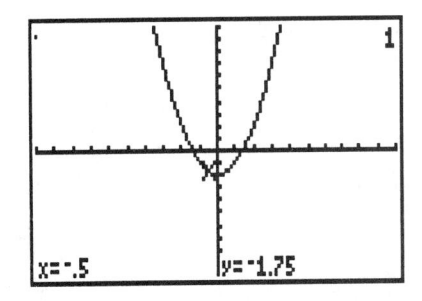

[2nd] [CALC] [1 [: value]]

[EXIT] [F1 [EVAL]] [3] [ENTER]

[3] [ENTER]

[2nd] [CALC] [1 [: value]]

[EXIT] [F1 [EVAL]] [(-)] [1]

[(-)] [1] [÷] [2] [ENTER]

[÷] [2] [ENTER]

The three different values of the function are: $f(0) = -2$, $f(3) = 7$, and $f\left(-\frac{1}{2}\right) = -1.75$. The three points read from the calculator screens are: $(0, -2)$, $(3, 7)$, and $\left(-\frac{1}{2}, -1.75\right)$.

LABORATORY 8: FUNCTIONS

EXERCISES

1. Determine if the following equations are functions or are just relations. Sketch the graph of each of the following, showing how you determined if they were functions or not. For each function, determine the interval where the function increases, decreases, or remains constant.

(a) $y = 3x^2 - 5$
(b) $y = \pm\sqrt{2x}$
(c) $y = 2x + x^2$
(d) $x^2 - 2y^2 - 3 = 0$
(e) $2x^2 = 3y^2 - 4$

2. State the *domain* and *range* of each of the following. Explain why they are or are not functions. Sketch the graphs if need be.

(a) (0, 2), (1, 4), (2, 8), (3, 16), (4, 32).
(b) $y = \sqrt{x-5}$
(c) $y = |x - 4|$
(d) $y = x^3 - 3x^2$

3. Find the roots (if possible) and evaluate the function (if possible) at the specified value of the independent variable.

(a) $f(x) = 2x - 3$, for $f(-2), f(0), f(2/3)$
(b) $f(x) = \sqrt{x+8} + 2$, for $f(-8), f(3), f(17/2)$
(c) $f(x) = |x + 2| + |x - 2|$, for $f(-1), f(0), f(9)$.

4. The distance traveled by a freely falling body is a function of the elapsed time t,

$$y(t) = v_0 t + \frac{1}{2} g t^2$$

where v_0 is the initial velocity and g is the acceleration due to gravity (32.2 ft/s²). If v_0 is 65.0 ft/s, set up a table and find $y(10.0), y(15.0),$ and $y(20.0)$. Graph the function and use the graph to determine the values of y.

5. **Peanut Production.** For 1984 through 1991, the peanut production, P (in billions of pounds), in the United States is shown in the table. The table also shows the average yield, y (in pounds per acre). Was the peanut production a function of average yield? Explain.

Year	1984	1985	1986	1987	1988	1989	1990	1991
Average yield, y	2880	2805	2299	2335	2440	2440	1895	2596
Production, P	4.44	4.13	3.71	3.60	4.00	4.00	3.62	5.09

6. The function $P(d) = 1 + (d/33)$ gives the pressure, in atmospheres (atm), at a depth d in the sea (d is in feet). Note that $P(0) = 1$ atm, $P(33) = 2$ atm. Find the pressure at 12 feet, 20 feet, 30 feet, and 105 feet. Set the viewing rectangle to reflect **xMin:** 0, **xMax:** 110, **xScl:** 10, **yMin:** 0.7, **yMax:** 12, and **yScl:** 1.

LABORATORY 9: POLYNOMIALS

Purpose:

The purpose of this laboratory is to understand how to graph and find the roots of equations of degree higher than two using the TI-83/82 and TI-85/86 calculators.

Analytic Approach:

Let $a_0, a_1, a_2, ..., a_n$ be real numbers, and n is non-negative, then a **polynomial in x** is expressed in the form

$$a_n x^n + \cdots + a_2 x^2 + a_1 x + a_0$$

where $a_n \neq 0$ and a_0 is the constant term. The polynomial is of **degree** n, and the number, a_n, is called the **leading coefficient**. A polynomial written with the powers of x in *descending order* is in **standard form**. In the polynomial $3x^5 + 2x^3 - 5x - 3$, the degree of the polynomial is 5 while the leading coefficient is 3 and the constant term, a_0, is -3.

A **term** is either a constant, variable, or an expression that can be written as a product of constants and variables. For example a^2, $3x$, 3, and 0 are each called a **term**. Polynomials or expressions containing one term are referred to as **monomials**. The number 3 and certain variables such as e and π are called **constant polynomials**. The number 0 is sometimes referred to as the **zero polynomial**.

The expression $x^2 - 3$ is a polynomial that has two terms. It is the difference of the two terms x^2 and 3. An expression or polynomial that contains two terms is called a **binomial**.

The expression $z^2 + xz - \pi$ is a three termed *polynomial in two variables*, z and x. It is called a **trinomial**. The Greek letter π is *not* a variable because it has a *constant value*. However, π is a term.

The **degree of a polynomial** is the **largest sum of the exponents of any of its monomial terms** (provided like terms have been combined). The *degree of a monomial in more than one variable* is the **sum of the exponents** of the variables. The *degree* of the monomial $3x^2 y^7$ is $2 + 7$ or 9. In the polynomial $4x^2 y - 3x^3 y^4 + 2xy^4$, the sum of the powers of the middle term yields the greatest sum 7. The *degree* of this trinomial is 7.

An expression such as $\dfrac{a^2 + 4}{2}$ is a polynomial, while the expression $\dfrac{a^2 + 4}{a - 1}$ is *not* a polynomial. The constant divisor of 2 (the denominator 2) in $\dfrac{a^2 + 4}{2}$ allows for the rewriting of the polynomial as, $\dfrac{1}{2} a^2 + \dfrac{4}{2}$, which is the same as, $\dfrac{1}{2} a^2 + 2$. The expression $\dfrac{a^2 + 4}{a - 1}$ is really the quotient of two polynomials and is known as a *rational expression*.

LABORATORY 9: POLYNOMIALS

Numerical Evaluation of Polynomials

Many times it is necessary to evaluate a polynomial (assuming it is not a constant) for different numerical values of the variable(s). In place of the variable(s) insert the given numerical value that the variable(s) represent. If the same variable occurs more than once, replace it with the same number each time the variable occurs.

For the problems in this Laboratory, set the **Float** to three decimal places.

Problem 1: Evaluate each polynomial using the specified values for the variable.

(a) $\frac{1}{3}(a^3 + ab - 3b^2)$; when $a = -1$ and $b = 2$.

(b) $m^2 - 2mn + n^3 - t^4$; when $m = -2$ and $n = 0.3$ and $t = -2.2$.

The keystrokes and the screens are the same on all four TI calculators. Therefore, one screen will be shown for *both* parts (a) and (b).

Graphing Calculator Solution

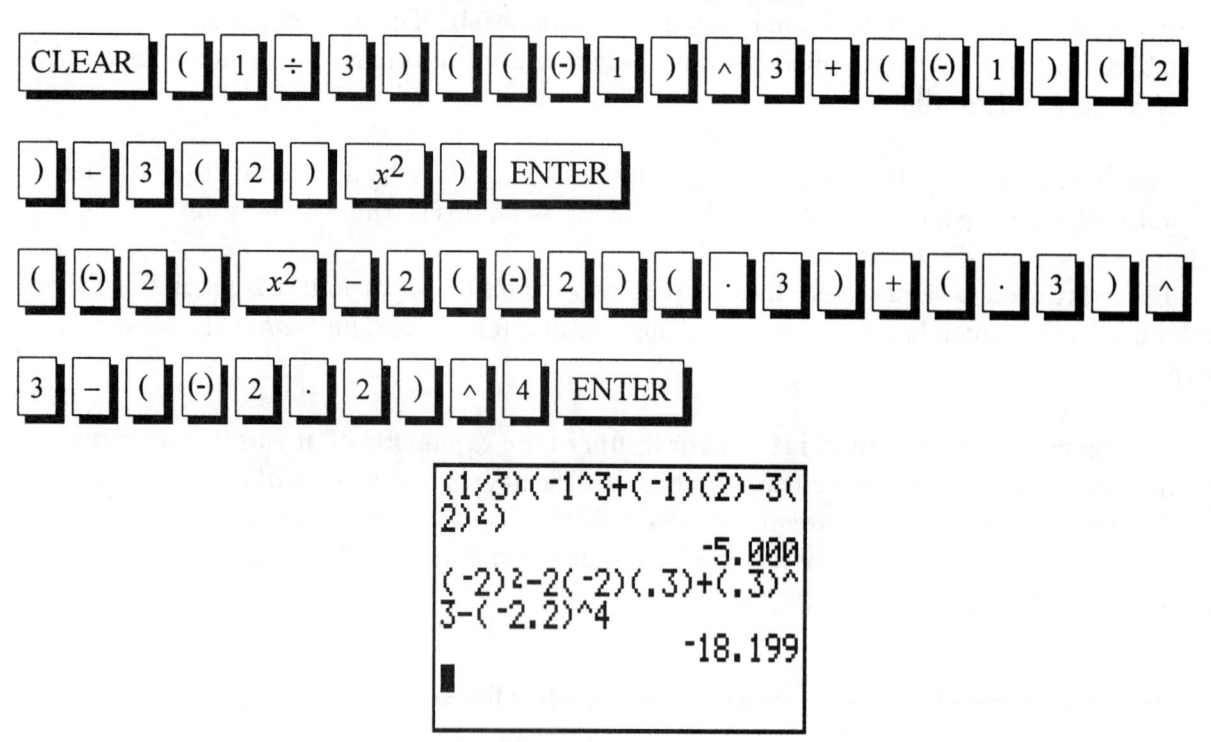

[CLEAR] [(] [1] [÷] [3] [)] [(] [(] [(-)] [1] [)] [^] [3] [+] [(] [(-)] [1] [)] [(] [2]
[)] [−] [3] [(] [2] [)] [x^2] [)] [ENTER]

[(] [(-)] [2] [)] [x^2] [−] [2] [(] [(-)] [2] [)] [(] [.] [3] [)] [+] [(] [.] [3] [)] [^]
[3] [−] [(] [(-)] [2] [.] [2] [)] [^] [4] [ENTER]

```
(1/3)(-1^3+(-1)(2)-3(
2)²)
                -5.000
(-2)²-2(-2)(.3)+(.3)^
3-(-2.2)^4
               -18.199
```

The value of the polynomial in part (a) is -5.000. How many terms does the polynomial in part (a) have and name them? What is the degree of the polynomial, $\frac{1}{3}(a^3 + ab - 3b^2)$?

LABORATORY 9: POLYNOMIALS

The result from the substitution of -2 for *m*, 0.3 for *n*, and -2.2 for *t*, in $m^2 - 2mn + n^3 - t^4$ in part (b) is -18.199. How many terms does the polynomial have? What is the degree of the polynomial expression?

Problem 2: Evaluate the polynomial expression $y^4 + 2 - 5x^2 + 3x^3y + 9xy$ when

(a) $x = -1$ and $y = 2$ (b) $x = -4$ and $y = \frac{2}{3}$.

This problem can be evaluated on the **TI-83/82** and the **TI-85/86** in the same manner as was *Problem 1*.

If an expression is to be used in successive evaluations, it is easier to enter the expression *once* and use the **SOLVER** feature to do the evaluation. Since the polynomial expression is to be evaluated twice using different values for *x* and *y*, the **SOLVER** feature on the **TI-85/86** will be shown. Although the **TI-83** and **TI-82** have a **SOLVER** or **SOLVE** feature, they can only work with one variable at a time.

Part (a)

On the **TI-83/82**, evaluate the polynomial as you practiced in *Problem 1*. On the **TI-85/86** enter the polynomial expression exactly as it appears in the problem by pressing the following keys.

TI-85 Graphing Calculator Solution

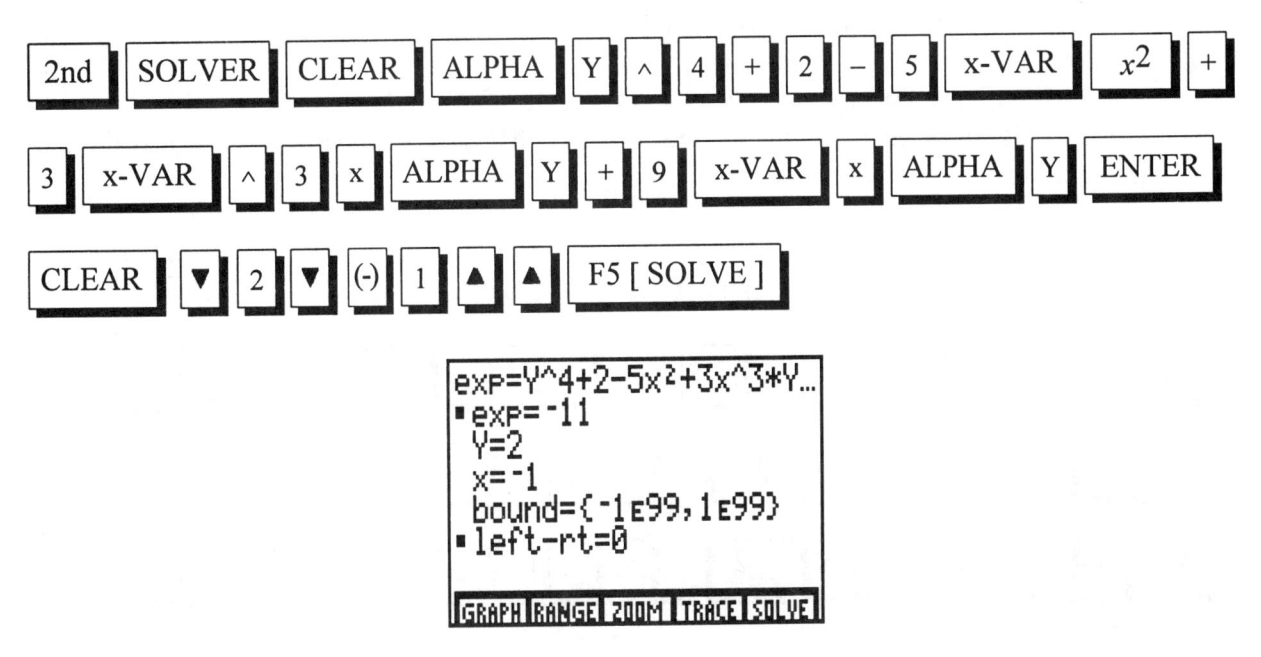

| 2nd | SOLVER | CLEAR | ALPHA | Y | ^ | 4 | + | 2 | - | 5 | x-VAR | x^2 | + |

| 3 | x-VAR | ^ | 3 | x | ALPHA | Y | + | 9 | x-VAR | x | ALPHA | Y | ENTER |

| CLEAR | ▼ | 2 | ▼ | (-) | 1 | ▲ | ▲ | F5 [SOLVE] |

```
exp=Y^4+2-5x²+3x^3*Y...
■exp=-11
 Y=2
 x=-1
 bound={-1E99,1E99}
■left-rt=0

GRAPH RANGE ZOOM TRACE SOLVE
```

The evaluation of the polynomial expression $y^4 + 2 - 5x^2 + 3x^3y + 9xy$, for $x = -1$ and $y = 2$ is -11.

LABORATORY 9: POLYNOMIALS

Part (b)

Since the expression has already been entered, you need only change the values for *x* and *y*.

TI-85 Graphing Calculator Solution

[CLEAR] [▼] [2] [÷] [3] [▼] [(-)] [4] [▲] [▲] [F5 [SOLVE]]

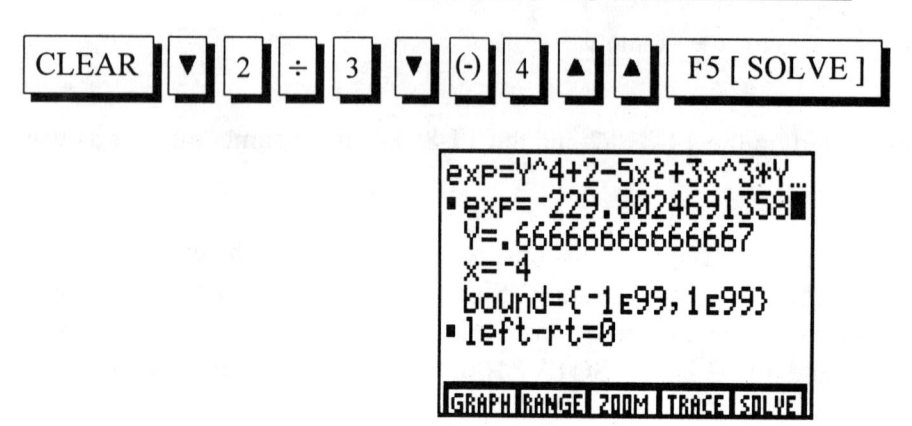

```
exp=Y^4+2-5x²+3x^3*Y...
•exp=-229.8024691358
 Y=.66666666666667
 x=-4
 bound={-1E99,1E99}
•left-rt=0
```
[GRAPH][RANGE][ZOOM][TRACE][SOLVE]

The result of the evaluation of the expression using -4 for *x* and $\frac{2}{3}$ for *y* in part (b) is -229.802.

Problem 3: When a ball is thrown upward from the ground with an initial velocity of 64 feet per second, its height *y* after *t* seconds is given by

$$y(t) = -16t^2 + 64t$$

Find its height (a) after 1.5 seconds, (b) after 2 seconds, and (c) after 3.5 seconds. **TI-85/86** users can put the expression in **SOLVER** to evaluate *or* follow the given keystrokes below.

Graphing Calculator Solution

[CLEAR] [(-)] [1] [6] [(] [1] [.] [5] [)] [x^2] [+] [6] [4] [(] [1] [.] [5] [)] [ENTER]

[(-)] [1] [6] [(] [2] [)] [x^2] [+] [6] [4] [(] [2] [)] [ENTER]

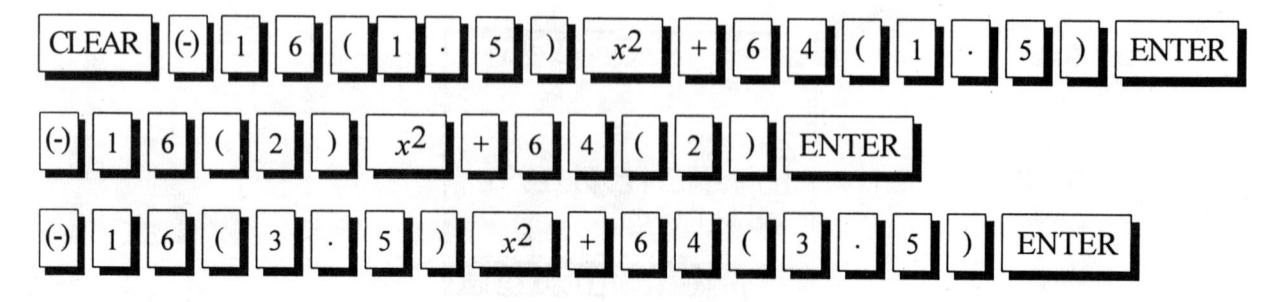

[(-)] [1] [6] [(] [3] [.] [5] [)] [x^2] [+] [6] [4] [(] [3] [.] [5] [)] [ENTER]

LABORATORY 9: POLYNOMIALS

Graphing Calculator Screen

```
-16(1.5)²+64(1.5)
                60.000
-16(2)²+64(2)
                64.000
-16(3.5)²+64(3.5)
                28.000
■
```

The height of the ball (a) after 1.5 seconds was 60 feet, (b) after 2 seconds the ball was 64 feet above the ground, and (c) the height of the ball after 3.5 seconds was 28 feet.

Interesting questions arise from *Problem 2*. Questions such as: "Find the time at which the ball reaches its maximum height" and "At what time does the ball hit the ground?". If the polynomial function, $y(t)$ is set equal to zero ($-16t^2 + 64t = 0$), and the resulting equation is then solved for t, the solution(s) can be found. The solution(s) to the polynomial equation is(are) the time(s) when the height of the ball is zero, i.e., the ball is at ground level.

The maximum point (highest point on the curve) that the graph of $y(t) = -16t^2 + 64t$ reaches is the point at which the ball stops rising and starts to fall.

Problem 4: Graph $y(t) = -16t^2 + 64t$

    (a) Determine the two roots of $y(t)$, and
    (b) determine the point on the graph where the ball stops rising and starts to fall.

Looking back at the results of Problem 3, we see that the graph reached a height of 64 feet when *t* was equal to 2 seconds. Set the viewing rectangle to **xMin: 0, yMin: -15, yMax: 70,** and a **yScl: 5**.

TI-83 Graphing Calculator Solution   TI-85 Graphing Calculator Solution

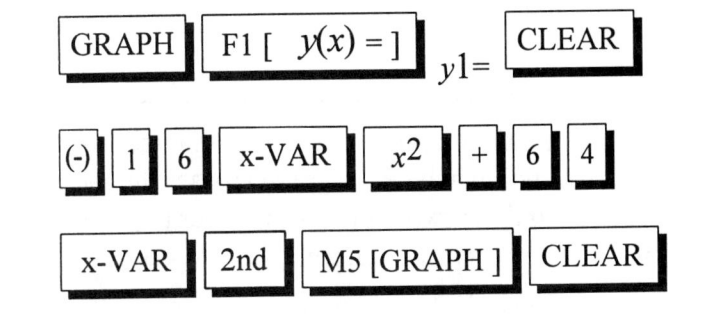

| Y= | CLEAR | \Y₁= | (-) | 1 | 6 | | GRAPH | F1 [y(x) =] | y1= | CLEAR |

| X,T,θ,n | x² | + | 6 | 4 | | (-) | 1 | 6 | x-VAR | x² | + | 6 | 4 |

| X,T,θ,n | GRAPH | | x-VAR | 2nd | M5 [GRAPH] | CLEAR |

LABORATORY 9: POLYNOMIALS

Graphing Calculator Screen

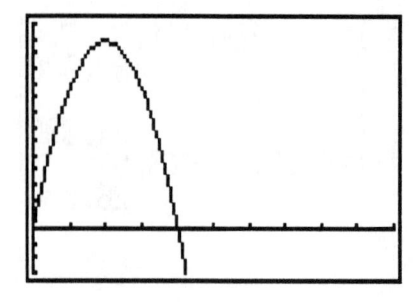

From the screen showing the graphical solution, you get a much clearer picture of the projectile of the ball as well as the points at which the ball is at ground level.

To determine the point at which the ball stops rising and starts to fall, find the highest point that the graph reaches. Press TRACE and move along the graph to approximately $x = 2$. Now press the following keys to find the actual y value (height).

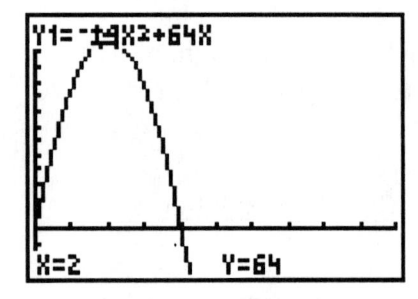

TI-83 Graphing Calculator Solution   TI-85 Graphing Calculator Solution

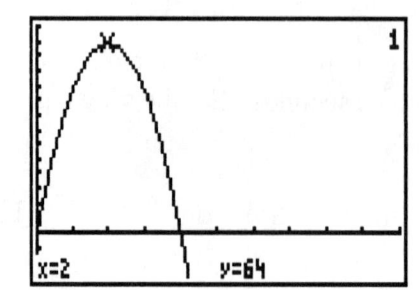

| 2nd | CALC | 1 [: value] | | GRAPH | MORE | MORE |

| 2 | ENTER | | F1 [EVAL] | 2 | ENTER |

It took the ball 2 seconds to reach its maximum height of 64 feet above the ground. Upon examining the graph, it is apparent that there are two places where the curve crosses the x-axis. The two points where the graph crosses the x-axis are the two roots or zeros. Press the following keys to find the two roots. Remember to press **F1** on the **TI-86** when determining the root. Then follow the keystroke pattern on the **TI-83** to set the lower and upper bounds and a guess.

Press ENTER after each.

LABORATORY 9: POLYNOMIALS

TI-83 Graphing Calculator Solution

[2nd] [CALC] [2 [: zero]]

set the bounds and a guess pressing

[ENTER] after each.

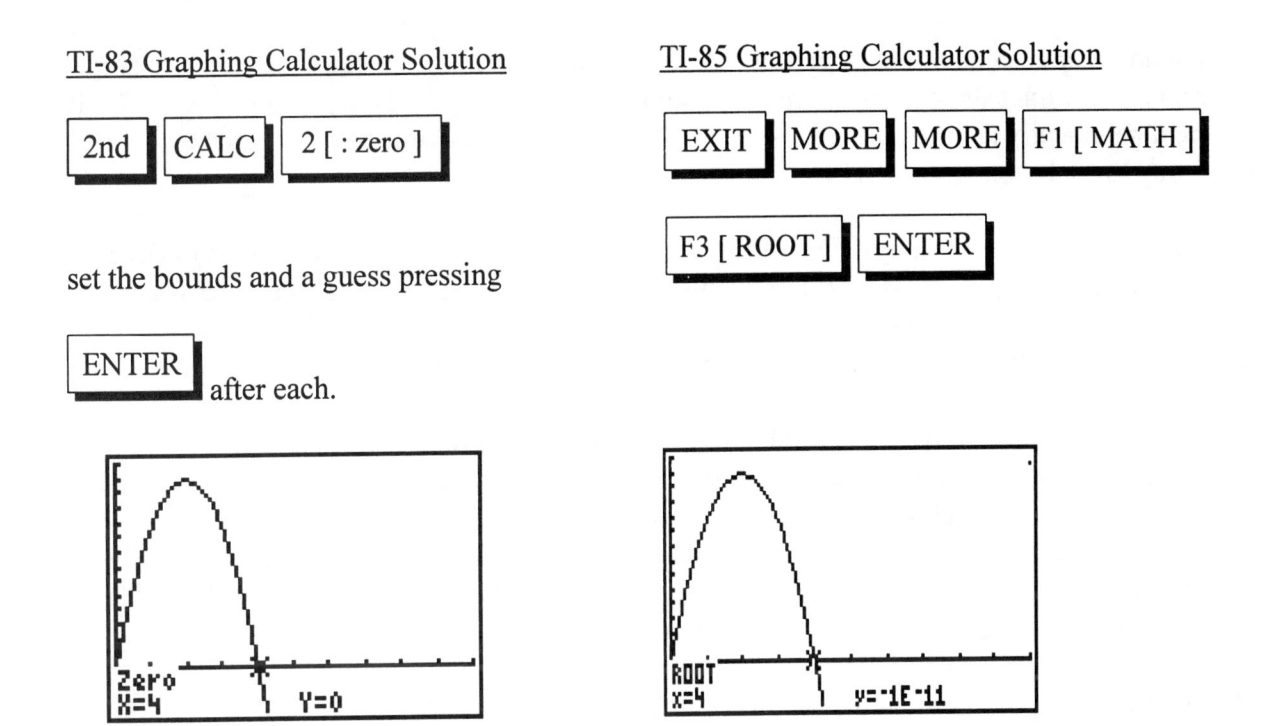

TI-85 Graphing Calculator Solution

[EXIT] [MORE] [MORE] [F1 [MATH]]

[F3 [ROOT]] [ENTER]

The two roots are 0 and 4. The roots define what happens to the ball in relation to the ground. At the point (0, 0) the ball has not been thrown. At the point (4, 0) the ball has returned to the ground. Any point below the x-axis is meaningless, since the ball comes to rest upon hitting the ground.

Graphing Polynomials of Degree Higher Than Two

In this section, we turn our attention to polynomials of degree greater than two. A polynomial function of nth degree has the form

$$f(x) = a_n x^n + a_1 x^{n-1} + \cdots + a_{n-1} x + a_0, \quad a_n \neq 0.$$

A **polynomial of degree n** has n and *only n roots*. Two or more of the roots may be equal; however, there can *never be more than n roots*.

The roots of $f(x) = a_n x^n + a_1 x^{n-1} + \cdots + a_{n-1} x + a_0$ can be found by setting $f(x) = 0$ and solving for x. This is called the **Fundamental Theorem of Algebra.**

The factors of a polynomial function can be found by applying the **Factor Theorem:** If r is a root of $f(x) = 0$, then $x - r$ is a factor of $f(x)$. Conversely, if $x - r$ is a factor of $f(x)$, then r is a root of $f(x) = 0$. The roots of a polynomial may not be **real**. These roots are called **complex roots**. If one complex root is found, there must exist another complex root, called its **conjugate.** Complex roots occur in conjugate pairs.

LABORATORY 9: POLYNOMIALS

In a polynomial equation of **degree two**, there are **at most 2 real roots**. The two roots are either *both real numbers* or they are *both complex numbers*. Why? The equation $x^2 - x - 12 = 0$ can be algebraically factored into $(x - 4)$ and $(x + 3) = 0$. Each factor is then set equal to zero, obtaining the solutions, $x = 4$ and $x = -3$.

For a polynomial equation of **degree three**, the roots can be (a) all three real roots or (b) one real root with 2 complex roots. All three roots can not be complex, why? In a similar way, $x^3 - 7x - 6 = 0$, can be factored into $(x +1)(x -3)(x +2) = 0$. By setting each factor equal to zero, the three roots of the equation can be stated as: $x = -1$, $x = 3$, and $x = -2$.

There can also exist what is referred to as **multiplicity of roots**. *Multiplicity of roots* is defined as the *same root* appearing *more than once*. For example, in the polynomial equation $(x - 1)(x + 2)^3 = 0$, set each factor equal to 0 and solve for x. The polynomial equation is of degree **four** and, therefore, has **four roots**. Setting the first factor $x - 1 = 0$, the root $x = 1$ is determined. When the second factor of the polynomial equation $(x + 2)^3$ is set equal to 0, we find the root $x = -2$ is a root of multiplicity three.

Problem 5:   Graph $f(x) = x^4 - 5x^2 + 4$. Use the graph of $f(x)$ to state the roots.
There should be four roots. Why?

Remember to set the viewing rectangle to the standard setting. **Remember** to press **F1** on the **TI-86** to determine the root. Then follow the keystroke pattern on the **TI-83** to set the bounds and a guess. The keystrokes for only one root will be demonstrated.

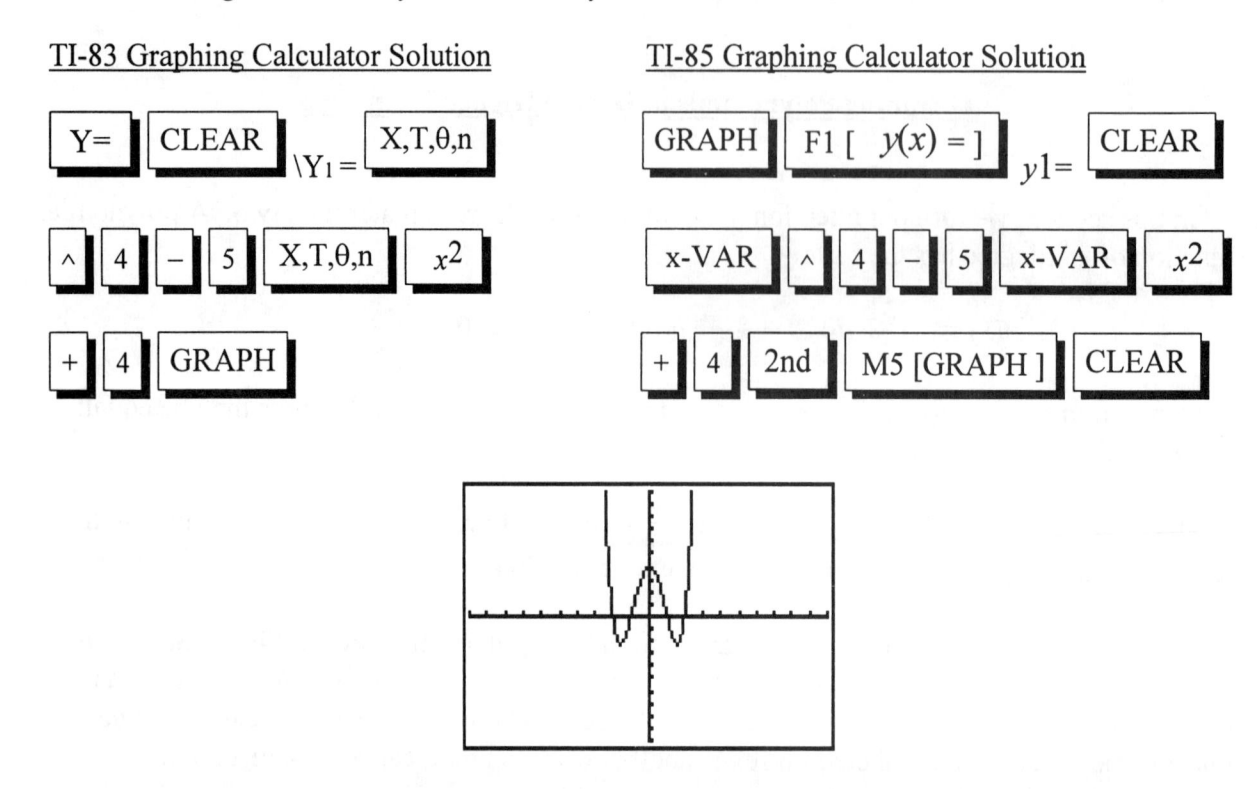

TI-83 Graphing Calculator Solution   TI-85 Graphing Calculator Solution

| Y= | CLEAR | \Y₁= | X,T,θ,n | | GRAPH | F1 [y(x) =] | y1= | CLEAR |

| ^ | 4 | – | 5 | X,T,θ,n | x² | | x-VAR | ^ | 4 | – | 5 | x-VAR | x² |

| + | 4 | GRAPH | | + | 4 | 2nd | M5 [GRAPH] | CLEAR |

There are four points shown on the screen above where the graph crosses the *x*-axis.

LABORATORY 9: POLYNOMIALS

TI-83 Graphing Calculator Solution

[2nd] [CALC] [2 [: zero]] [▶]

[▶] [▶] [▶] [ENTER] [▶] [ENTER]

[ENTER]

TI-85 Graphing Calculator Solution

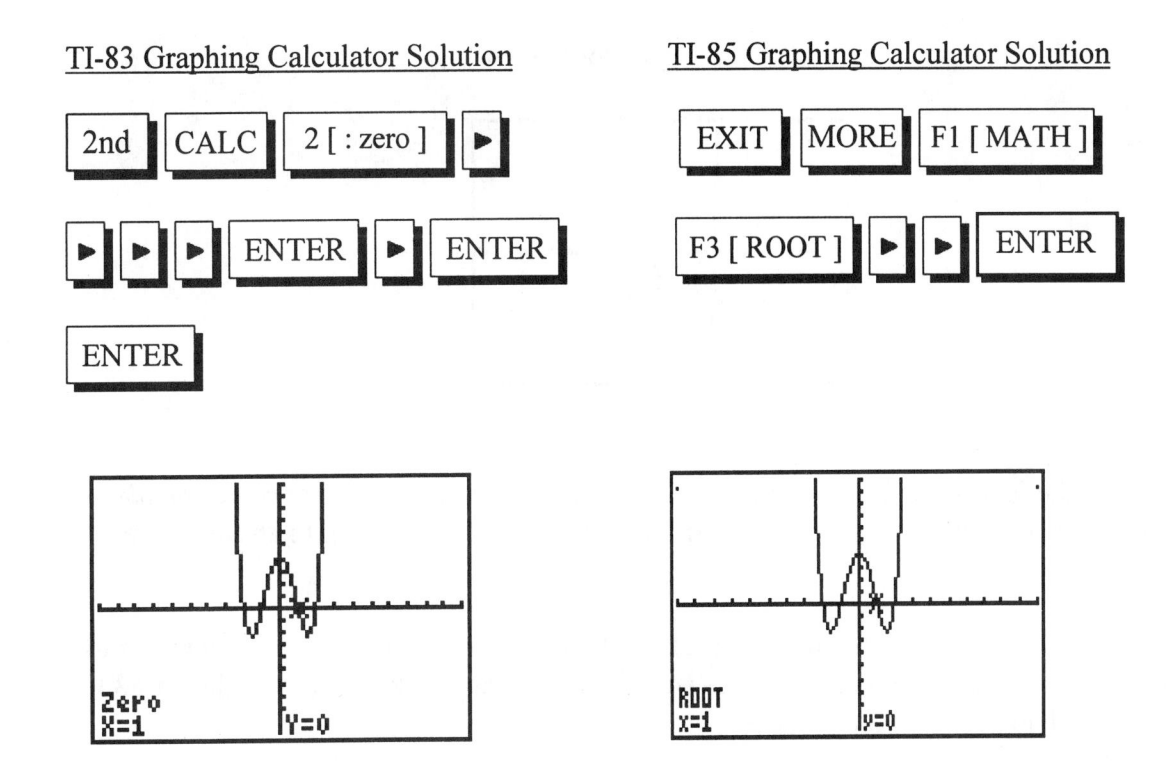

[EXIT] [MORE] [F1 [MATH]]

[F3 [ROOT]] [▶] [▶] [ENTER]

Remember, on the **TI-83, 82,** and **TI-86** you must reset the upper and lower bounds three more times to find the three remaining roots of the polynomial function. On the **TI-85**, use the arrow keys to move along the graph to, *one at a time*, define the three remaining roots.

After performing the proper keystrokes, the four roots of $f(x) = x^4 - 5x^2 + 4$ are located at the points (-2,0), (-1,0), (1,0), (2,0). For this function all the roots are real. **Remember, only real roots can be found on the graph!**

Problem 6: Graph $f(x) = (x + 3)(x - 1)^3$. Determine the real roots of $f(x)$. The degree of the polynomial function is 4 (1 + 3 = 4).

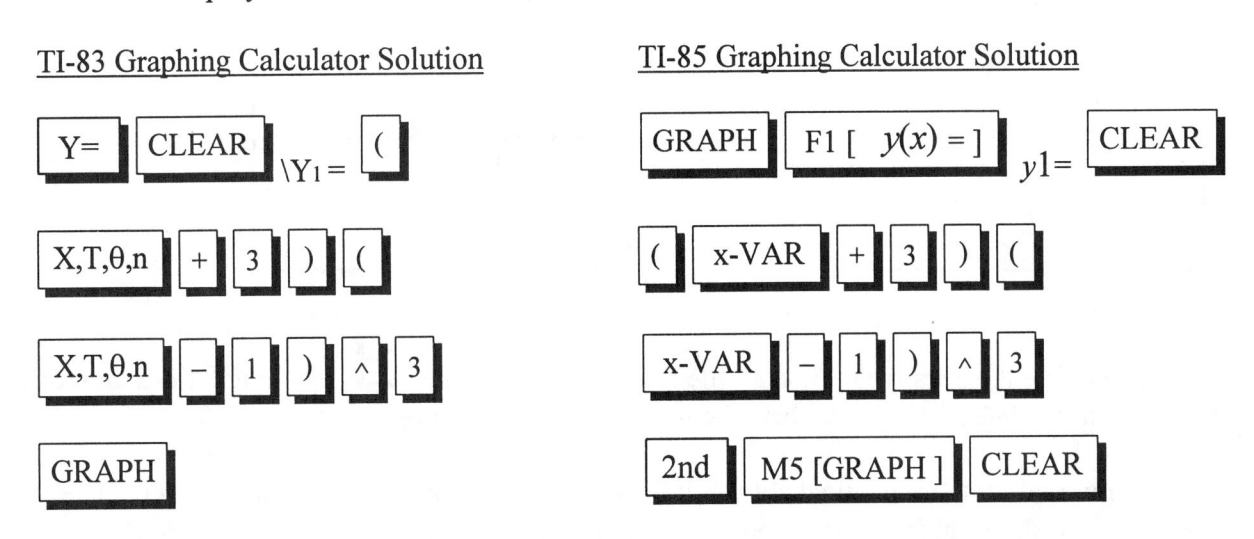

TI-83 Graphing Calculator Solution

[Y=] [CLEAR] \Y₁= [(]

[X,T,θ,n] [+] [3] [)] [(]

[X,T,θ,n] [−] [1] [)] [^] [3]

[GRAPH]

TI-85 Graphing Calculator Solution

[GRAPH] [F1 [y(x) =]] y1= [CLEAR]

[(] [x-VAR] [+] [3] [)] [(]

[x-VAR] [−] [1] [)] [^] [3]

[2nd] [M5 [GRAPH]] [CLEAR]

LABORATORY 9: POLYNOMIALS

Graphing Calculator Screen

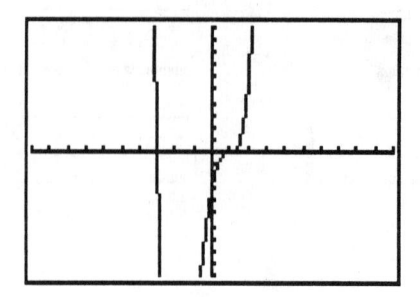

Why aren't the four roots shown on the graph? Does this mean that two roots are *complex* or does it indicate *multiplicity of roots*? Look carefully at $f(x) = (x + 3)(x - 1)^3$ to answer the two questions.

Use the following keystrokes to find the roots of the function graphed above. **Remember** to press **F1** to determine the root on the **TI-86.** Then follow the keystroke pattern on the **TI-83** to set the lower and upper bound and a guess.

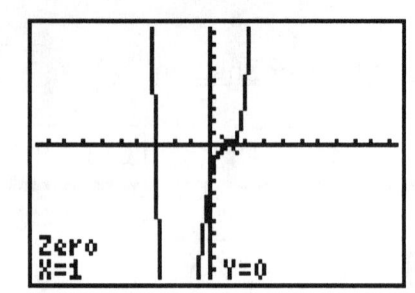

TI-83 Graphing Calculator Solution   TI-85 Graphing Calculator Solution

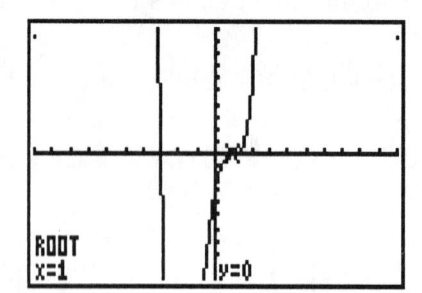

| 2nd | CALC | 2 [: zero] |

| ◄ | ► | ENTER | ► | ► | ► |

| ENTER | ENTER |

| EXIT | MORE | F1 [MATH] |

| F3 [ROOT] | ► | ► | ENTER |

To determine the negative root on all the calculators, press the ◄ key enough times to move the cursor to the **left** of the point where the graph intersects the *x*-axis.

The four real roots are pictured above. It happens that the root $x = 1$ is a multiple root. The other real root shown above is at the point (-3, 0) and is written as $x = -3$.

110

LABORATORY 9: POLYNOMIALS

Problem 7: Graph $f(x) = x^3 + 3x^2 - 4x + 5$.
        (a) How many roots should there be? and
        (b) State all the **real** roots.

Reset the value of **yMax** to 20 on all the TI calculators.

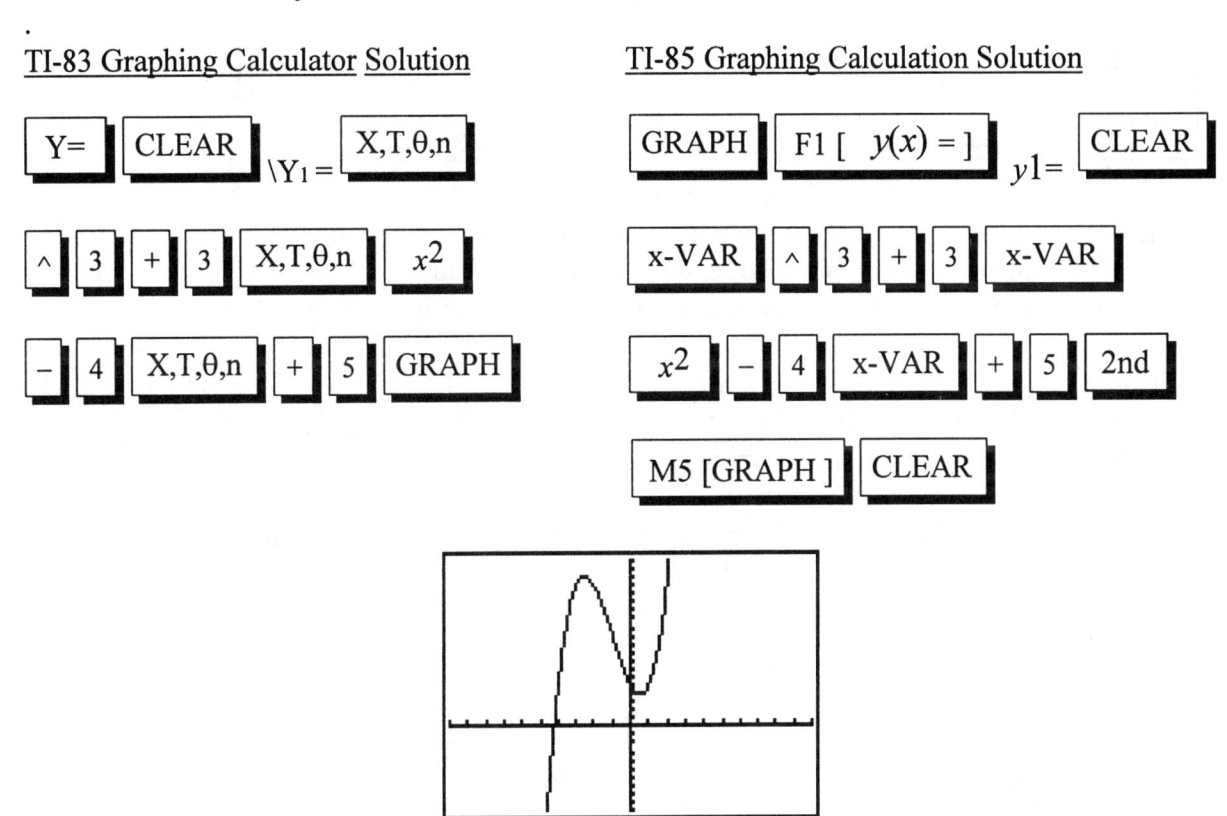

TI-83 Graphing Calculator Solution

| Y= | CLEAR | \Y₁= | X,T,θ,n |

| ^ | 3 | + | 3 | X,T,θ,n | x² |

| – | 4 | X,T,θ,n | + | 5 | GRAPH |

TI-85 Graphing Calculation Solution

| GRAPH | F1 [y(x) =] | y1= | CLEAR |

| x-VAR | ^ | 3 | + | 3 | x-VAR |

| x² | – | 4 | x-VAR | + | 5 | 2nd |

| M5 [GRAPH] | CLEAR |

To determine the one real root of the polynomial function (the graph only crosses the *x*-axis once), press the following keys. **Remember** to press **F1** to determine the root on the **TI-86**. Then follow the keystroke pattern on the **TI-83** to set the lower and upper bound and a guess.

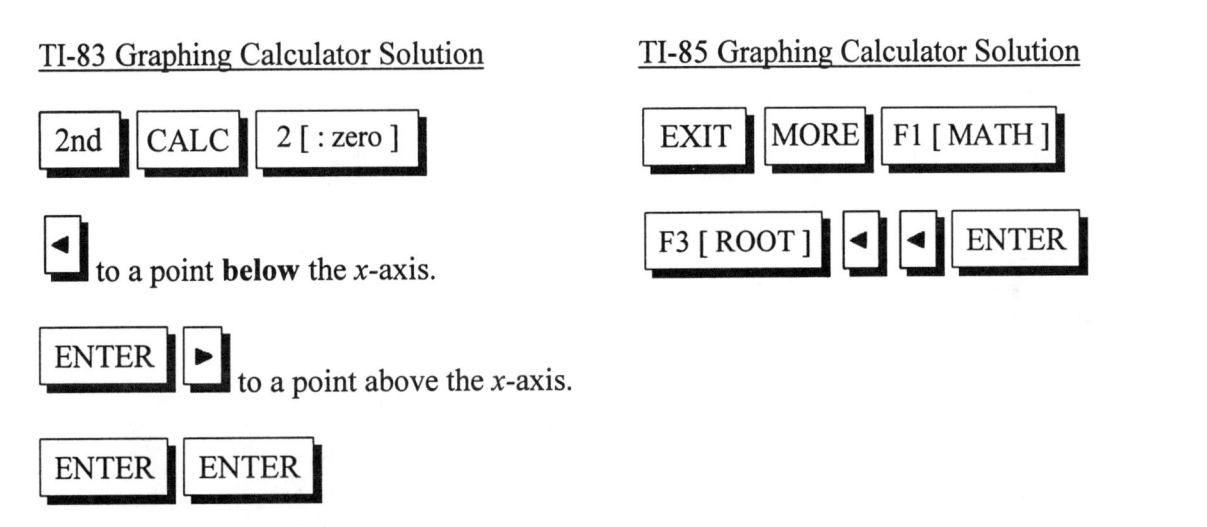

TI-83 Graphing Calculator Solution

| 2nd | CALC | 2 [: zero] |

◄ to a point **below** the *x*-axis.

| ENTER | ► | to a point above the *x*-axis.

| ENTER | ENTER |

TI-85 Graphing Calculator Solution

| EXIT | MORE | F1 [MATH] |

| F3 [ROOT] | ◄ | ◄ | ENTER |

111

LABORATORY 9: POLYNOMIALS

TI-83 Graphing Calculator Screen   TI-85 Graphing Calculator Screen

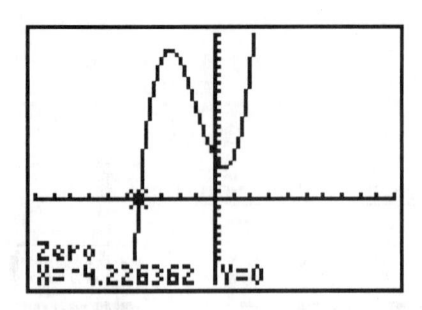

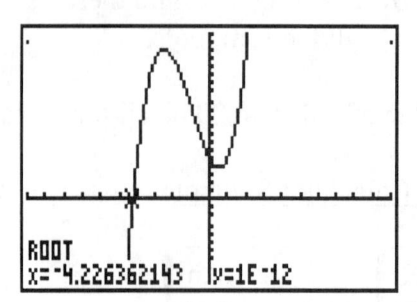

There is only one **real** root to this function (-4.226, 0). The remaining two roots are **complex** and cannot be found on the graph. The remaining two roots can be found algebraically. Refer to your text for help in doing the algebra necessary to find the two remaining roots.

Problem 8:   Find all the roots of $f(x) = x^3 + 8$. For **TI-83 and 82 users**, get pen and paper and *algebraically* find the roots. One root of $f(x)$ is real and can be found by graphing the function. Complex roots cannot be found on graphs.

TI-85 and TI-86 users can find both the complex and real roots, using the **POLY** feature. The **Poly** feature will solve any polynomial equation of nth degree. **All** the roots *both* real and complex are determined. The **TI-83 and TI-82** do not have a **POLY** feature.

TI-85 Graphing Calculator Solution

| 2nd | POLY | 3 | ENTER | 1 | ▼ | 0 | ▼ | 0 | ▼ | 8 |

TI-85 Graphing Calculator Screen

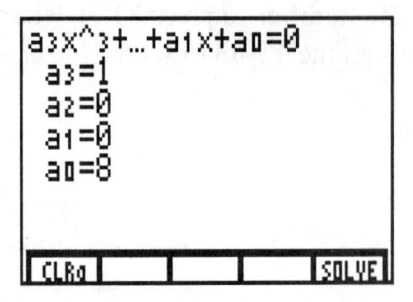

```
a3x^3+...+a1x+a0=0
 a3=1
 a2=0
 a1=0
 a0=8


 CLRq                         SOLVE
```

F5 [SOLVE]

LABORATORY 9: POLYNOMIALS

TI-85 Graphing Calculator Screen

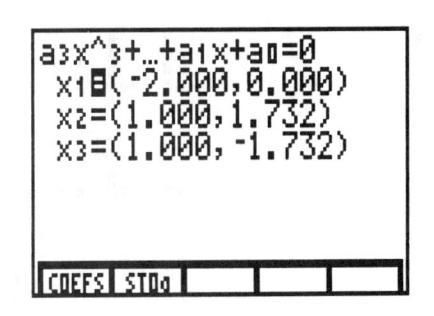

```
a3x^3+...+a1x+a0=0
 x1∎(-2.000,0.000)
 x2=(1.000,1.732)
 x3=(1.000,-1.732)



 COEFS STOa
```

The three roots of $f(x) = x^3 + 8$ are (-2, 0), the real root, and the complex roots (1.000, 1.732) and (1.000, -1.732). The complex roots should be written as $1 + 1.732i$ and $1 - 1.732i$, where i represents the *imaginary* number $\sqrt{-1}$.

Problem 9: A company that produces CD players estimates that the profit for selling a particular model is given by

$$P = -66x^3 + 4230x^2 - 310{,}000$$

where P is the profit in dollars and x is the advertising expense (in $10,000's). Using this model, what would the smaller of two advertising amounts that yield a profit of $1,500,000 be?

The substitution of 1,500,000 for P into $P = -66x^3 + 4230x^2 - 310{,}000$ must be done. Then the equations must be set equal to zero before using the **SOLVER** or **SOLVE** features on the **TI-83** and **TI-82** calculators.

The **TI-82 SOLVE** feature is found on the same key in the **MATH** menu as on the **TI-83**. Then **enter the expression** followed by a **coma**. Press [X,T,θ] followed by a **coma**. Press **0** followed by a **right parenthesis**. Finally, press [ENTER].

TI-83 Graphing Calculator Solution

[MATH] [0[: Solver ...]] [▲]

[CLEAR] [(-)] [6] [6] [X,T,θ,n]

[^] [3] [+] [4] [2] [3] [0]

TI-85 Graphing Calculator Solution

[2nd] [SOLVER] [CLEAR] [ALPHA] [P]

[ALPHA] [=] [(-)] [6] [6] [x-VAR] [^]

[3] [+] [4] [2] [3] [0] [x-VAR] [x^2]

LABORATORY 9: POLYNOMIALS

<u>TI-83 Graphing Calculator Solution</u>

[X,T,θ,n] [x²] [−] [1] [8] [1]

[0] [0] [0] [0] [ENTER]

[ALPHA] [SOLVE]

```
-66X^3+4230X²...=0
·X=27.303439667...
 bound={-1E99,■...
·left-rt=0
```

<u>TI-85 Graphing Calculator Solution</u>

[−] [3] [1] [0] [0] [0] [0] [ENTER]

[1] [5] [0] [0] [0] [0] [0] [ENTER]

[F5 [SOLVE]]

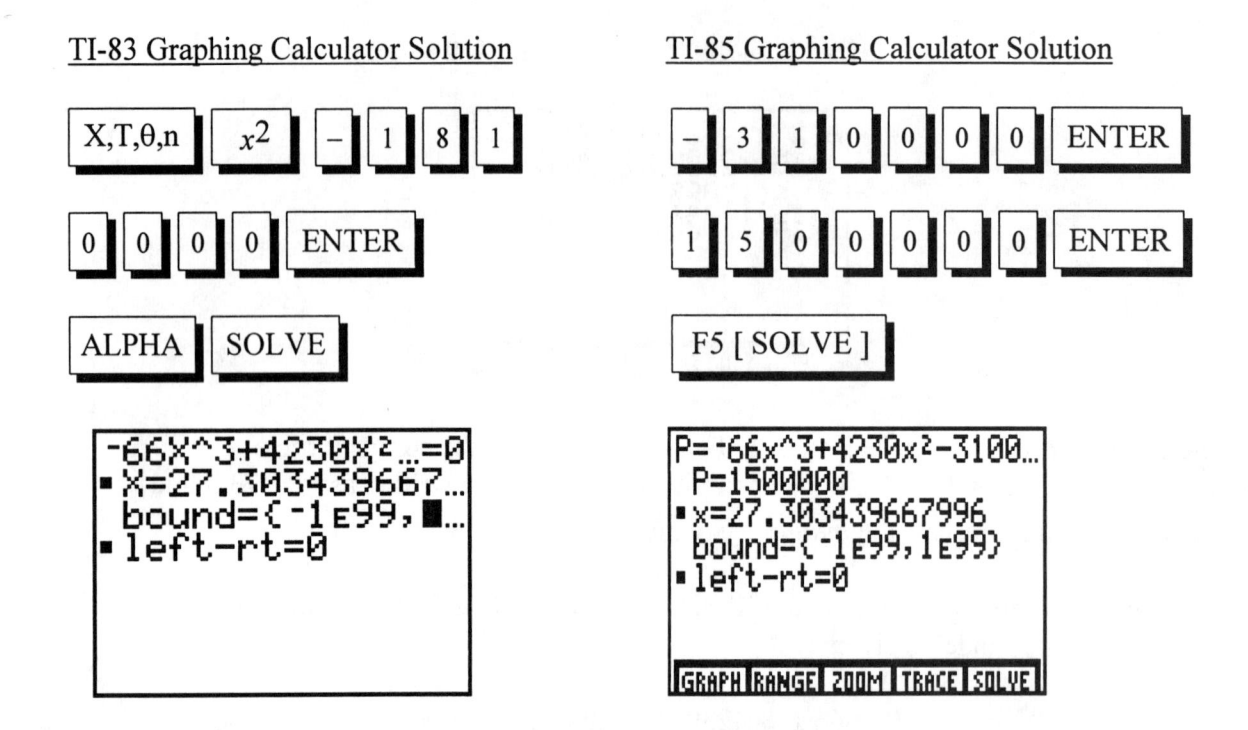

```
P=-66x^3+4230x²-3100...
 P=1500000
·x=27.303439667996
 bound={-1E99,1E99}
·left-rt=0
```
[GRAPH][RANGE][ZOOM][TRACE][SOLVE]

The company would have to spend a minimum of $27,303.44 in advertising costs to make a profit of $1,500,000.

The **SOLVER** and **SOLVE** features found on the TI calculators are not the really the best choice in determining the solution of an equation that has more than one possible root. Obtaining the correct root of an equation with more than one solution is dependent upon how close the cursor is to that root on the graph. If you are using a **TI-85** or a **TI-86** with the **SOLVER** feature, you can use [F1 [GRAPH]] to examine the characteristics of the graph and its possible roots. When using the **TI-83** and **TI-82** it is easier to just use **GRAPH** from the beginning. This type of problem will be revisited again in a later laboratory.

LABORATORY 9: POLYNOMIALS

EXERCISES

1. Classify each of the following polynomials as either monomial, binomial, or a trinomial.

   (a) $4x + 3$ (b) $3x^2 + 2 - x$ (c) $8abc^2$ (d) $x^3 - 6$

2. Find the degree of each of the following polynomials.

   (a) $x^4 + 4x^5$ (b) $ab - ab^2 + 3a^2b^5$ (c) $abc + a^2bc$

3. Evaluate the following polynomials using $x = -2, y = 3, z = 0.25$.

   (a) $x^2y - 3xy^2 + 5$ (b) $x^2z^3 + 4 - 2zy^2$ (c) $3zx^2 + 0.5x^3y^2z$ (d) $\frac{3}{4}xy^3 + \frac{1}{3}y^3z^4 - x$

4. A rectangular box has dimensions $x + 2$, $x - 1$, and $2x + 2$. Find the polynomial that represents the volume of the box. (Drawing a picture of the box and labeling the dimensions (sides) is helpful). Determine the volume of the box when $x = 3$ inches, $x = 5.5$ inches, and $x = 9$ inches.

5. Graph $f(x) = (x + 1)^2(x - 2)^2$. What is the degree of the polynomial function? How many roots should f have? Name all the roots and state if there exists a multiplicity of roots.

6. Graph $f(x) = x^4 - x^3 - 2x^2$. There should be four real roots. One of the roots has *multiplicity* of 2 (which is called a **double root**). Use your knowledge of algebra to determine which root is the double root.

7. Find all the real and complex roots of the following polynomial functions. TI-83 and TI-82 users may have to use paper and pencil to solve if graphing the function does not yield *all* the roots. TI-85 and TI-86 users can either **GRAPH** or use **POLY**.

   (a) $f(x) = x^3 - 3x^2 + 3x + 2$

   (b) $f(x) = x^4 - 4x + 2$

8. The cost, in cents per kilometer, of operating an automobile at speed s, in kilometers per hour, is approximated by the polynomial

$$C(s) = 0.002s^2 - 0.21s + 15.$$

How much does it cost to operate at 80 kilometers per hour? Find the value of s when $C(s) = 0$.

LABORATORY 9: POLYNOMIALS

9. Find the diameter (d) in mils to the nearest hundredth of a copper wire conductor whose resistance (R) is 1.013 ohms and whose length is 3631.5 feet. (Formula: $R = \dfrac{KL}{d^2}$, where K is 10.4 for copper wire). Find the diameter of a copper wire whose resistance is 2.344 ohms and whose length is 3011.6 feet.

10. A square of land is 162.25 m² in area. What is the length of a side to the nearest hundredth? (Use the formula $A = s^2$.)

11. A company that manufactures exercise equipment estimates that the profit for selling exercise bikes is given by

$$P = -35 x^2 + 2030 x - 215{,}000$$

where P is the profit in dollars and x is the advertising expense (in 10,000's of dollars). According to this model, find the smaller of two advertising amounts that yield a profit of $631,000. A profit of $754,000. No profit at all ($P = 0$).

12. While circling at the top of a 1600 foot cliff, a helicopter pilot accidentally drops a pen out of his hand. The height of the pen after t seconds is given by the polynomial function

$$h(t) = -16 t^2 + 1600$$

When will the pen hit the ground? What is the height of the pen after 1.4 seconds?

13. The area of a rectangular garden is 90 square inches. The length of the garden is 3 inches longer than twice its width. Find the dimensions of the garden.

14. The center of a nearby town is 40 square miles. The center of the town is in the shape of a rectangle whose length is 2 miles less than triple its width. Find the dimensions of the center of the town.

LABORATORY 10: RATIONAL EXPRESSIONS

Purpose:

The purpose of this laboratory is to learn how to evaluate and solve rational expressions on the TI-83/82 and TI-85/86 calculators.

Analytic Approach:

Before defining a rational expression, recall that every integer **N** can be expressed as a **rational number**, $\frac{N}{1}$. A *rational number* when written in the form $\frac{a}{b}$, where *a* and *b* are real numbers and $b \neq 0$, is called a **fraction**. Remember that *a* is called the *numerator* of the fraction and *b* is the *denominator* of the fraction.

A fraction that has a polynomial for both its numerator and its denominator is called a **rational expression**. Some examples of rational expressions are:

$$\frac{2}{5}, \frac{x-1}{x}, \frac{y^2+y-1}{y-2}, \frac{x^2-xz+z^2}{x^2+xz-z^2}.$$

Before beginning the problems in this laboratory set the **Float** to reflect three decimal places. Also, set the viewing rectangle to the standard setting.

Evaluating Rational Expressions

To evaluate a rational expression for given values of the variables, each time the variable appears, replace the variable with its given value.

Problem 1: Evaluate each of the following for the given values of the variable:

(a) $\dfrac{a^2+b^2}{a-b^3}$; for $a = -3$ and $b = -\dfrac{2}{3}$.

(b) $\dfrac{x-ax+y^2}{ax^2-xy}$; for $a = 2$, $x = -\pi$, $y = \sqrt{3}$.

To evaluate each of the rational expressions using all of the TI calculators, the *substitution of the numerical values* can be done. However, the **SOLVER** feature on the **TI-85** and **TI-86** will be used in this problem to determine the answer. The **SOLVER** feature on the **TI-85** and **TI-86** can be used with **either** an *expression* or an *equation*. The **SOLVER** feature on the **TI-83** and the **SOLVE** feature on the **TI-82** are not the method of choice when evaluating expressions. **Remember** that when using the TI-82 you may have to add a left parenthesis after the root key.

LABORATORY 10: RATIONAL EXPRESSIONS

Part (a)

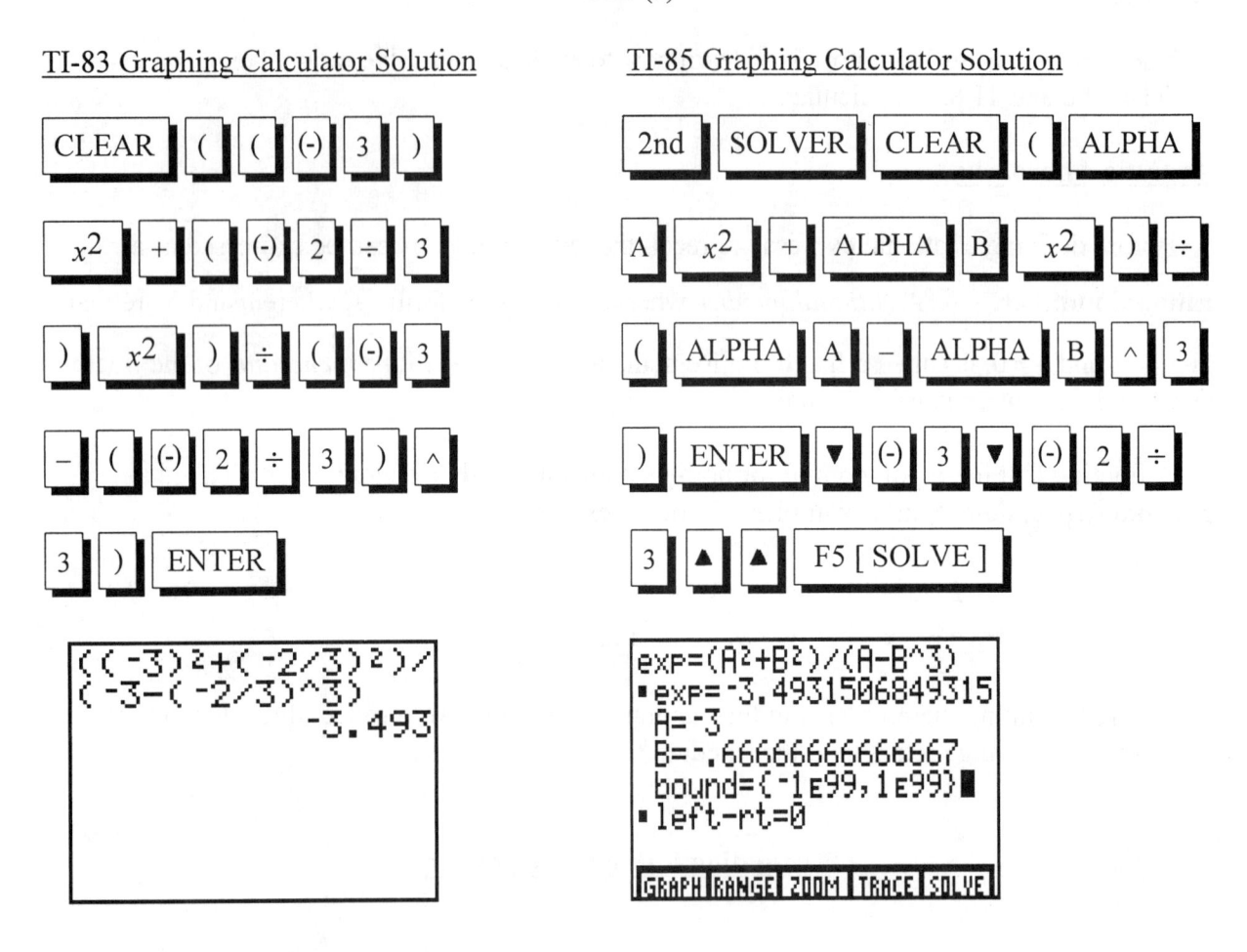

TI-83 Graphing Calculator Solution

| CLEAR | (| (| (-) | 3 |) |

| x^2 | + | (| (-) | 2 | ÷ | 3 |

|) | x^2 |) | ÷ | (| (-) | 3 |

| − | (| (-) | 2 | ÷ | 3 |) | ^ |

| 3 |) | ENTER |

```
((-3)²+(-2/3)²)/
(-3-(-2/3)^3)
              -3.493
```

TI-85 Graphing Calculator Solution

| 2nd | SOLVER | CLEAR | (| ALPHA |

| A | x^2 | + | ALPHA | B | x^2 |) | ÷ |

| (| ALPHA | A | − | ALPHA | B | ^ | 3 |

|) | ENTER | ▼ | (-) | 3 | ▼ | (-) | 2 | ÷ |

| 3 | ▲ | ▲ | F5 [SOLVE] |

```
exp=(A²+B²)/(A-B^3)
•exp=-3.4931506849315
 A=-3
 B=-.66666666666667
 bound={-1E99,1E99}
•left-rt=0
```
| GRAPH | RANGE | ZOOM | TRACE | SOLVE |

The answer to part (a) is -3.493.

Part (b)

TI-83 Graphing Calculator Solution

| CLEAR | (| (| (-) | 2nd | π |

|) | − | (| 2 | x | (| (-) | 2nd |

| π |) |) | + | (| 2nd | √ |

| 3 |) |) | x^2 |) | ÷ | (| (|

TI-85 Graphing Calculator Solution

| ▲ | CLEAR | (| x-VAR | − | ALPHA |

| A | x | x-VAR | + | ALPHA | Y | x^2 |

|) | ÷ | (| ALPHA | A | x | x-VAR |

| x^2 | − | x-VAR | x | ALPHA | Y |) |

118

LABORATORY 10: RATIONAL EXPRESSIONS

TI-83 Graphing Calculator Solution

[2] [x] [(] [(-)] [2nd] [π] [)]

[x²] [)] [-] [(] [(-)] [2nd]

[π] [)] [x] [(] [2nd] [√] [3]

[)] [)] [)] [ENTER]

```
((-π)-(2*(-π))+(
√(3))²)/((2*(-π)
)²-(-π)*(√(3)))
              .244
```

TI-85 Graphing Calculator Solution

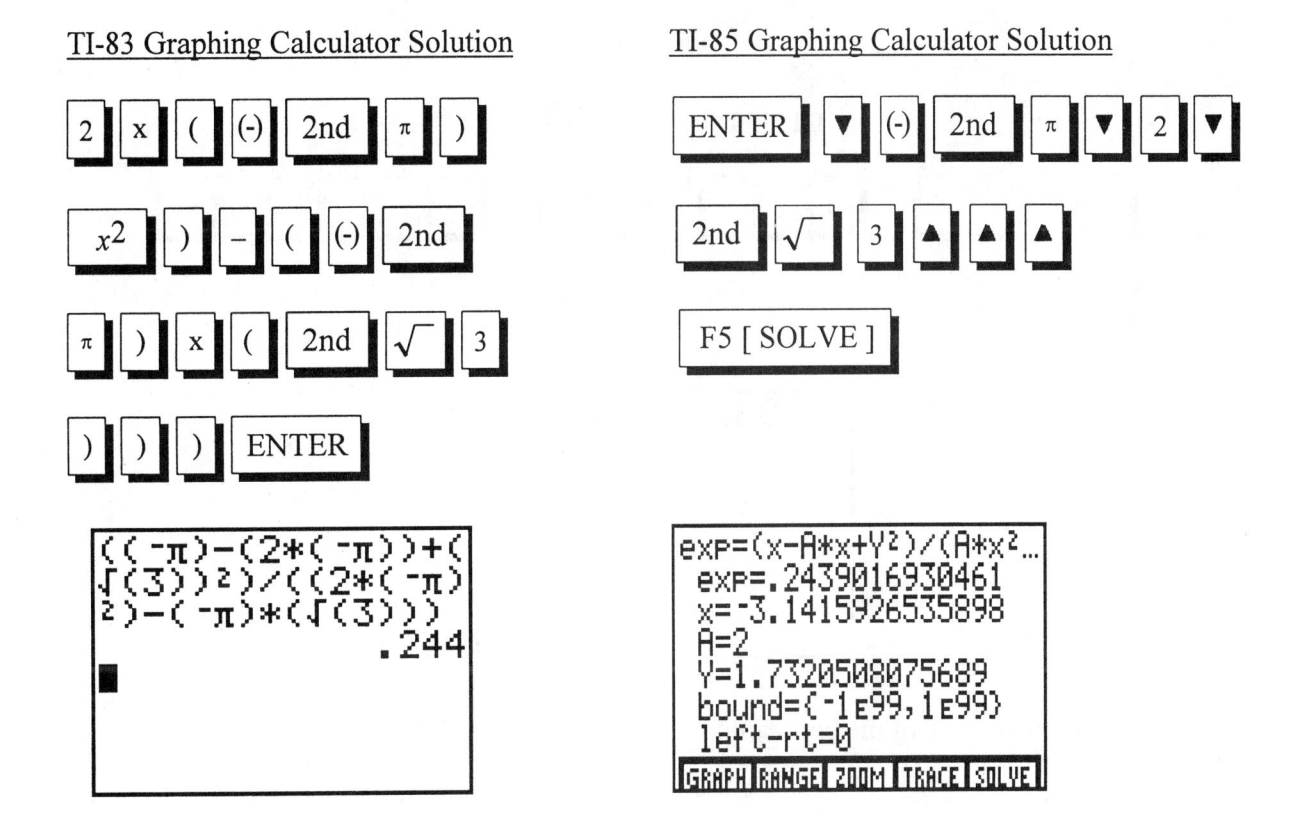

[ENTER] [▼] [(-)] [2nd] [π] [▼] [2] [▼]

[2nd] [√] [3] [▲] [▲] [▲]

[F5 [SOLVE]]

```
exp=(x-A*x+Y²)/(A*x²...
 exp=.2439016930461
 x=-3.1415926535898
 A=2
 Y=1.7320508075689
 bound={-1E99,1E99}
 left-rt=0
GRAPH RANGE ZOOM TRACE SOLVE
```

The answer to part (b) is .244.

Evaluating Rational Functions

Rational expressions can be used to describe functions on restricted domains. The function $f(x) = \dfrac{2x+1}{x^2-3}$ is a **rational function** on the restricted domain $x \neq \pm\sqrt{3}$.

Problem 2: The cost (*C*, in thousands of dollars) of eliminating *x*% of pollutants from a river is given by $C(x) = \dfrac{5x}{150-x}$.

(a) What is the cost to remove 83% of the pollutants from the river?
(b) What is the cost to remove 97% of the pollutants from the river?

Although the fastest way to evaluate the rational function is to use **SOLVER** on the **TI-85/86**, the *substitution of numbers* in place of the variables will be shown.

LABORATORY 10: RATIONAL EXPRESSIONS

Part (a)

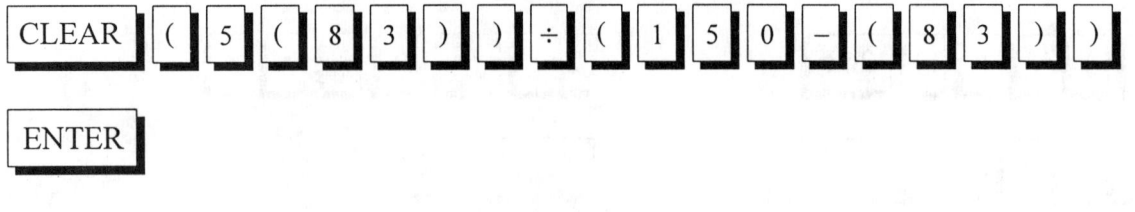

Graphing Calculator Solution

`CLEAR` `(` `5` `(` `8` `3` `)` `)` `÷` `(` `1` `5` `0` `-` `(` `8` `3` `)` `)`

`ENTER`

```
(5(83))/(150-(83
))
            6.194
```

The cost to remove 83% of the pollutants from the river is $6,194.

Part (b)

Graphing Calculator Solution

Press `2nd` `ENTRY`. Press the `◄` key four times and change the number 83 to `9` `7`. Then press the `◄` key thirteen times and change the remaining 83 to `9` `7` `ENTER`.

```
(5(83))/(150-(83
))
            6.194
(5(97))/(150-(97
))
            9.151
```

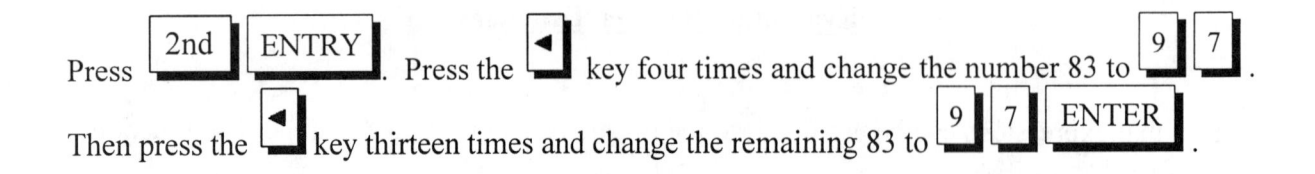

The cost to remove 97% of the pollutants from the river is $9,151.

LABORATORY 10: RATIONAL EXPRESSIONS

Domain of a Rational Expression

Rational expressions and functions often have *restricted domains*. Since *division by zero is not allowed*, any numerical replacement of the variable(s) that causes the *denominator* of a rational expression or function to equal zero is called an **excluded value of the expression.** The **domain of a rational expression or function** excludes numbers that make the polynomial expression in the denominator 0.

Set the denominator of a rational expression equal to zero and solve to determine the *excluded value(s)*. Then state the restricted domain.

Problem 3: Find the excluded value of the expression and state the domain.

(a) $\dfrac{3x+2}{2x-1}$ (b) $\dfrac{x^2-x+3}{3x+4}$ (c) $\dfrac{x^2+3x-1}{x^2-9}$

For both parts (a) and (b), use the **SOLVER** feature with the **TI-83**, **TI-85**, and **TI-86**. The **SOLVER** feature does not round off the answers.

TI-82 users must make some keystroke adjustments when using the **SOLVE** feature. The **SOLVE** feature is found on the same key in the **MATH** menu. The **rational expression** is then entered followed by a **coma**. Then the [X,T,θ] key is pressed followed by a **coma**. Press **zero** followed by a **right parenthesis**. Finally, press [ENTER].

Part (a)

TI-83 Graphing Calculator Solution

[MATH] [0[: Solver ...]] [▲]

[CLEAR] [2] [X,T,θ,n] [−] [1]

[ENTER] [ALPHA] [SOLVE]

TI-85 Graphing Calculator Solution

[2nd] [SOLVER] [CLEAR] [2] [x-VAR]

[−] [1] [ALPHA] [=] [0] [ENTER]

[F5 [SOLVE]]

LABORATORY 10: RATIONAL EXPRESSIONS

TI-83 Graphing Calculator Screen

```
2X-1=0
■X=.5
 bound={-1E99,1…
■left-rt=0
```

TI-85 Graphing Calculator Screen

```
2x-1=0
■x=.5
 bound={-1E99,1E99}
■left-rt=0
```
GRAPH RANGE ZOOM TRACE SOLVE

The excluded value of the variable in part (a) is .5. The domain of the expression is all real numbers *except* .5. When .5 is substituted for *x* in the denominator of the rational expression $\frac{3x+2}{2x-1}$, the rational expression is undefined.

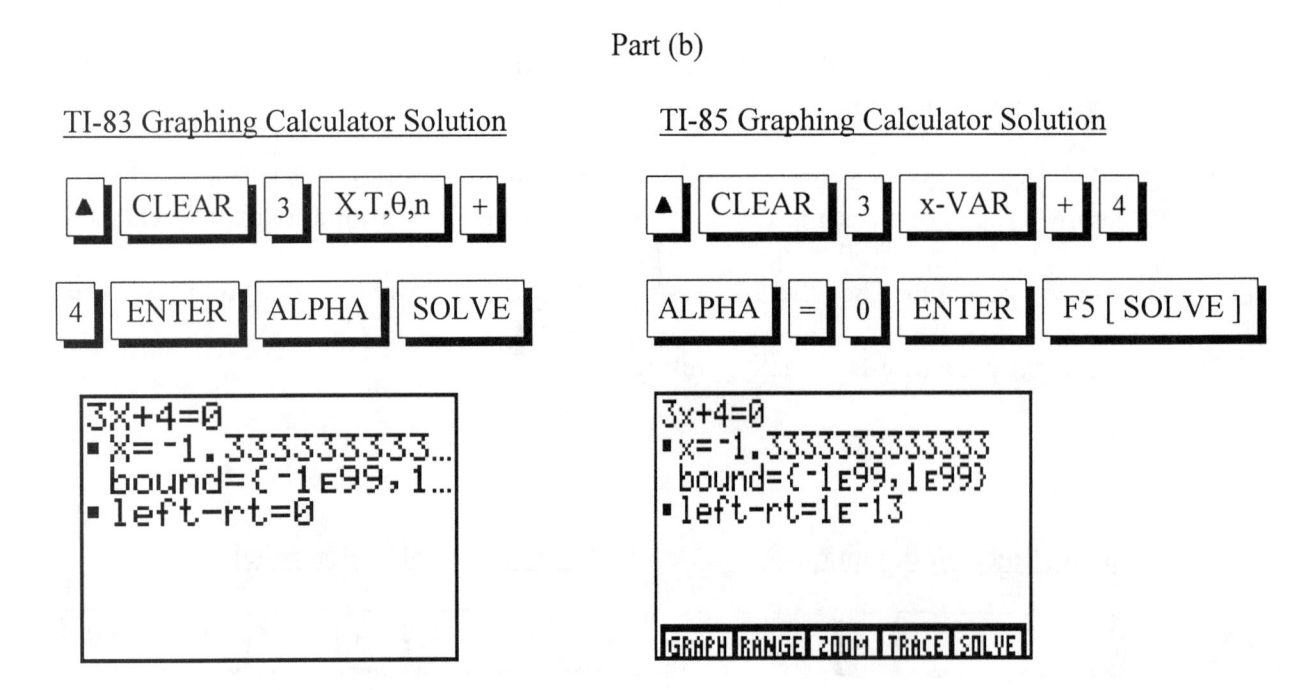

Part (b)

TI-83 Graphing Calculator Solution

[▲] [CLEAR] [3] [X,T,θ,n] [+]

[4] [ENTER] [ALPHA] [SOLVE]

TI-85 Graphing Calculator Solution

[▲] [CLEAR] [3] [x-VAR] [+] [4]

[ALPHA] [=] [0] [ENTER] [F5 [SOLVE]]

```
3X+4=0
■X=-1.333333333…
 bound={-1E99,1…
■left-rt=0
```

```
3x+4=0
■x=-1.333333333333
 bound={-1E99,1E99}
■left-rt=1E-13
```
GRAPH RANGE ZOOM TRACE SOLVE

The domain of the rational expression $\frac{x^2-x+3}{3x+4}$ is all real numbers *except* -1.333 or -4/3.

Part (c)

The polynomial $x^2 - 9$ can be graphed and the *zeroes* or *roots* of the expression can be defined. The keystrokes for entering the polynomial are given along with the calculator screen showing the graph. Review **Laboratory 8: Polynomials** for the appropriate keystrokes for finding the values of the zeroes (on the **TI-82**) or roots (on the **TI-86**).

LABORATORY 10: RATIONAL EXPRESSIONS

Before beginning part (c), **CLEAR** all existing graphs from the **GRAPH** menu.

TI-83 Graphing Calculator Solution

| Y= | \Y₁= | X,T,θ,n |

| x² | – | 9 | GRAPH |

TI-85 Graphing Calculator Solution

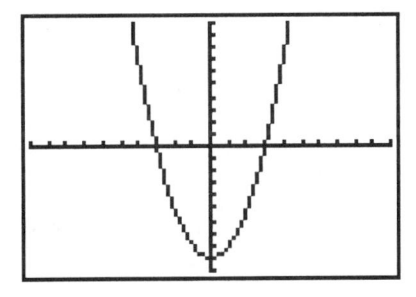

| GRAPH | F1 [y(x) =] | y1= |

| x-VAR | x² | – | 9 | 2nd |

| M5 [GRAPH] | CLEAR |

The screen clearly shows the graph crossing the x-axis in two places, (-3, 0) and (3, 0). The excluded values of the rational expression $\dfrac{x^2 + 3x - 1}{x^2 - 9}$ are -3 and 3. Therefore, $x \ne -3$ and $x \ne 3$ in the denominator of the rational expression.

The Greatest Common Divisor

Recall that in addition and subtraction of fractions, the denominators must be alike. The same principle is true when adding or subtracting rational expressions. Although the graphing calculators (with the exception of the **TI-92**) can not perform the algebraic symbolization necessary to combine rational expressions, the **TI-83, TI-85**, and **TI-86** calculators can make finding the common denominators easier through the use of the **gcd** and **lcm** features.

Changing fractions to like denominators is the same as finding **equivalent fractions**. We can rename rational expressions in equivalent forms through the use of the **Fundamental Rule of Rational Expressions** which states that:

> If a, b, and c are polynomials, then $\dfrac{ac}{bc} = \dfrac{a}{b}$, provided $b \ne 0$ and $c \ne 0$.

LABORATORY 10: RATIONAL EXPRESSIONS

Since *c* is a common polynomial factor in both the numerator and denominator of the rational expression, *c,* as the common polynomial factor, can be eliminated from both the numerator and denominator of the rational expression.

One way of finding equivalent fractions, which is referred to as *"reducing fractions"*, is to find the **greatest common divisor**. To find the *greatest common divisor* (commonly referred to as the **gcd**), factor both the numerator and denominator **completely**. The largest common factor is then removed. In the example $\frac{4}{8} = \frac{1}{2}$, 4 is the *gcd*.

The **TI-82** does *not* have a *gcd* feature.

Problem 4: Find the **gcd** of each of the following numbers. Only the *gcd* of two numbers at a time can be determined.

    (a) 32, 214   (b) 12, 1104   (c) 13, 65, 235

Parts (a) and (b)

TI-83 Graphing Calculator Solution

| CLEAR | MATH | ▶ |

| 9[: gcd( ] | 3 | 2 | , | 2 | 1 |

| 4 |) | ENTER | MATH | ▶ |

| 9[: gcd( ] | 1 | 2 | , | 1 | 1 |

| 0 | 4 |) | ENTER |

```
gcd(32,214)
           2.000
gcd(12,1104)
          12.000
■
```

TI-85 Graphing Calculator Solution

| CLEAR | 2nd | MATH | F5[MISC] |

| F5 [gcd] | 3 | 2 | , | 2 | 1 | 4 |) |

| ENTER | F5 [gcd] | 1 | 2 | , | 1 |

| 1 | 0 | 4 |) | ENTER |

```
gcd(32,214)
           2.000
gcd(12,1104)
          12.000
■
```
```
NUM  PROB  ANGLE  HYP  MISC
sum  prod   seq   lcm  gcd
```

The greatest common divisor in part (a) is 2. The *gcd* of 12 and 1104 in part (b) is 12.

LABORATORY 10: RATIONAL EXPRESSIONS

Part (c)

To find the *gcd* of three numbers, find the *gcd* of the first two numbers. Use the *gcd* of the first two numbers along with the third number to find the *gcd* of all three numbers.

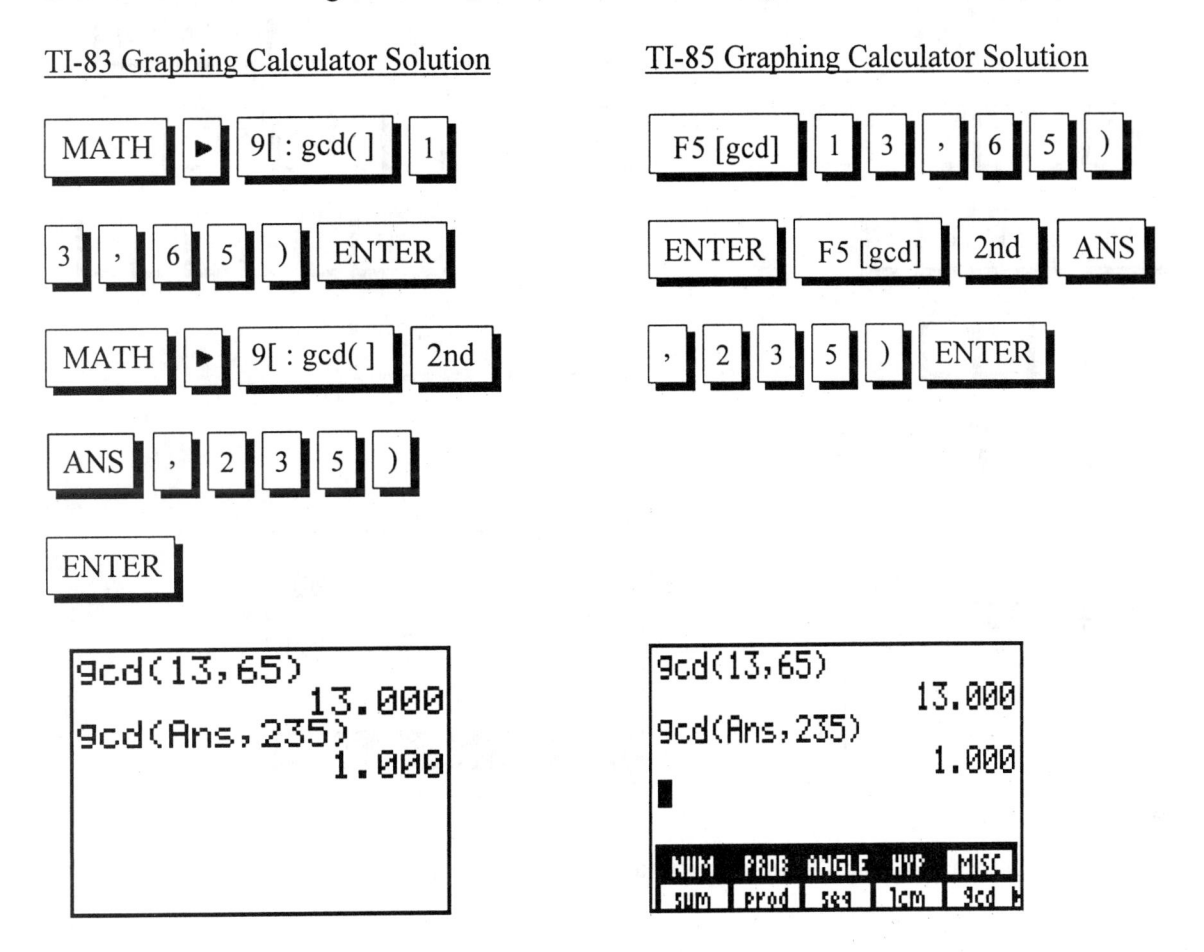

TI-83 Graphing Calculator Solution

| MATH | ▶ | 9[: gcd( ] | 1 |

| 3 | , | 6 | 5 |) | ENTER |

| MATH | ▶ | 9[: gcd( ] | 2nd |

| ANS | , | 2 | 3 | 5 |) |

| ENTER |

```
gcd(13,65)
            13.000
gcd(Ans,235)
            1.000
```

TI-85 Graphing Calculator Solution

| F5 [gcd] | 1 | 3 | , | 6 | 5 |) |

| ENTER | F5 [gcd] | 2nd | ANS |

| , | 2 | 3 | 5 |) | ENTER |

```
gcd(13,65)
            13.000
gcd(Ans,235)
            1.000
```

The *gcd* of 13, 65, and 235 is 1. The largest factor that 13, 65, and 235 have in common is therefore 1.

The Least Common Multiple

Another way of finding equivalent fractions is to determine the **least common multiple** (commonly referred to as the **lcm**). The *least common multiple* is the product of the prime factors of each number making sure not to repeat the common factors. To find the least common multiple, factor each number **completely**.

Problem 5: Find the *lcm* of each of the following numbers. Again, only the lcm of two numbers at a time can be found. The **TI-82** does *not* have a *lcm* feature.

(a) 23, 35 (b) 14, 1032 (c) 20, 45, 375

LABORATORY 10: RATIONAL EXPRESSIONS

Parts (a) and (b)

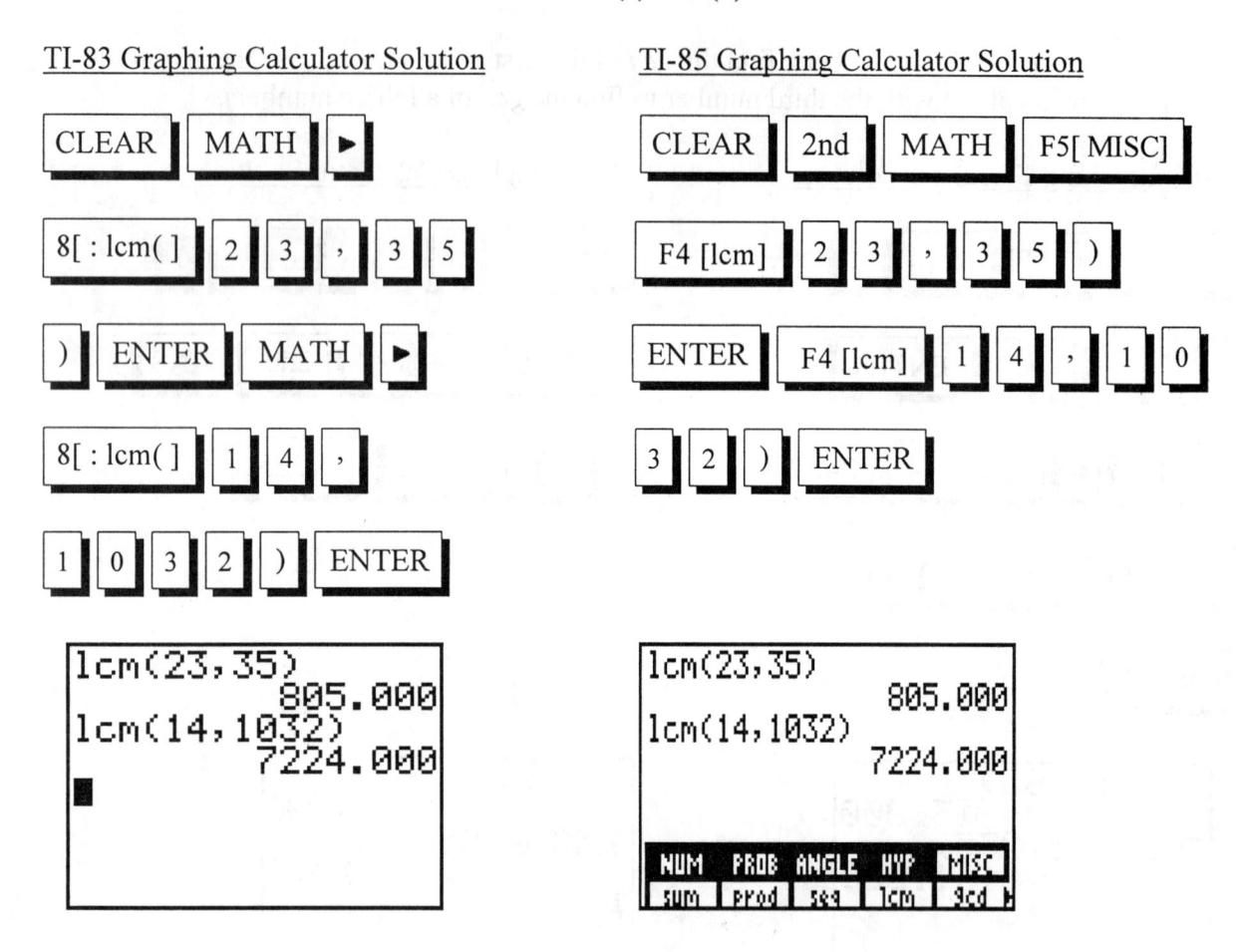

TI-83 Graphing Calculator Solution

| CLEAR | MATH | ▶ |

| 8[: lcm( ] | 2 | 3 | , | 3 | 5 |

|) | ENTER | MATH | ▶ |

| 8[: lcm( ] | 1 | 4 | , |

| 1 | 0 | 3 | 2 |) | ENTER |

```
lcm(23,35)
            805.000
lcm(14,1032)
           7224.000
■
```

TI-85 Graphing Calculator Solution

| CLEAR | 2nd | MATH | F5[MISC] |

| F4 [lcm] | 2 | 3 | , | 3 | 5 |) |

| ENTER | F4 [lcm] | 1 | 4 | , | 1 | 0 |

| 3 | 2 |) | ENTER |

```
lcm(23,35)
            805.000
lcm(14,1032)
           7224.000

NUM  PROB  ANGLE  HYP  MISC
sum  prod  seq   lcm  gcd ▶
```

The least common multiple of 23 and 35 in part (a) is 805. The *lcm* of 14 and 1032 is 7224.

Part (c)

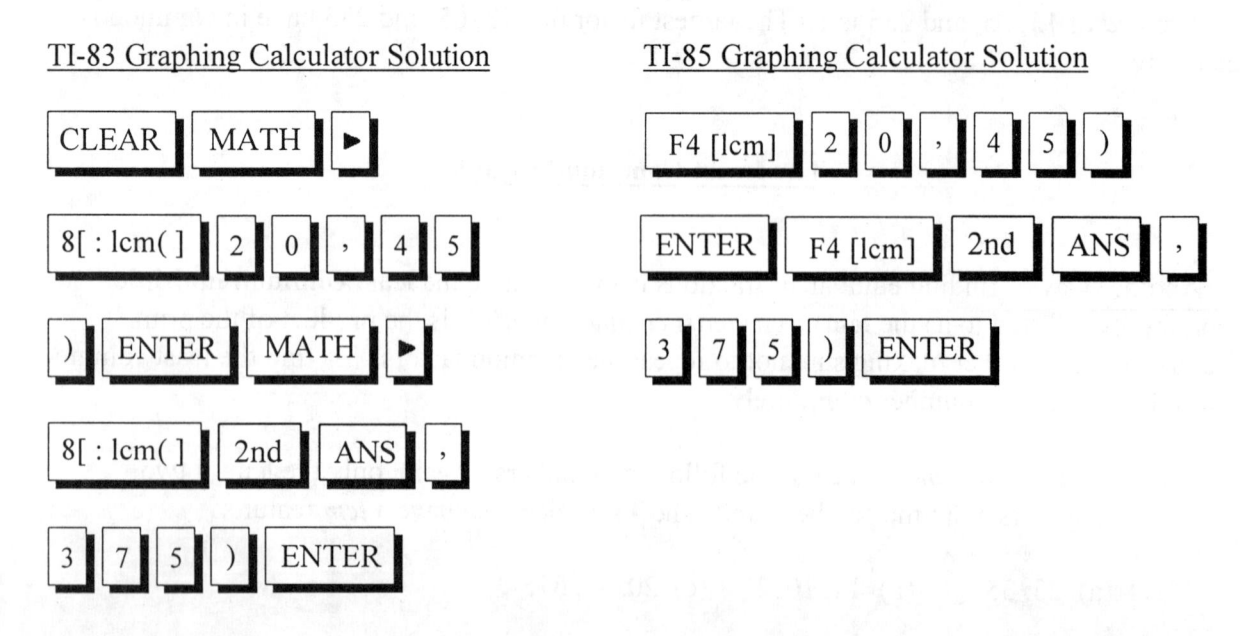

TI-83 Graphing Calculator Solution

| CLEAR | MATH | ▶ |

| 8[: lcm( ] | 2 | 0 | , | 4 | 5 |

|) | ENTER | MATH | ▶ |

| 8[: lcm( ] | 2nd | ANS | , |

| 3 | 7 | 5 |) | ENTER |

TI-85 Graphing Calculator Solution

| F4 [lcm] | 2 | 0 | , | 4 | 5 |) |

| ENTER | F4 [lcm] | 2nd | ANS | , |

| 3 | 7 | 5 |) | ENTER |

LABORATORY 10: RATIONAL EXPRESSIONS

TI-83 Graphing Calculator Screen	TI-85 Graphing Calculator Screen

```
lcm(20,45)
           180.000
lcm(Ans,375)
          4500.000
```

```
lcm(20,45)
           180.000
lcm(Ans,375)
          4500.000
```

The least common multiple of 20, 45, and 375 is 4500.

As *Problems 3 and 4* have shown, finding the *gcd* and *lcm* of large numbers is not difficult when using the **TI-83**, **TI-85**, and **TI-86** calculators. Using pen and paper to determine the prime factorization of large numbers is both long and tedious **and** still yields the same results.

LABORATORY 10: RATIONAL EXPRESSIONS

EXERCISES

1. Evaluate each of the following rational expressions when $x = -2$ and $y = 4$.

   (a) $\dfrac{3x}{6y}$
   (b) $\dfrac{x^2 - 2xy}{xy^2 + x^4}$
   (c) $\dfrac{x^3 - 2xy}{x^2} - \dfrac{x^2 y + y^3}{y^3 - x}$

2. Find the restricted domain of the rational functions.

   (a) $f(x) = \dfrac{3x + 8}{3x + 2}$
   (b) $f(x) = \dfrac{x^2 - 7}{-6x + 5}$
   (c) $f(x) = \dfrac{4x + 7}{x^2 - 16}$

3. The cost (C, in tens of thousands of dollars) to inoculate $x\%$ of the number of herds of cattle is described by
$$C = \dfrac{110x}{230 - x}.$$

   (a) What will it cost to inoculate 38% of the number of herds of cattle?
   (b) What is the domain of the rational expression?
   (c) What will it cost to inoculate 67% of the number of herds of cattle?
   (d) What is happening to the cost of inoculating the herds of cattle as the value of x approaches 100 %? How can you interpret this observation?

4. The total revenue from the sale of a new book is approximated by the rational function
$R(x) = \dfrac{500x^2}{x^2 + 6}$ where x represents the number of years since the publication of the book.
$R(x)$ is the total revenue in millions of dollars.

   (a) Find the total revenue after the first year.
   (b) Find the total revenue after the third year.
   (c) Find the total revenue between the first and the third year.

5. Find the *lcm* for each of the following.

   (a) 6, 105
   (b) 20, 30, 45
   (c) 15, 23, 107

6. Find the *gcd* for each of the following.

   (a) 9, 48
   (b) 25, 65, 120
   (c) 13, 36, 105

LABORATORY 11: RATIONAL EQUATIONS

Purpose:

The purpose of this laboratory is to learn how to evaluate and solve rational equations on the TI-83/82 and TI-85/86 calculators.

Analytic Approach:

An equation that contains one or more rational expressions is often called a **fractional equation**. Another name for a **fractional equation** is a **rational equation**. Some examples of rational or fractional equations are:

$$\frac{2}{3} + \frac{1}{2} = \frac{1}{x}, \quad \frac{x-2}{x+3} = \frac{6}{x-4}, \quad \frac{x^2}{x^3+1} = \frac{x}{6}.$$

There are many different ways to use the TI calculators to solve rational equations. On the **TI-83** and **TI-85/86** calculators, we will first use the **SOLVER** feature. The **TI-82** has a **SOLVE** feature that can be used to solve equations. *Secondly*, we will **GRAPH** the equations on all four TI calculators and use the graph to determine the solution.

Problem 1 will demonstrate the use of the **SOLVER** feature on the **TI-83**, **TI-85**, and **TI-86** and the **SOLVE** feature on the **TI-82**, to find the solution to a rational equation.

Problem 1: Solve the rational equation $\dfrac{x^2}{x-2} = \dfrac{4}{x-2}$, $x \neq 2$.

Remember to set the equation equal to zero when using the TI-83 and TI-82. There are some keystrokes that must be adjusted when using the **TI-82**. The **SOLVE** feature is found on the same key in the **MATH** menu as on the **TI-83**. The **rational expression** is then entered followed by a **coma**. Then press $\boxed{\text{X,T,}\theta}$ followed by another **coma**. Press **0** and then a **right parenthesis**. Finally, press $\boxed{\text{ENTER}}$.

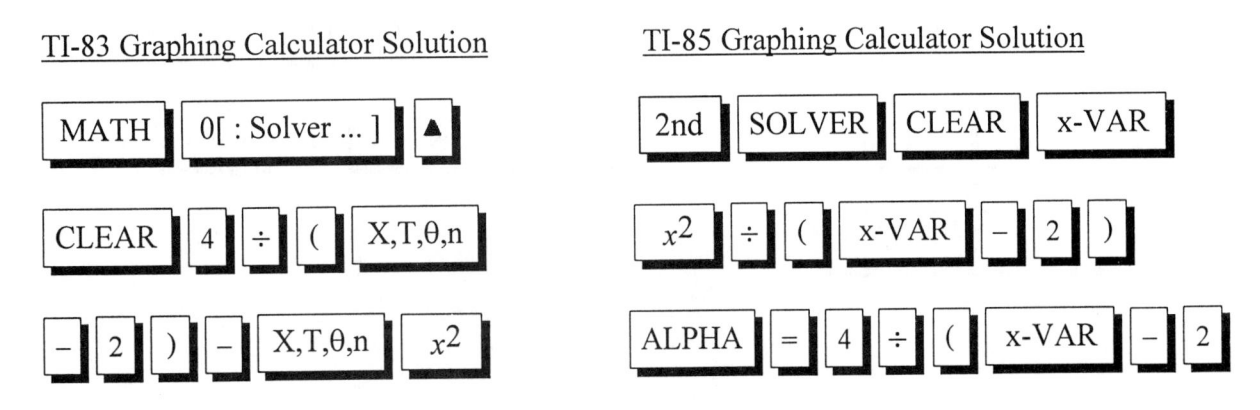

TI-83 Graphing Calculator Solution

$\boxed{\text{MATH}}$ $\boxed{0[: \text{Solver} \ldots]}$ $\boxed{\blacktriangle}$

$\boxed{\text{CLEAR}}$ $\boxed{4}$ $\boxed{\div}$ $\boxed{(}$ $\boxed{\text{X,T,}\theta\text{,n}}$

$\boxed{-}$ $\boxed{2}$ $\boxed{)}$ $\boxed{-}$ $\boxed{\text{X,T,}\theta\text{,n}}$ $\boxed{x^2}$

TI-85 Graphing Calculator Solution

$\boxed{\text{2nd}}$ $\boxed{\text{SOLVER}}$ $\boxed{\text{CLEAR}}$ $\boxed{\text{x-VAR}}$

$\boxed{x^2}$ $\boxed{\div}$ $\boxed{(}$ $\boxed{\text{x-VAR}}$ $\boxed{-}$ $\boxed{2}$ $\boxed{)}$

$\boxed{\text{ALPHA}}$ $\boxed{=}$ $\boxed{4}$ $\boxed{\div}$ $\boxed{(}$ $\boxed{\text{x-VAR}}$ $\boxed{-}$ $\boxed{2}$

LABORATORY 11: RATIONAL EQUATIONS

TI-83 Graphing Calculator Solution

| ÷ | (| X,T,θ,n | − | 2 |) |

| ENTER | ALPHA | SOLVE |

```
4/(X-2)-X²/(X...=0
■X= -2
 bound={-1E99,1...
■left-rt=0
```

TI-85 Graphing Calculator Solution

|) | ENTER | F5 [SOLVE] |

```
x²/(x-2)=4/(x-2)
■x= -2
 bound={-1E99,1E99}
■left-rt=0
```
GRAPH RANGE ZOOM TRACE SOLVE

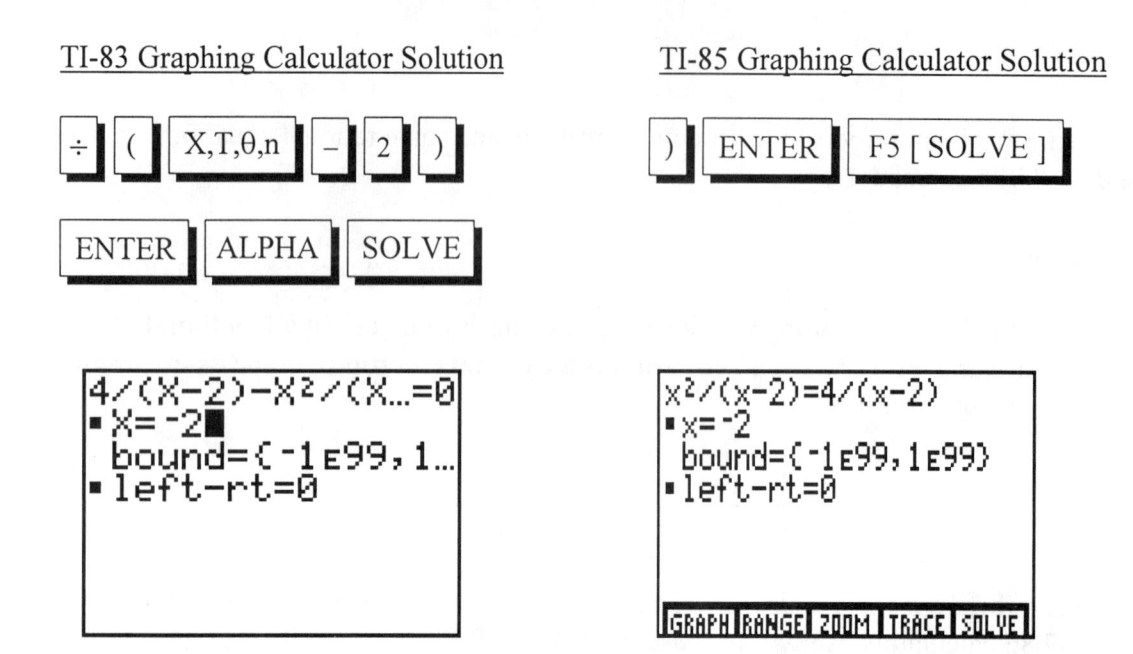

The solution to the rational equation is -2. Be sure to check that when -2 is substituted for *x*, the fractions are equivalent. Why can't the value of *x* in the denominator be positive two as is stated in the beginning of the problem?

Problem 2: What happens when you put the rational equation, $\frac{2}{3} - \frac{5}{6} = \frac{1}{x}, x \neq 0$,

in **SOLVER** and try to solve for *x*? Remember to make the proper keystroke adjustments when using the **SOLVE** feature on the **TI-82**.

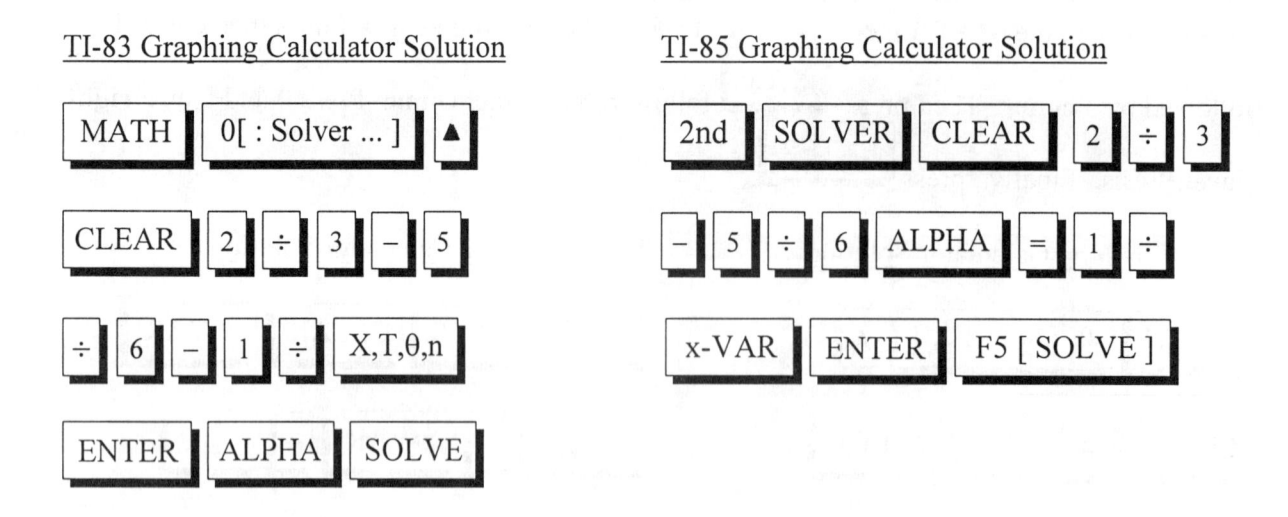

TI-83 Graphing Calculator Solution

| MATH | 0[: Solver ...] | ▲ |

| CLEAR | 2 | ÷ | 3 | − | 5 |

| ÷ | 6 | − | 1 | ÷ | X,T,θ,n |

| ENTER | ALPHA | SOLVE |

TI-85 Graphing Calculator Solution

| 2nd | SOLVER | CLEAR | 2 | ÷ | 3 |

| − | 5 | ÷ | 6 | ALPHA | = | 1 | ÷ |

| x-VAR | ENTER | F5 [SOLVE] |

LABORATORY 11: RATIONAL EQUATIONS

TI-83 Graphing Calculator Screen

TI-85 Graphing Calculator Screen

```
2/3-5/6-1/X=0
•X=-6.000000000...
 bound={-1E99,1...
•left-rt=0
```

```
2/3-5/6=1/x
•x=-6.0000000000002
 bound={-1E99,1E99}
•left-rt=0
```
GRAPH RANGE ZOOM TRACE SOLVE

The solution to the rational equation $\frac{2}{3} - \frac{5}{6} = \frac{1}{x}$ is -6.

All the TI calculators use a numerical analysis method for determining solutions to equations in **SOLVER** and **SOLVE**. This is why you *may have* received an **ERROR Message**. The **ERROR Message**: "**NO SIGN CHANGE**" may have appeared on **all four** calculators. It depends on how close your initial *x* value is to the correct answer as to whether you get a solution or an error message. To illustrate this point, try the same problem with an initial guess of $x = 10$.

The best way *to avoid* the possible appearance of an error message is to *clear the fractions* from both sides of the equation by multiplying by the *lcm*. The *lcm* for the rational equation $\frac{2}{3} - \frac{5}{6} = \frac{1}{x}$ is 6*x*. Multiply both sides of the equation by 6*x*, obtaining the equation $4x - 5x = 6$. Now solve the equation $4x - 5x = 6$ using **SOLVER** on the **TI-83, 85,** or **TI-86** calculator. Make the proper keystroke adjustments when using the **SOLVE** feature on the **TI-82**.

When using the **SOLVER** or **SOLVE** features on the TI calculators to solve rational equations, it is almost always better to clear the denominators first. Once the equation has been rewritten without any denominators, enter the equation in **SOLVER** or **SOLVE** and find the solution.

Another method for solving rational equations is to **GRAPH** the equation. The **GRAPH** feature can be used with all TI calculators.

Remember to always **CLEAR** any existing graphs **and** to set the viewing rectangle to the standard setting.

Problem 3: Find the solution to the following rational equation by graphing the equation.

$$-\frac{1}{3} - \frac{5}{4x} = \frac{3}{4} - \frac{1}{6x}, \ x \neq 0$$

Be sure to set the equation equal to zero before using **GRAPH** with all four calculators.

LABORATORY 11: RATIONAL EQUATIONS

TI-83 Graphing Calculator Solution

[Y=] [CLEAR] \Y₁= [(-)] [(]
[1] [÷] [3] [)] [−] [(] [5] [÷]
[(] [4] [X,T,θ,n] [)] [)] [−] [(]
[3] [÷] [4] [)] [+] [(] [1]
[÷] [(] [6] [X,T,θ,n] [)] [)]
[GRAPH]

TI-85 Graphing Calculator Solution

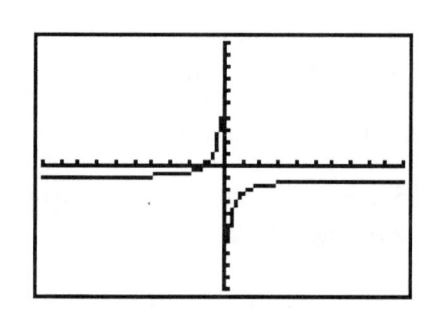

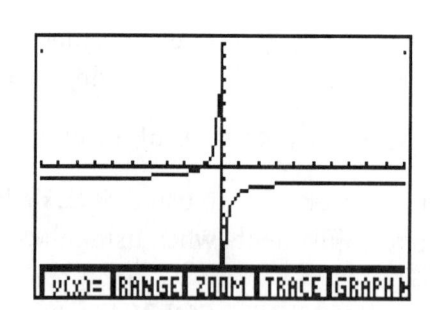

[GRAPH] [F1 [y(x) =]] y1= [CLEAR]
[(-)] [(] [1] [÷] [3] [)] [−] [(] [5] [÷] [(]
[4] [x-VAR] [)] [)] [−] [(] [3] [÷] [4]
[)] [+] [(] [1] [÷] [(] [6] [x-VAR] [)]
[)] [2nd] [M5 [GRAPH]]

Before finding the solution to the equation, notice the **discontinuity** of the graph. This *break* in the graph occurs at $x = 0$ on the graph of $-\frac{1}{3} - \frac{5}{4x} = \frac{3}{4} - \frac{1}{6x}$. We saw that $x = 0$ is a **vertical asymptote** for the graph $-\frac{1}{3} - \frac{5}{4x} = \frac{3}{4} - \frac{1}{6x}$ because the graph of $-\frac{1}{3} - \frac{5}{4x} = \frac{3}{4} - \frac{1}{6x}$ gets closer and closer to the *vertical line* $x = 0$ even though it does not touch it. Remember that this is one of the reasons for stating in the beginning of the problem *the excluded value of the variable*.

In later courses such as College Algebra or Precalculus, you will learn more about *asymptotes*. Besides *vertical asymptotes* there are also *horizontal* and *slant asymptotes*.

Use the following keystrokes to determine the solution to the rational equation.

LABORATORY 11: RATIONAL EQUATIONS

TI-83 Graphing Calculator Solution

| 2nd | | CALC | | 2 [: zero] |

Use the ◀ key to set the lower bound.

| ENTER |

Use the ▶ key to set the upper bound. This should not be positive.

| ENTER | | ENTER |

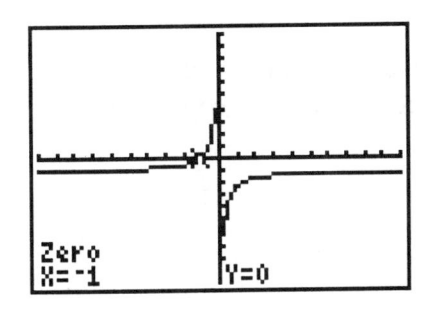

TI-85 Graphing Calculator Solution

| MORE | | F1 [MATH] | | F3 [ROOT] |

Use the ◀ key to move the cursor toward the x intercept.

| ENTER |

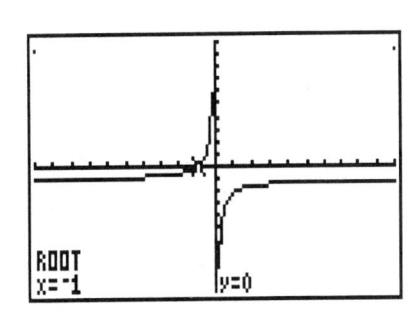

The keystrokes used in finding the root on the **TI-86** are the same as those for the **TI-85** *until* you go to press the root key itself. On the **TI-86** the root key is found on **F1** of the same menu. Then follow the keystrokes for the **TI-83** to set the bounds.

The *root or zero* of the rational equation, $-\dfrac{1}{3} - \dfrac{5}{4x} = \dfrac{3}{4} - \dfrac{1}{6x}$, is -1.

The following problem is *only* for users of the **TI-85** and **TI-86**, since it uses a feature of **SOLVER** that the **TI-83** and **TI-82** do *not* have.

Problem 4: Use **SOLVER** and the | F1 [GRAPH] | feature found on the **TI-85 and TI-86** to aid in finding the solution to the rational equation:

$$-\dfrac{1}{3} - \dfrac{5}{4x} = \dfrac{3}{4} - \dfrac{1}{6x}, \quad x \neq 0.$$

LABORATORY 11: RATIONAL EQUATIONS

TI-85 and TI-86 Graphing Calculator Solution

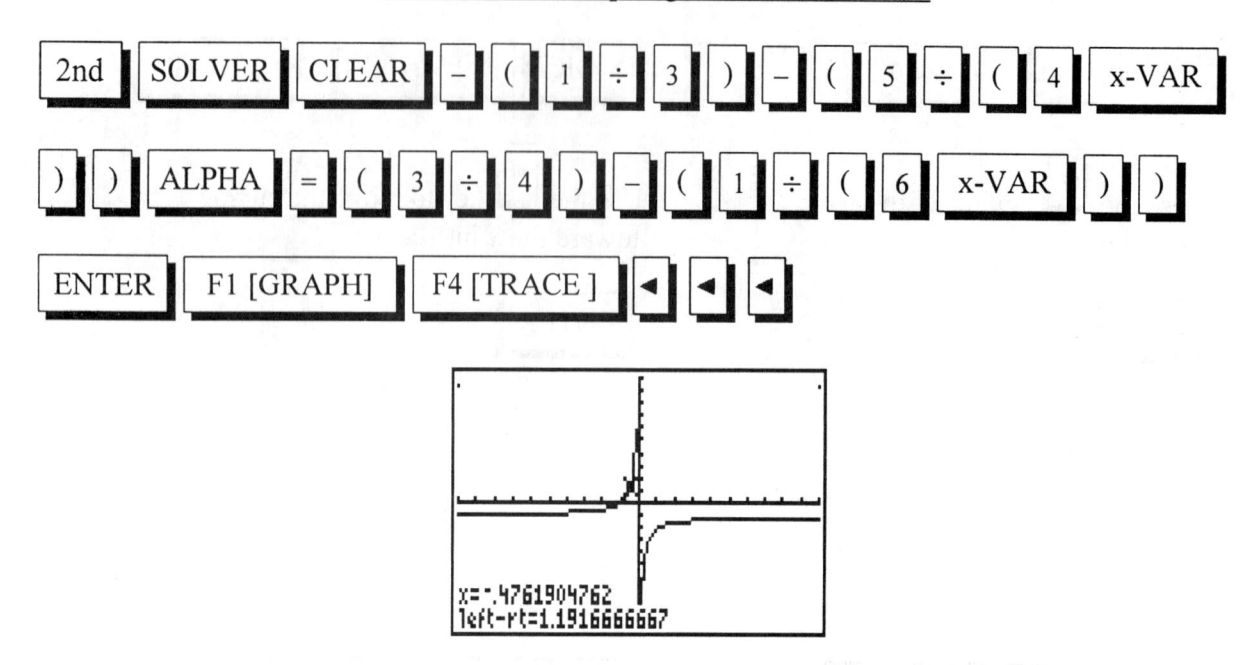

| 2nd | SOLVER | CLEAR | – | (| 1 | ÷ | 3 |) | – | (| 5 | ÷ | (| 4 | x-VAR |

|) |) | ALPHA | = | (| 3 | ÷ | 4 |) | – | (| 1 | ÷ | (| 6 | x-VAR |) |) |

| ENTER | F1 [GRAPH] | F4 [TRACE] | ◄ | ◄ | ◄ |

TRACE is used to move the cursor toward the point of intersection on the *x*-axis which is the *root* of the rational equation. Before graphing the equation make sure you have used the CLEAR key to remove any existing values for *x*.

TI-85 and TI-86 Graphing Calculator Solution

| EXIT | F5 [SOLVE] |

Notice that the answer, -1, is the same as in *Problem 3*. Why did this result in the same answer?

Problem 5: Graph the rational equation $\dfrac{2x-14}{x^2+3x-28} + \dfrac{2-x}{4-x} - \dfrac{x+3}{x+7} = 0$. State the solution to the rational equation.

134

LABORATORY 11: RATIONAL EQUATIONS

SOLVER is not the best choice in determining the solution of an equation that has more than one possible root. If you are using a **TI-85/86** with the **SOLVER** feature, you can use F1 [GRAPH] to examine the characteristics of the graph and its possible roots. When using the **TI-83** and **TI-82** it is easier to just use **GRAPH** from the beginning.

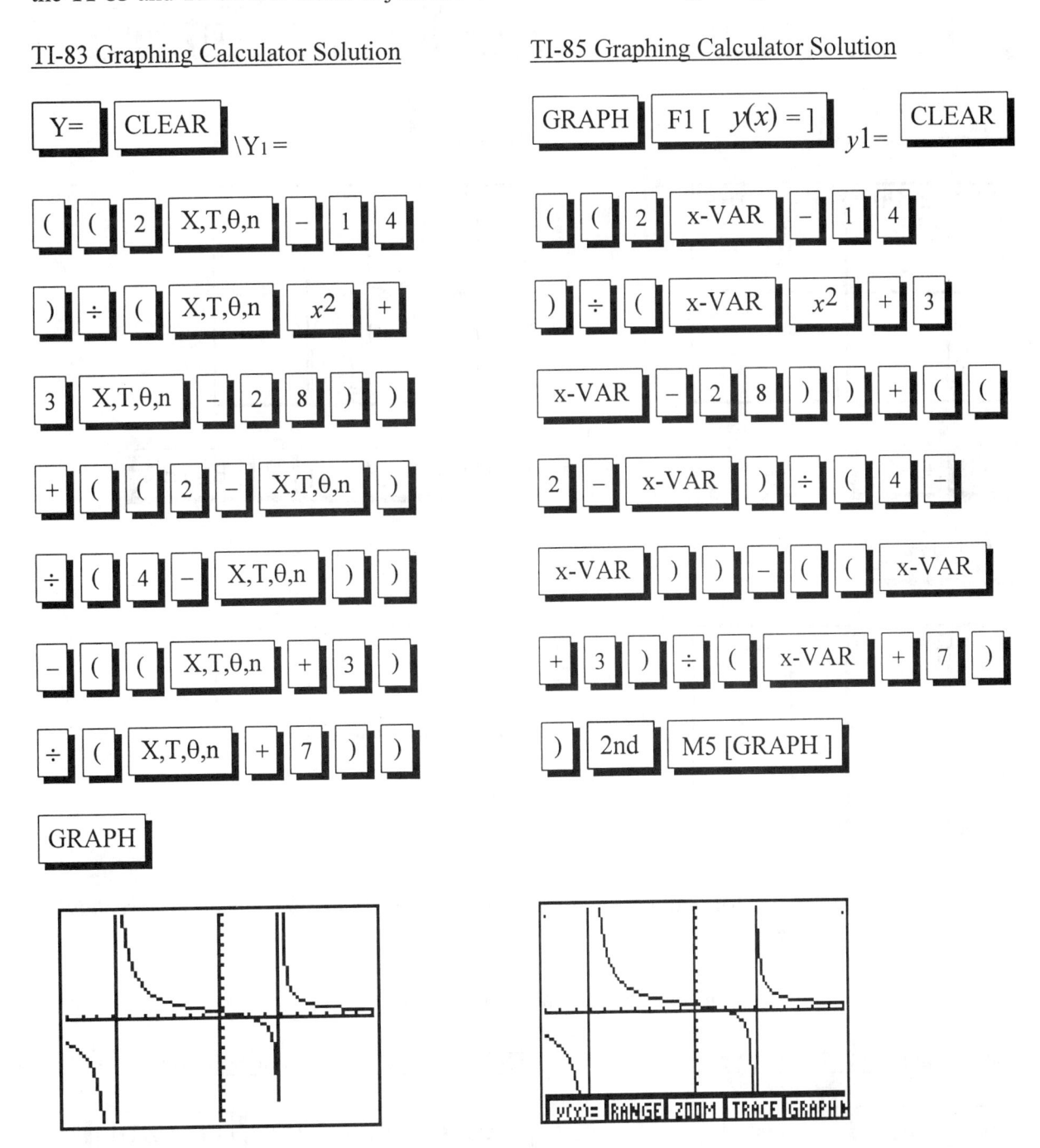

TI-83 Graphing Calculator Solution

Y= | CLEAR \Y₁=

((2 X,T,θ,n – 1 4

) ÷ (X,T,θ,n x² +

3 X,T,θ,n – 2 8))

+ ((2 – X,T,θ,n)

÷ (4 – X,T,θ,n))

– ((X,T,θ,n + 3)

÷ (X,T,θ,n + 7))

GRAPH

TI-85 Graphing Calculator Solution

GRAPH | F1 [y(x) =] y1= CLEAR

((2 x-VAR – 1 4

) ÷ (x-VAR x² + 3

x-VAR – 2 8)) + ((

2 – x-VAR) ÷ (4 –

x-VAR)) – ((x-VAR

+ 3) ÷ (x-VAR + 7)

) 2nd M5 [GRAPH]

There are two apparent **vertical asymptotes**. To determine if the two "breaks" in the graph actually do appear at the *vertical asymptotes* $x = -7$ and $x = 4$ examine the value of y at each

135

LABORATORY 11: RATIONAL EQUATIONS

point. When $x = -7$ there should not be a value for y. When $x = 4$ there should not be a value for y. The following keystrokes will confirm the previous two statements.

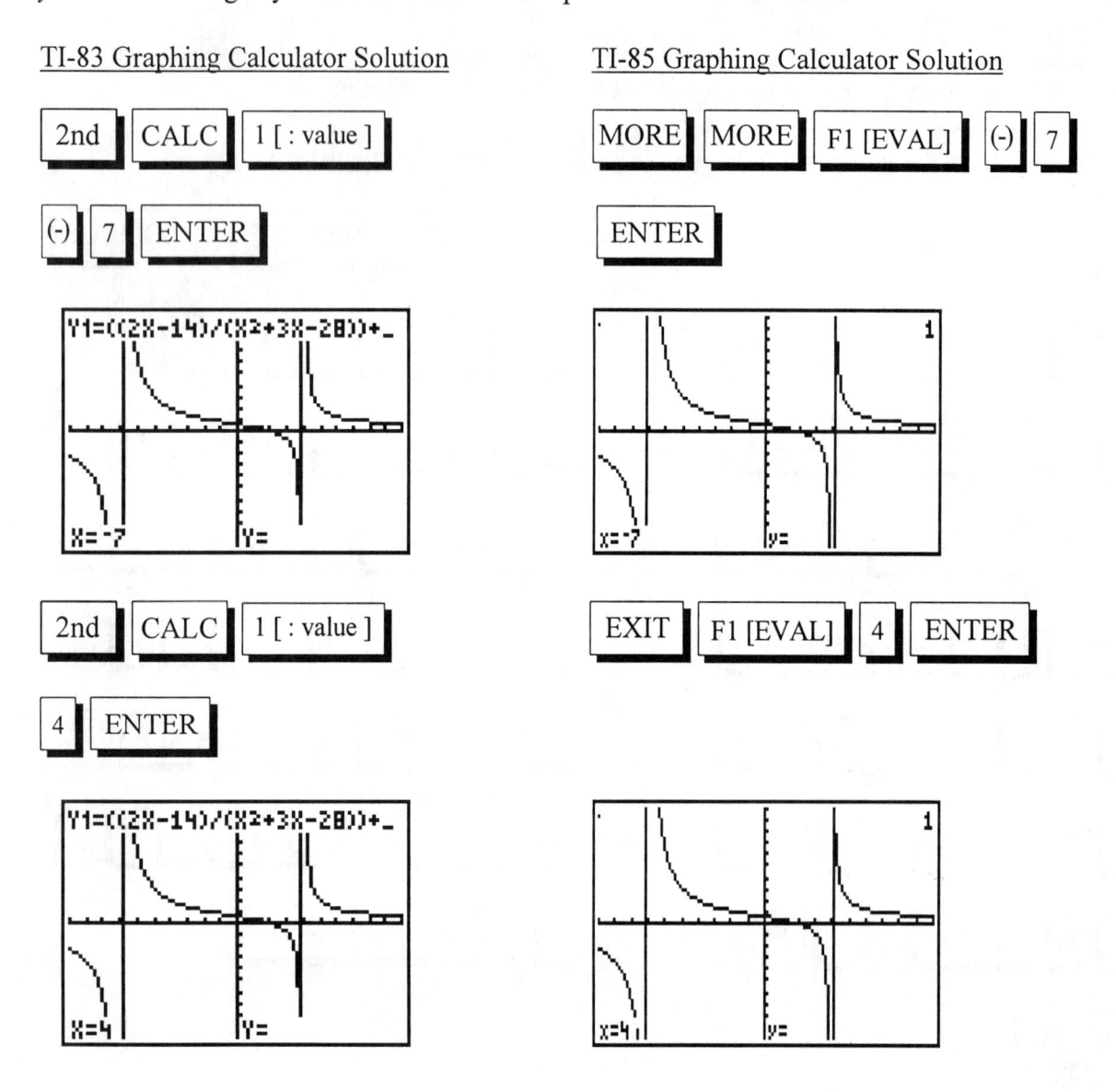

TI-83 Graphing Calculator Solution

[2nd] [CALC] [1 [: value]]

[(-)] [7] [ENTER]

TI-85 Graphing Calculator Solution

[MORE] [MORE] [F1 [EVAL]] [(-)] [7]

[ENTER]

[2nd] [CALC] [1 [: value]]

[4] [ENTER]

[EXIT] [F1 [EVAL]] [4] [ENTER]

The two missing values for y in the above two screens confirm the existence of the *vertical asymptotes*, $x = -7$ and $x = 4$.

Now determine the one *root* or *zero* of the rational equation.

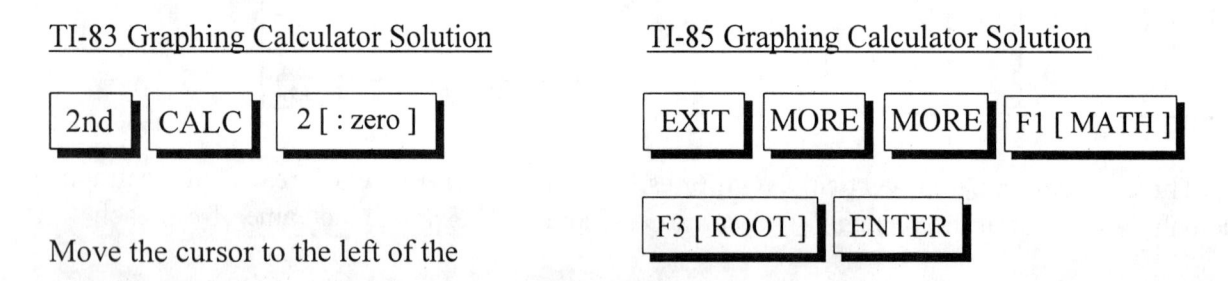

TI-83 Graphing Calculator Solution

[2nd] [CALC] [2 [: zero]]

Move the cursor to the left of the

TI-85 Graphing Calculator Solution

[EXIT] [MORE] [MORE] [F1 [MATH]]

[F3 [ROOT]] [ENTER]

LABORATORY 11: RATIONAL EQUATIONS

TI-83 Graphing Calculator Solution

root by pressing the ◄ key.

To set the lower bound, press

[ENTER].

We must now set the upper limit by pressing the ► key until the cursor is to the right of the zero. Then press

[ENTER] [ENTER].

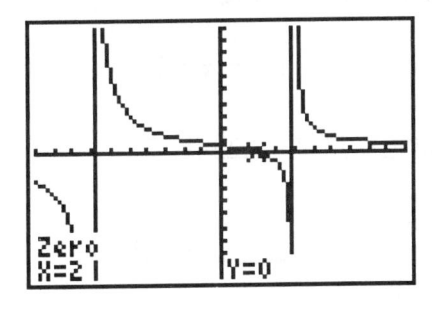

TI-85 Graphing Calculator Solution

Remember to press the **F1** key on the **TI-86**, and then, follow the keystroke pattern on the **TI-83**

for setting the upper and lower bounds.

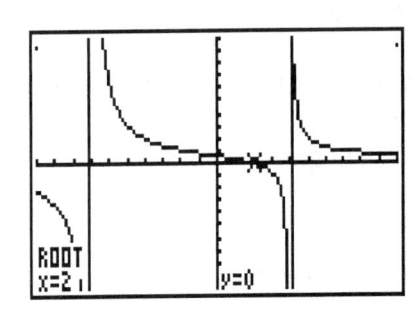

The root of the rational equation is 2.

Problem 6: The average monthly temperature in Anytown, USA., can be found by using the model

$$T = \frac{-191(x-30)}{x^2 - 16.5x + 114}$$

where T is measured in degrees Fahrenheit and $x = 1, 2, ..., 12$ represents the months of the year. What was the average monthly temperature for July?

On all the TI calculators, set the viewing rectangle for **xMin**: 0; **xMax**: 12; **xScl**: 1; **yMin**: 0; **yMax** 110; and **yScl**: 10.

TI-83 Graphing Calculator Solution

[Y=] [CLEAR] \Y₁= [(] [(-)] [1]

[9] [1] [(] [X,T,θ,n] [−] [3] [0]

TI-85 Graphing Calculator Solution

[GRAPH] [F1 [y(x) =]] y1= [CLEAR]

[(] [(-)] [1] [9] [1] [(] [x-VAR] [−] [3]

137

LABORATORY 11: RATIONAL EQUATIONS

TI-83 Graphing Calculator Solution

[ )] [ )] [÷] [( ] [X,T,θ,n] [x^2]

[−] [1] [6] [.] [5] [X,T,θ,n]

[+] [1] [1] [4] [ )] [GRAPH]

TI-85 Graphing Calculator Solution

[0] [ )] [ )] [÷] [( ] [x-VAR] [x^2] [−]

[1] [6] [.] [5] [x-VAR] [+] [1] [1] [4]

[ )] [2nd] [M5 [GRAPH]]

July is the seventh month of the year. When evaluating for *x*, be sure to use *x* = 7.

TI-83 Graphing Calculator Solution

[2nd] [CALC] [1 [: value]]

[7] [ENTER]

TI-85 Graphing Calculator Solution

[MORE] [MORE] [F1 [EVAL]] [7]

[ENTER]

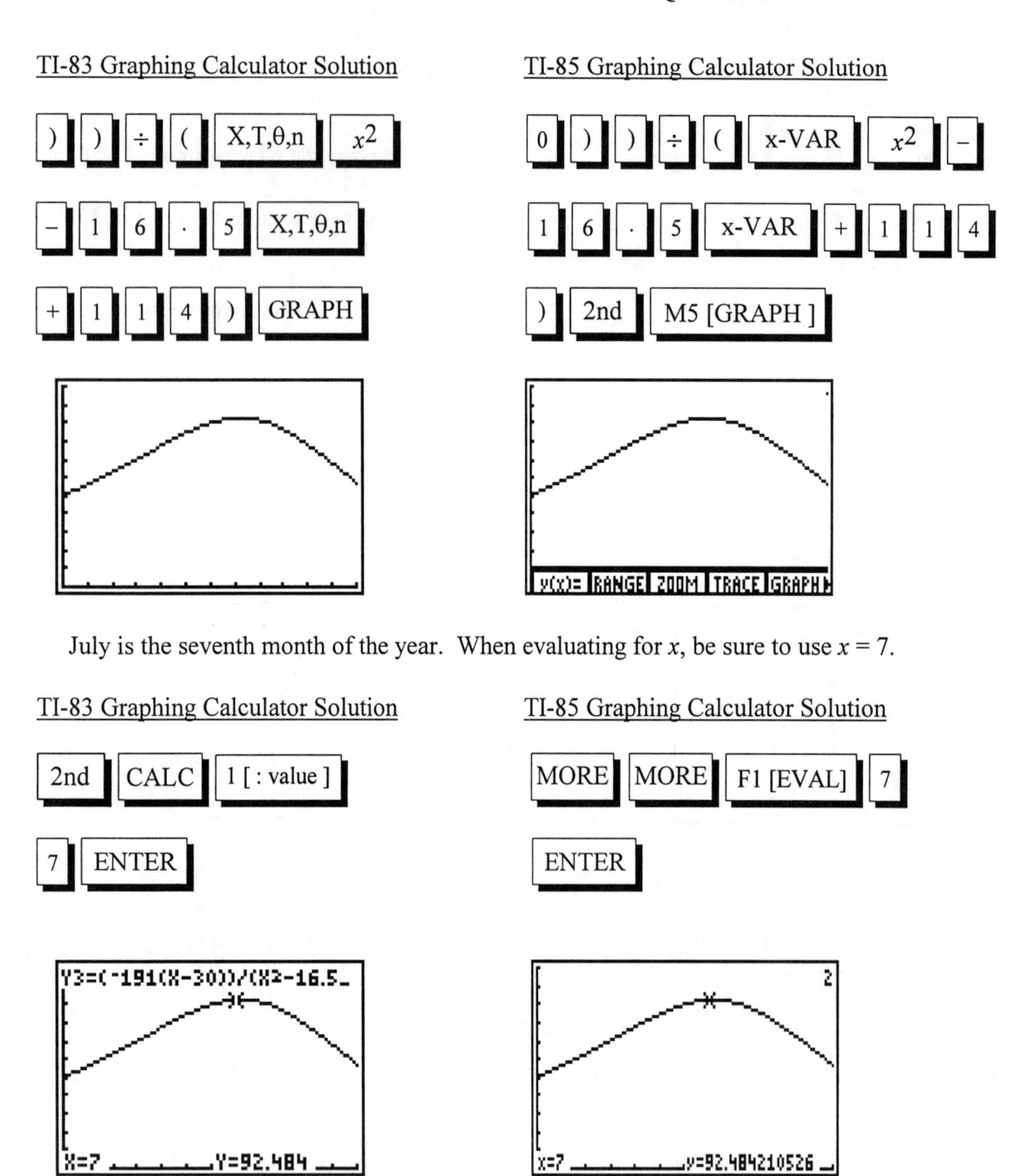

The average monthly temperature during the month of July in Anytown, USA is 92.5°.

138

LABORATORY 11: RATIONAL EQUATIONS

EXERCISES

1. Solve the rational equations. Use the **SOLVER** or **SOLVE** feature on the calculators.

   (a) $\dfrac{5x}{2x+1} + 3 = \dfrac{x}{2x+1}$, $x \neq ?$

   (b) $\dfrac{2x}{x+3} - 4 = \dfrac{5x}{x+3}$, $x \neq ?$

2. Graph each of the following rational equations and state the root or zero of the equation.

   (a) $\dfrac{2}{x+4} + \dfrac{2x-1}{x^2+2x-8} = \dfrac{1}{x-2}$

   (b) $\dfrac{3x}{x+2} + \dfrac{72}{x^3+8} = \dfrac{24}{x^2-2x+4}$

3. The formula $\dfrac{1}{R} = \dfrac{1}{r_1} + \dfrac{1}{r_2}$ gives the resistance R of two resistors r_1 and r_2 connected in parallel. What is R if r_1 is 4 ohms and r_2 is 3 more than half of r_1? What is r_1 if R is 25 ohms and r_2 is 3 ohms?

4. In x years from 1990, the population (P, in thousands) of a small town will be

$$P = 30 - \dfrac{4}{x+1}.$$

   What will the population of the town be in 1999? When will the population of the town be 29,000?

5. A shipyard that orders x gallons of gasoline pays a price per gallon of G dollars, described by

$$G = 0.53 + \dfrac{0.85}{x}.$$

   How many gallons of gasoline should the shipyard order if it wishes to pay $0.55/gallon? What is the price per gallon if the shipyard orders 465 gallons of gas?

6. For the years 1985 to 1995, the total revenue from special events at a Sports Arena, P (in millions of dollars) can be described by the model

$$P = \dfrac{69(x^2 + 3x + 30)}{-x^3 + 16x^2 - 15x + 240}$$

   where $x = 0$ represents 1985. In what year did the total revenue approximate $16,000,000? What was the revenue in 1991? In 1994?

LABORATORY 11: RATIONAL EQUATIONS

7. According to Boyle's law, the pressure exerted by a gas is inversely proportional to the volume, as long as the temperature stays the same. If a gas exerts a pressure of 1150 pounds per square inch when the volume is 3 cubic feet, find the volume when the pressure is 900 pounds per square inch.

8. The intensity of light, in foot-candles, that is x feet from its source is given by the rational function
$$I(x) = \frac{320}{x^2}.$$

How far away is the source if the intensity of light is 4 foot-candles? What is the intensity of the light if you are 10 feet away? 8.3 feet away? 3 feet away?

LABORATORY 12: RADICALS AND EXPONENTS

Purpose:

The purpose of this laboratory is to use the laws of exponents to simplify expressions having integral exponents, simplify radical expressions, add, subtract, multiply, and divide radicals, and solve radical equations using the TI-83/82 and TI-85/86 calculators.

Analytic Approach:

The TI-83/82 and TI-85/86 calculators can*not* show the algebra steps necessary to evaluate or solve expressions and equations. It is imperative, therefore, that you understand the basic laws of exponents.

If a number is to be multiplied by itself repeatedly, the product can be written using **exponential notation**. *Exponential notation* is a form of shorthand. For example, 4 x 4 x 4 is written as 4^3. This is read as 4 cubed, or simply, 4 to the third power. The number 4 is called the **base,** while the number 3 is called the **power** or **exponent.** An *exponent* tells how many times to multiply the *base* by itself. The value of the expression, 64, remains the same no matter how it is written. *Exponents* are not limited to just integer values. **Any real number** can be an exponent.

> **CAUTION**: x^n *does not mean to multiply n by x!*

An exponent applies only to the symbol directly before it. For example, in the expression $2x^5$, the *exponent* 5 refers to the number of times the *base x* is to be multiplied by itself. Once x has been multiplied by itself 5 times, the product is then multiplied by 2. In the expression $(2x)^5$, *because of the parentheses,* both the 2 *and* the x are multiplied by themselves 5 times, yielding $2^5 x^5$ in exponential form, or $32x^5$.

Before beginning each problem in this laboratory, set the **Float** to reflect three decimal places. Also, set the viewing rectangle to the standard setting.

Problem 1: Evaluate the following expressions:

(a) $3^2 + 5$
(b) $-3^2 + 5$
(c) $4(-3)^2 + 5$

Graphing Calculator Solution

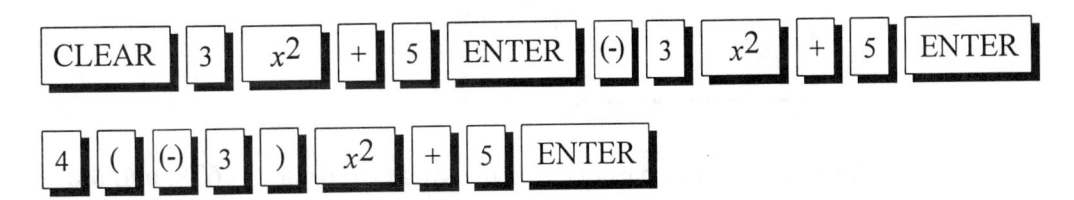

| CLEAR | 3 | x^2 | + | 5 | ENTER | (-) | 3 | x^2 | + | 5 | ENTER |

| 4 | (| (-) | 3 |) | x^2 | + | 5 | ENTER |

LABORATORY 12: RADICALS AND EXPONENTS

Graphing Calculator Screen

```
3²+5
            14.000
-3²+5
            -4.000
4(-3)²+5
            41.000
```

The product in part (a) is 14, the product in part (b) is -4. Part (c), $4(-3)^2 + 5$, yields a product of 41.

An exponent of 0 or 1 has a special meaning. For any number x, x^1 **is defined to be** x, *and* for any nonzero number x, x^0 **is defined to be 1**.

Problem 2: Evaluate the following expressions.

(a) $(345)^0$
(b) $(678)^1$
(c) $(432)^1 + (6934)^0$

TI-83 and TI-85 Graphing Calculator Solution

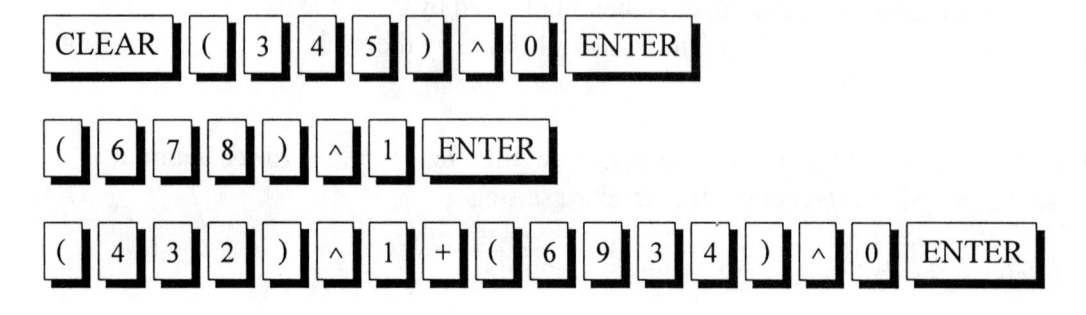

[CLEAR] [(] [3] [4] [5] [)] [^] [0] [ENTER]

[(] [6] [7] [8] [)] [^] [1] [ENTER]

[(] [4] [3] [2] [)] [^] [1] [+] [(] [6] [9] [3] [4] [)] [^] [0] [ENTER]

```
(345)^0
            1.000
(678)^1
          678.000
(432)^1+(6934)^0
          433.000
```

The value of the exponential expression in part (a) is 1; the value of $(678)^1$ is 678; and the value of the expression $(432)^1 + (6934)^0$ is 433.

LABORATORY 12: RADICALS AND EXPONENTS

The Laws of Exponents

A *summary* of the **Laws of Exponents** is stated in each of the problems below. The summary of the laws of exponents should be used in conjunction with your textbook.

Exponent Law 1. When multiplying like bases, add the exponents.

Problem 3: Show that the statement $(4^2)(4^3) = 4^5$ is true.

Graphing Calculator Solution

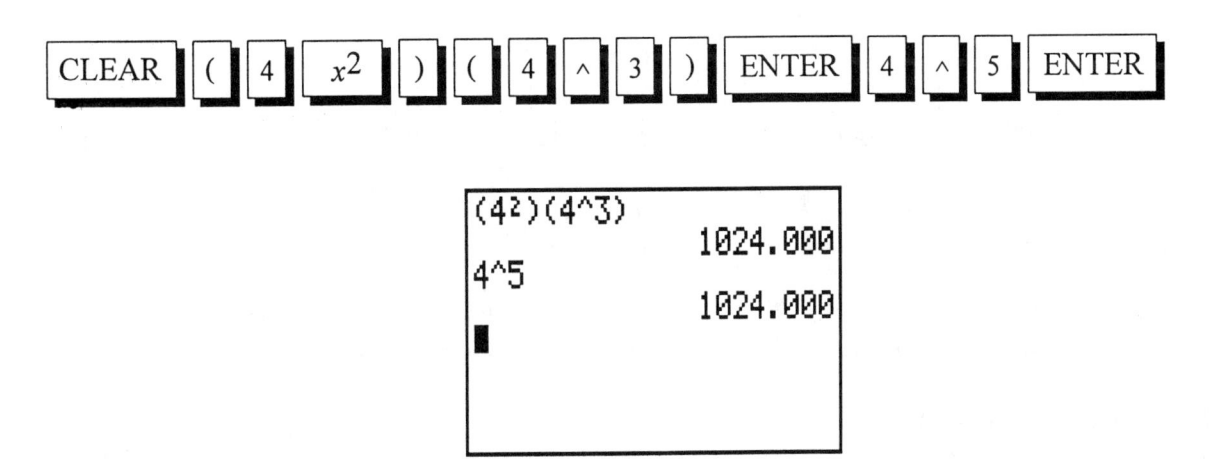

| CLEAR | (| 4 | x2 |) | (| 4 | ^ | 3 |) | ENTER | 4 | ^ | 5 | ENTER |

```
(4²)(4^3)
              1024.000
4^5
              1024.000
```

Note that the result of $(4^2)(4^3)$ and the result of 4^5 is the same, 1024.

Exponent Law 2. When dividing with like bases, subtract the exponents. If this results in a negative exponent, rewrite your answer with a positive exponent

---

Problem 4: Show that the following statements are true:

(a) $\dfrac{(5^6)}{(5^3)} = 5^3$
b) $\dfrac{(7^2)}{(7^5)} = \dfrac{1}{7^3}$

Part (a)

Graphing Calculator Solution

| CLEAR | 5 | ^ | 6 | ÷ | 5 | ^ | 3 | ENTER | 5 | ^ | 3 | ENTER |

LABORATORY 12: RADICALS AND EXPONENTS

Graphing Calculator Screen

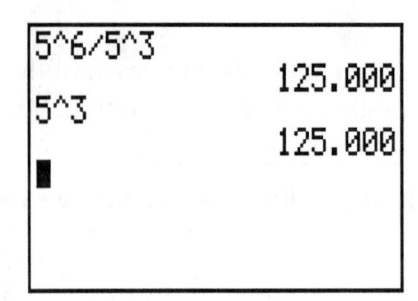

```
5^6/5^3
              125.000
5^3
              125.000
```

The two sides of the statement $\dfrac{(5^6)}{(5^3)} = 5^3$ are the same, 125.

Part (b)

TI-83 Graphing Calculator Solution

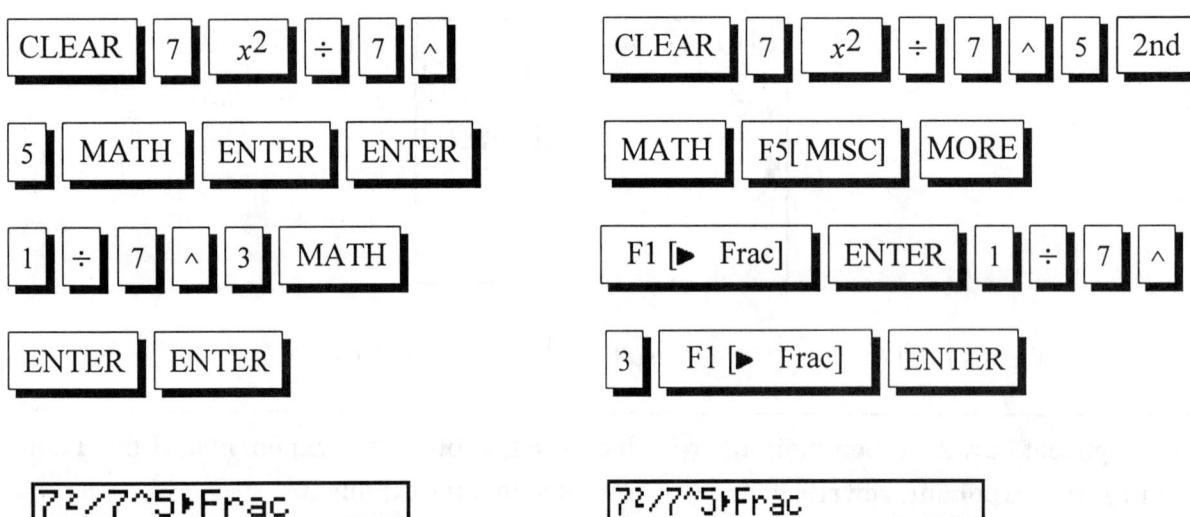

[CLEAR] [7] [x^2] [÷] [7] [^]

[5] [MATH] [ENTER] [ENTER]

[1] [÷] [7] [^] [3] [MATH]

[ENTER] [ENTER]

```
7²/7^5▶Frac
              1/343
1/7^3▶Frac
              1/343
```

TI-85 Graphing Calculator Solution

[CLEAR] [7] [x^2] [÷] [7] [^] [5] [2nd]

[MATH] [F5[MISC]] [MORE]

[F1 [▶ Frac]] [ENTER] [1] [÷] [7] [^]

[3] [F1 [▶ Frac]] [ENTER]

```
7²/7^5▶Frac
              1/343
1/7^3▶Frac
              1/343

NUM  PROB  ANGLE  HYP  MISC
▶Frac  %  pEval  ˣ√  eval
```

It does not matter how the statement $\dfrac{(7^2)}{(7^5)} = \dfrac{1}{7^3}$ is evaluated, the result is the same, $\dfrac{1}{343}$.

LABORATORY 12: RADICALS AND EXPONENTS

Exponent Law 3a. The Power Rule: When raising a power to a power, keep the base the same, and multiply the exponents.

Problem 5: Show that the statement $(15^2)^3 = 15^6$ is true.

Graphing Calculator Solution

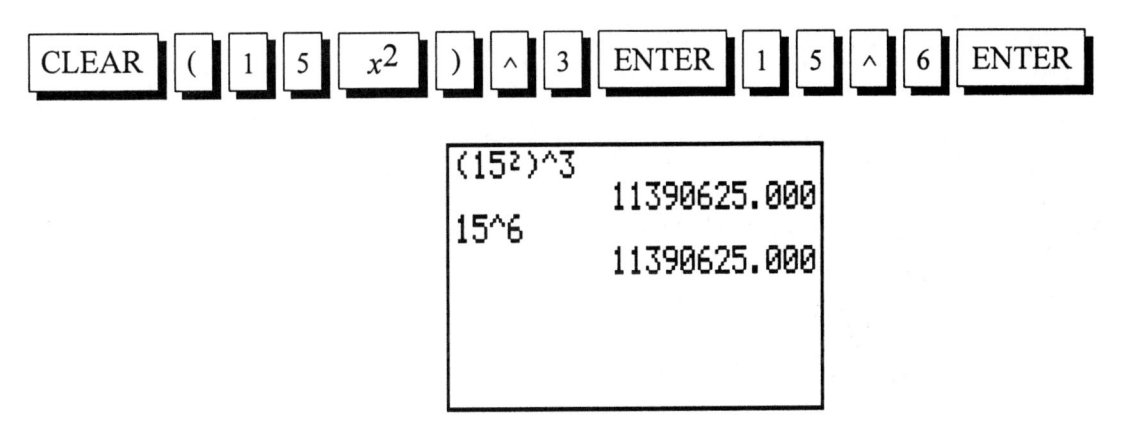

[CLEAR] [(] [1] [5] [x^2] [)] [^] [3] [ENTER] [1] [5] [^] [6] [ENTER]

```
(15²)^3
              11390625.000
15^6
              11390625.000
```

The same result of 11,390,625 is obtained when each side of the statement $(15^2)^3 = 15^6$ is evaluated.

Exponent Law 3b. Raising a Product to a Power. *To raise a product to the nth power, raise each factor to the nth power.*

Problem 6: Evaluate $(4x^3y^5)^2$ when $x = 1.5$ and $y = -2.4$.

Graphing Calculator Solution

[CLEAR] [(] [4] [(] [1] [.] [5] [)] [^] [3] [(] [(-)] [2] [.] [4] [)] [^] [5] [)] [x^2]

[ENTER]

```
(4(1.5)^3(-2.4)^5)²
              1155526.618
```

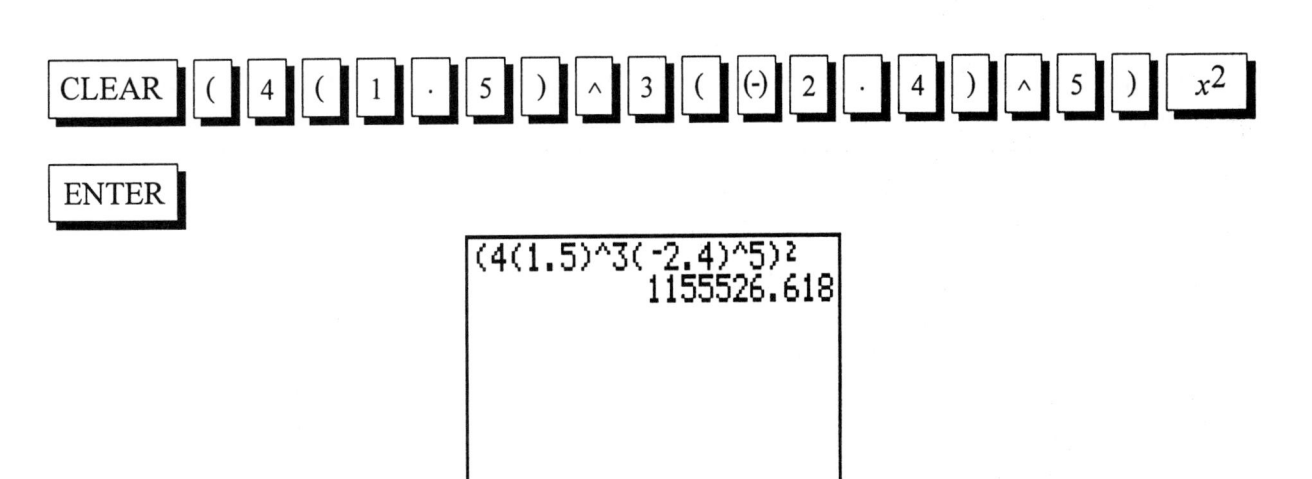

LABORATORY 12: RADICALS AND EXPONENTS

The evaluation of $(4x^3y^5)^2$ when $x = 1.5$ and $y = -2.4$ is 1155526.618. It is left to the student to determine if the evaluation of $4^2x^6y^{10}$ yields the same result of 1155526.618.

Exponent Law 3c. Raising a Quotient to a Power. *To raise a quotient to a power, raise **both the numerator and denominator** to the power.*

Problem 7: Show that the statement $\left(\dfrac{4^6}{2^2}\right)^3 = \dfrac{4^{18}}{2^6}$ is true.

Graphing Calculator Solution

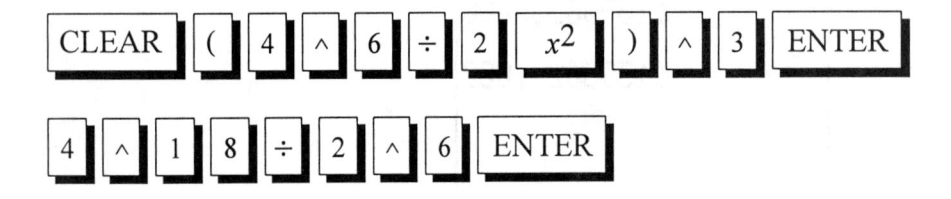

| CLEAR | (| 4 | ^ | 6 | ÷ | 2 | x^2 |) | ^ | 3 | ENTER |

| 4 | ^ | 1 | 8 | ÷ | 2 | ^ | 6 | ENTER |

```
(4^6/2²)^3
            1073741824.00
4^18/2^6
            1073741824.00
```

When the statement $\left(\dfrac{4^6}{2^2}\right)^3 = \dfrac{4^{18}}{2^6}$ is evaluated, the result is 1,073,741,824.

Exponent Law 4. When a base is raised to a negative power, invert the base and *change* the sign of the exponent.

Problem 8: Show that the statement $4^{-1} = \dfrac{1}{4}$ is true:

TI-83 Graphing Calculator Solution

| CLEAR | 4 | ^ | (-) | 1 |

| MATH | ENTER | ENTER |

TI-85 Graphing Calculator Solution

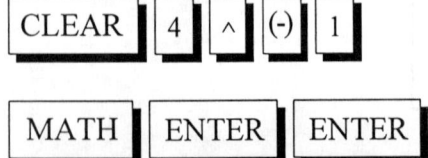

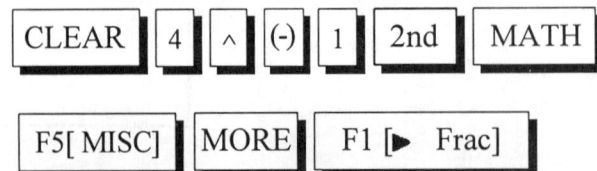

| CLEAR | 4 | ^ | (-) | 1 | 2nd | MATH |

| F5[MISC] | MORE | F1 [▶ Frac] |

LABORATORY 12: RADICALS AND EXPONENTS

TI-83 Graphing Calculator Solution   TI-85 Graphing Calculator Solution

ENTER

```
4^-1▶Frac
           1/4
```

```
4^-1▶Frac
           1/4
```

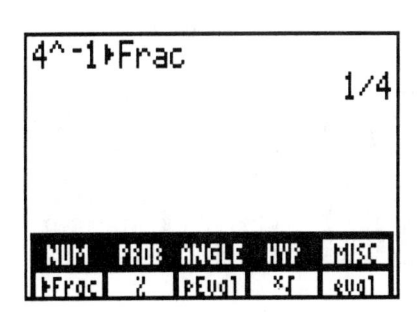

From the calculator screen 4^{-1} and $\frac{1}{4}$ are equal. The reciprocal of 4 is $\frac{1}{4}$.

Exponential expressions can be evaluated on all four TI calculators. The values for the variables must be stated.

Problem 9: Evaluate the expression $\frac{x^2 + 5y^3}{a - 6}$, when $x = -2$, $y = 2$, and $a = 0$.

Graphing Calculator Solution

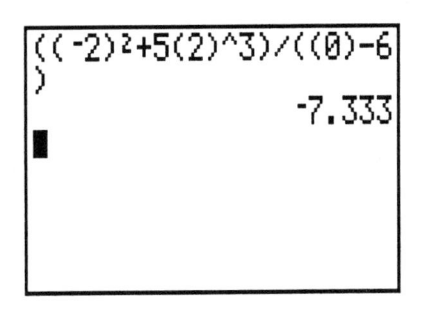

CLEAR (((-) 2) x^2 + 5 (2) ^ 3) ÷ ((0) − 6) ENTER

```
((-2)²+5(2)^3)/((0)-6
)
              -7.333
```

The result for the substitution of the variables into $\frac{x^2 + 5y^3}{a - 6}$ is -7.333 or $-7\frac{1}{3}$.

The **SOLVER** feature on the **TI-85/86**, unlike its counterpart on the **TI-83** and the **SOLVE** feature on the **TI-82**, can also be used, easily, to evaluate expressions in *more than one* variable.

LABORATORY 12: RADICALS AND EXPONENTS

Problem 10: Evaluate the expression $\dfrac{x^2 + 5y^3}{a - 6}$, using the **SOLVER** feature on the TI-85 and TI-86. Let $x = -2$, $y = 2$, and $a = 0$.

TI-85 Graphing Calculator Solution

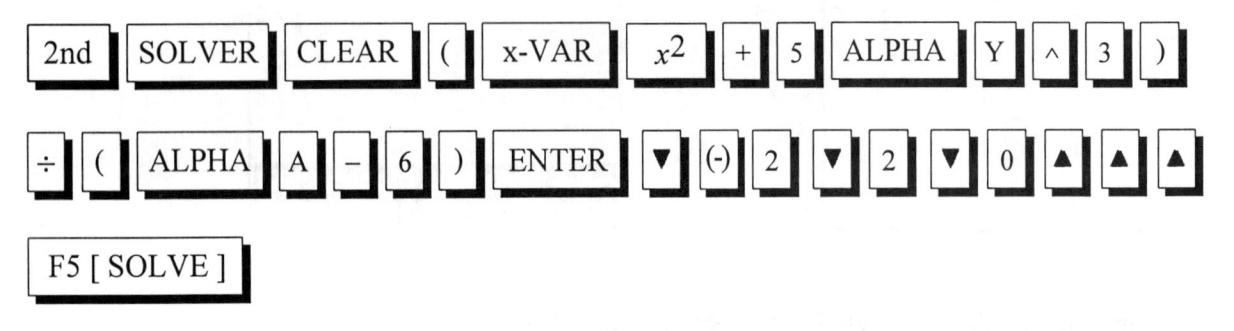

| 2nd | SOLVER | CLEAR | (| x-VAR | x² | + | 5 | ALPHA | Y | ^ | 3 |) |

| ÷ | (| ALPHA | A | − | 6 |) | ENTER | ▼ | (-) | 2 | ▼ | 2 | ▼ | 0 | ▲ | ▲ | ▲ |

| F5 [SOLVE] |

```
exp=(x²+5Y^3)/(A-6)
■exp=-7.3333333333333
  x=-2
  Y=2
  A=0
  bound={-1E99,1E99}
■left-rt=0
GRAPH RANGE ZOOM TRACE SOLVE
```

The same result -7.333 is obtained using a different approach.

Rational Exponents and Radicals

When a quantity is raised to a rational exponent, it can also be written as a **radical**. A **radical** is a *root* of a quantity. A radical is written in the following form: $\sqrt[n]{x}$, where *n* represents the **index** or **root** of the radical, $\sqrt{}$ is the **radical sign,** and *x* is the **radicand**. A common error made in the evaluation of a radical is confusing the *coefficient* of a radical with the *index* of the radical. In the example $4\sqrt{x} \neq \sqrt[4]{x}$, the *4 in front* of the square root is the *coefficient*, while the *4 in the radical sign* is the *index*.

The Properties of Radicals

> Property 1: **If *n* is a positive integer greater than 1 and $\sqrt[n]{b}$ is a real number, then $\sqrt[n]{b}$ can also be written as $b^{1/n}$. Therefore $b^{1/n} = \sqrt[n]{b}$.**

LABORATORY 12: RADICALS AND EXPONENTS

One example of a root that fits the above definition is the **square root** of a number. For any non-negative number b, **the principal square root** of b is its nonnegative root. The *principal square root* is written as $\sqrt{b}$. The *negative square root* is written as $-\sqrt{b}$.

Problem 11: Verify the statement $\sqrt{6} = 6^{\frac{1}{2}}$. When using a **TI-82** follow the **TI-85** keystroke sequence shown below.

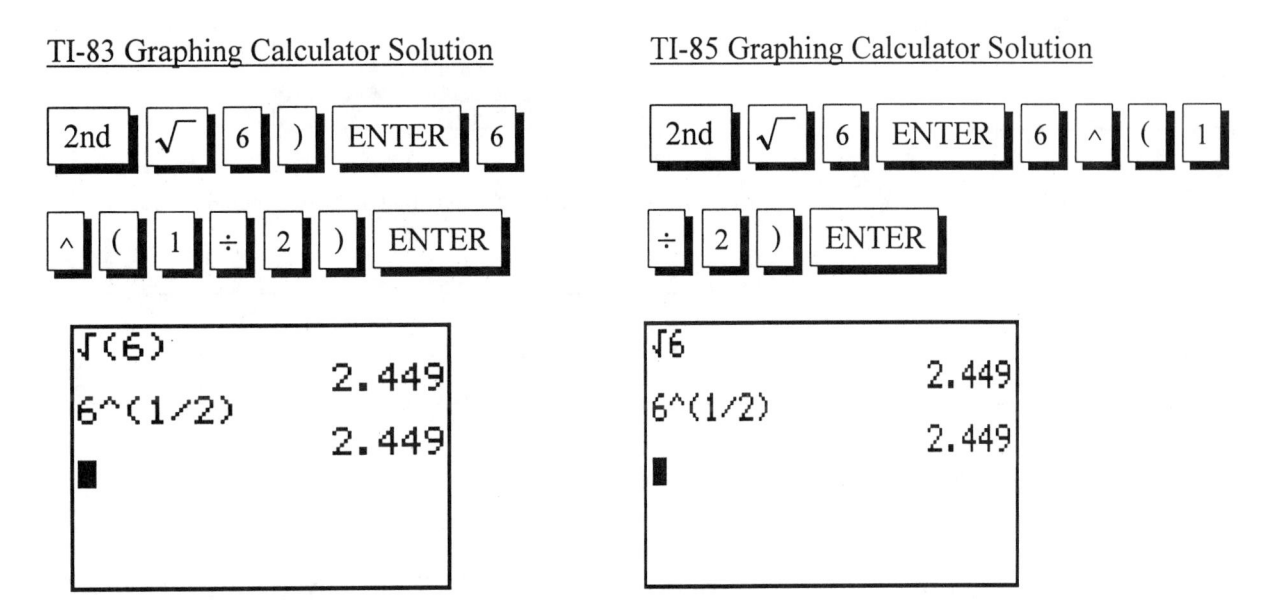

TI-83 Graphing Calculator Solution

| 2nd | √ | 6 |) | ENTER | 6 |

| ^ | (| 1 | ÷ | 2 |) | ENTER |

```
√(6)
        2.449
6^(1/2)
        2.449
```

TI-85 Graphing Calculator Solution

| 2nd | √ | 6 | ENTER | 6 | ^ | (| 1 |

| ÷ | 2 |) | ENTER |

```
√6
        2.449
6^(1/2)
        2.449
```

Only the principal root, 2.449, is stated. *Remember*, a negative sign must be placed before the radical sign if the negative answer is desired.

Property 2: If m and n are positive integers greater than 1 with $\dfrac{m}{n}$ in lowest terms, then $b^{m/n} = \sqrt[n]{b^m} = \left(\sqrt[n]{b}\right)^m$ as long as $\sqrt[n]{b}$ is a real number.

Problem 12: Verify the statement $\sqrt[3]{25^2} = 25^{\frac{2}{3}}$.

Remember that when using a **TI-82** that it may be necessary to add a left parenthesis after pressing the root key.

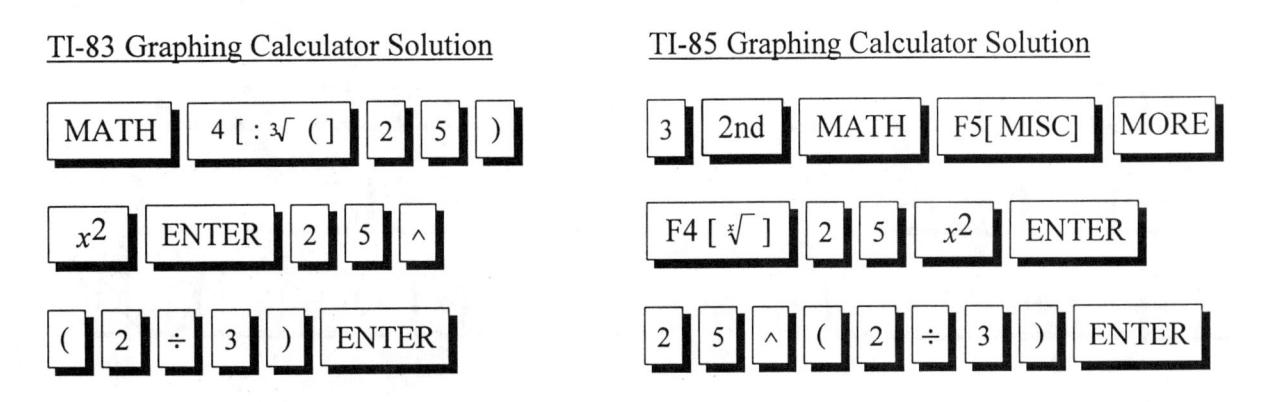

TI-83 Graphing Calculator Solution

| MATH | 4 [: ∛ ( ] | 2 | 5 |) |

| x^2 | ENTER | 2 | 5 | ^ |

| (| 2 | ÷ | 3 |) | ENTER |

TI-85 Graphing Calculator Solution

| 3 | 2nd | MATH | F5[MISC] | MORE |

| F4 [√] | 2 | 5 | x^2 | ENTER |

| 2 | 5 | ^ | (| 2 | ÷ | 3 |) | ENTER |

LABORATORY 12: RADICALS AND EXPONENTS

TI-83 Graphing Calculator Screen

```
³√(25)²
              8.550
25^(2/3)
              8.550
```

TI-85 Graphing Calculator Screen

```
3 ˣ√25²
              8.550
25^(2/3)
              8.550

NUM  PROB  ANGLE  HYP  MISC
▶Frac   ?   ▶Eval   ˣ√   eval
```

The cube root of 25 squared is the same as taking 25 to the two thirds power or 8.550.

Property 3: As long as $b^{m/n}$ is a nonzero real number, then $b^{-m/n} = \dfrac{1}{b^{m/n}}$.

Problem 13: Verify that $(-27)^{-2/3} = \dfrac{1}{\sqrt[3]{(-27)^2}} = \dfrac{1}{9}$.

The exponential expression, $(-27)^{-2/3}$, should be restated as, $((-27)^2)^{-\frac{1}{3}}$, when using the **TI-82** and **TI-85**. This form, $((-27)^2)^{-\frac{1}{3}}$, of the exponential expression allows the numerical algorithm to correctly evaluate this type of expression on the TI-85 and TI-82. **TI-82** users should follow the **TI-85** keystroke sequence to enter the expression, then revert to the **TI-83** keystroke sequence to write the answer as a fraction. Users of the **TI-86** can enter the expression $(-27)^{-2/3}$ in the same manner as on the **TI-83**.

When using a TI-82 to evaluate the radical side of the expression, remember to insert a left parenthesis before entering the radicand, -27. Then follow the **TI-83** keystroke sequence. Users of the **TI-86** should follow the **TI-85** keystroke sequence for entering the radical expression.

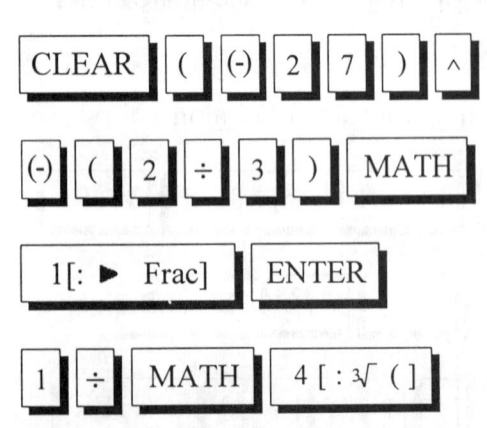

TI-83 Graphing Calculator Solution

| CLEAR | (| (-) | 2 | 7 |) | ^ |

| (-) | (| 2 | ÷ | 3 |) | MATH |

| 1[: ▶ Frac] | ENTER |

| 1 | ÷ | MATH | 4 [: ³√ ( ] |

TI-85 Graphing Calculator Solution

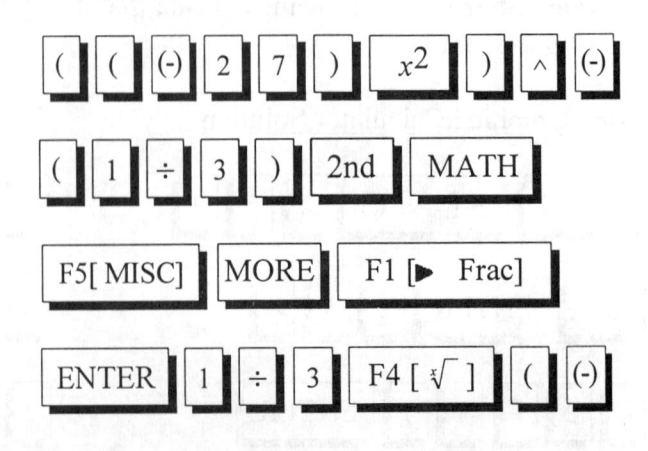

| (| (| (-) | 2 | 7 |) | x² |) | ^ | (-) |

| (| 1 | ÷ | 3 |) | 2nd | MATH |

| F5[MISC] | MORE | F1 [▶ Frac] |

| ENTER | 1 | ÷ | 3 | F4 [ˣ√] | (| (-) |

150

LABORATORY 12: RADICALS AND EXPONENTS

TI-83 Graphing Calculator Solution

[(-)] [2] [7] [)] [x^2] [MATH]

[1[: ▶ Frac]] [ENTER]

```
(-27)^-(2/3)▶Fra
c
                1/9
1/3√(-27)²▶Frac
                1/9
```

TI-85 Graphing Calculator Solution

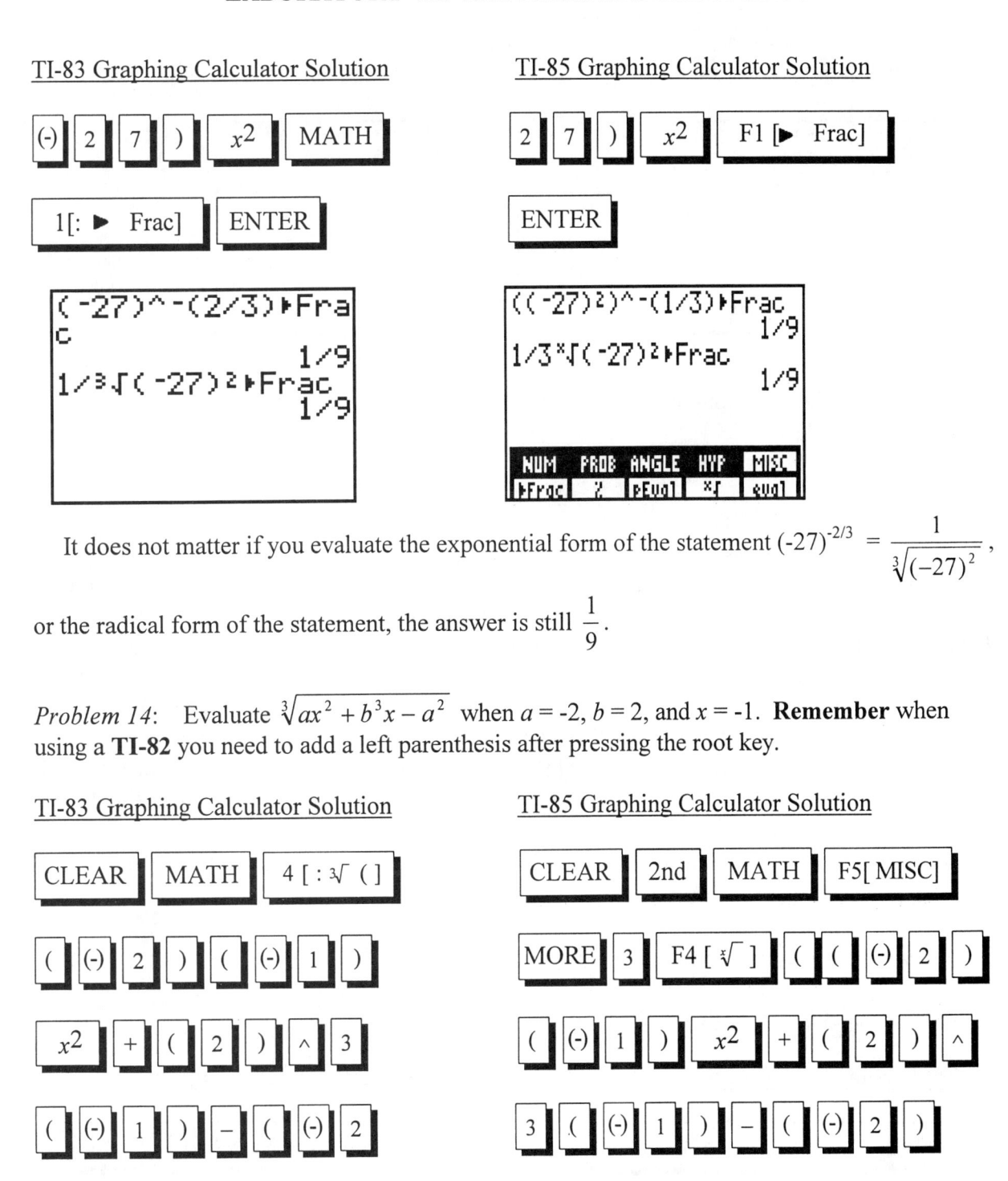

[2] [7] [)] [x^2] [F1 [▶ Frac]]

[ENTER]

```
((-27)²)^-(1/3)▶Frac
                1/9
1/3x√(-27)²▶Frac
                1/9
NUM  PROB ANGLE HYP  MISC
▶Frac  ?  ▶Eval  x√  eval
```

It does not matter if you evaluate the exponential form of the statement $(-27)^{-2/3} = \dfrac{1}{\sqrt[3]{(-27)^2}}$, or the radical form of the statement, the answer is still $\dfrac{1}{9}$.

Problem 14: Evaluate $\sqrt[3]{ax^2 + b^3x - a^2}$ when $a = -2$, $b = 2$, and $x = -1$. **Remember** when using a **TI-82** you need to add a left parenthesis after pressing the root key.

TI-83 Graphing Calculator Solution

[CLEAR] [MATH] [4 [: $\sqrt[3]{\ }$ ( ]]

[(] [(-)] [2] [)] [(] [(-)] [1] [)]

[x^2] [+] [(] [2] [)] [^] [3]

[(] [(-)] [1] [)] [-] [(] [(-)] [2]

[)] [x^2] [)] [ENTER]

TI-85 Graphing Calculator Solution

[CLEAR] [2nd] [MATH] [F5[MISC]]

[MORE] [3] [F4 [$\sqrt[x]{\ }$]] [(] [(] [(-)] [2] [)]

[(] [(-)] [1] [)] [x^2] [+] [(] [2] [)] [^]

[3] [(] [(-)] [1] [)] [-] [(] [(-)] [2] [)]

[x^2] [)] [ENTER]

151

LABORATORY 12: RADICALS AND EXPONENTS

TI-83 Graphing Calculator Screen

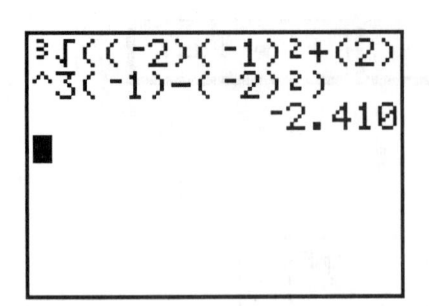

TI-85 Graphing Calculator Screen

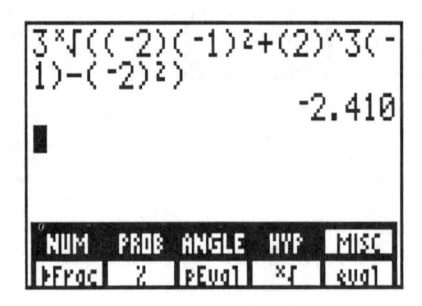

The result of the substitution of the given values for the variables in $\sqrt[3]{ax^2 + b^3x - a^2}$ is approximately -2.410.

Radical Equations

A **radical equation** contains *radicals* or *rational exponents*. To algebraically solve a **radical equation,** it is necessary to isolate the radical term to one side of the equal sign and then to *raise both sides* to a *power* that will eliminate the radical.

One method of choice in finding the solution to a radical equation using the **TI-83, 85**, and **TI-86** calculators is to use the **SOLVER** feature. The **TI-82** has a **SOLVE** feature and the keystrokes must be adjusted accordingly. On the **TI-83** and the **TI-82**, the radical equation *must be set equal to zero*. The equation does not have to be set equal to zero on the **TI-85** and **TI-86**.

Problem 15: Find the value of x that satisfies $\sqrt[3]{x + 8} = 5$. Remember to add a left parenthesis after the root key when using the **TI-82**.

If there already is an equation entered into **SOLVER** on the **TI-83**, then you must press the ▲ key in order to edit or delete the equation.

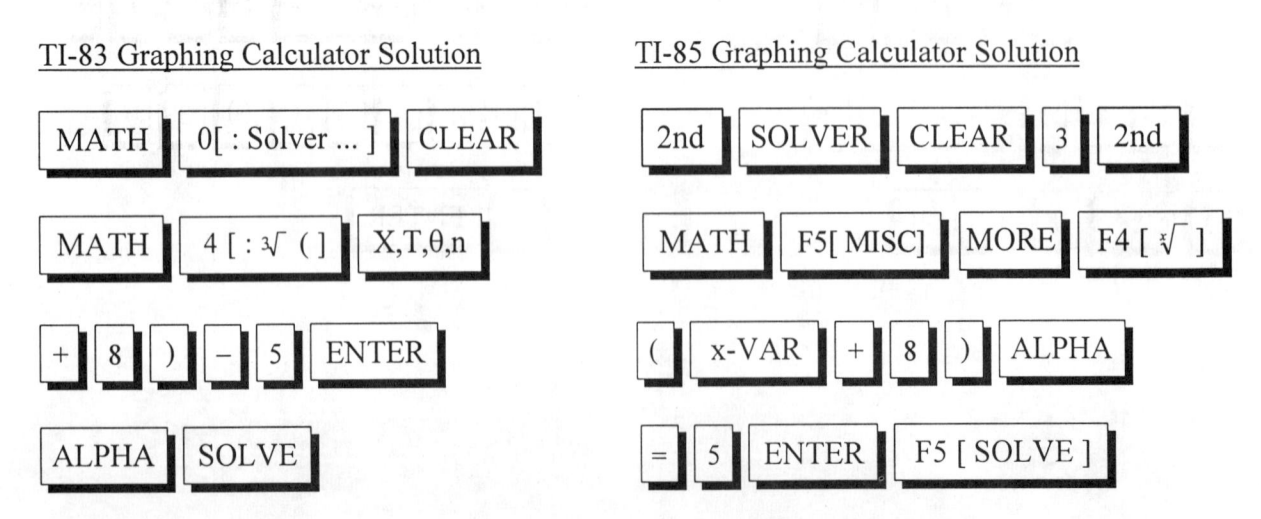

TI-83 Graphing Calculator Solution

| MATH | 0[: Solver ...] | CLEAR |

| MATH | 4 [: $\sqrt[3]{\ }$ ( )] | X,T,θ,n |

| + | 8 |) | − | 5 | ENTER |

| ALPHA | SOLVE |

TI-85 Graphing Calculator Solution

| 2nd | SOLVER | CLEAR | 3 | 2nd |

| MATH | F5[MISC] | MORE | F4 [$\sqrt[x]{\ }$] |

| (| x-VAR | + | 8 |) | ALPHA |

| = | 5 | ENTER | F5 [SOLVE] |

LABORATORY 12: RADICALS AND EXPONENTS

TI-83 Graphing Calculator Screen

```
³√(X+8)-5=0
 X=117.00000000…
 bound={-1E99,1…
```

TI-85 Graphing Calculator Screen

```
3ˣ√(x+8)=5
•x=117.00000000001
 bound={-1E99,1E99)
•left-rt=1E-13
```
[GRAPH|RANGE|ZOOM|TRACE|SOLVE]

For users of the **TI-82** adjust your keystrokes as follows. Access the **SOLVE** feature in the same manner as on the **TI-83**. Then **enter the expression** followed by a **coma**. Press [X,T,θ] followed by a **coma**. Press **0** followed by a **right parenthesis**. Finally, press [ENTER].

When $\sqrt[3]{x+8} = 5$ is solved for x, x has a value of 117. Always check your answer in the original radical equation. Use your TI calculators to help do the substitution of 117 for x in the original radical equation. Only one side of the equation needs to be evaluated to verify that the two sides of the equation equal 5.

Many times the squaring process, etc., needed to remove a radical yields what are called **extraneous roots**. *Extraneous roots* are discarded because they do not satisfy the original equation.

Problem 16:   Graphically solve the equation $\sqrt{x-2} + \sqrt{x} = 4$ for x.

Set the radical equation equal to zero before entering it onto **all the calculators**. **Remember** to add a left parenthesis after pressing the root key on the **TI-82**.

TI-83 Graphing Calculator Solution

[Y=] [CLEAR] \Y₁= [2nd]

[√] [X,T,θ,n] [−] [2] [)] [+]

[2nd] [√] [X,T,θ,n] [)] [−]

[4] [GRAPH] [2nd] [CALC]

[2 [: zero]]

TI-85 Graphing Calculator Solution

[GRAPH] [F1 [y(x) =]] y1= [CLEAR]

[2nd] [√] [(] [x-VAR] [−] [2] [)] [+]

[2nd] [√] [x-VAR] [−] [4] [2nd]

[M5 [GRAPH]] [MORE] [F1 [MATH]]

[F3 [ROOT]] [ENTER]

LABORATORY 12: RADICALS AND EXPONENTS

TI-83 Graphing Calculator Solution

Set the left and right bounds and a guess. Remember to press $\boxed{\text{ENTER}}$ after each.

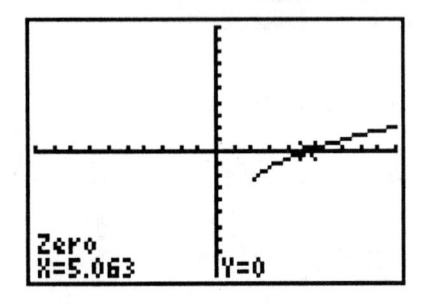

TI-85 Graphing Calculator Solution

The root function on the **TI-86** is located on **F1**. Remember to set the left and right bounds and a guess. Press $\boxed{\text{ENTER}}$ after each.

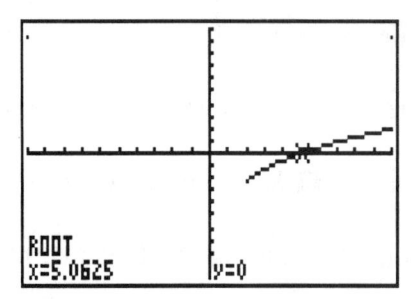

The solution to the radical equation $\sqrt{x-2} + \sqrt{x} = 4$ is 5.0625. Don't forget to check your answer by substituting 5.0625 for x in the original radical equation. When a graphical solution to a radical equation is found, the chance of finding any extraneous roots is diminished. It is always a good idea to check your answers in any case.

LABORATORY 12: RADICALS AND EXPONENTS

EXERCISES

1. Simplify the following exponential expressions. You may leave your answers in fraction form.

   (a) 27^8 (b) $(35)^0$ (c) $\left(\dfrac{3}{8}\right)^3$ (d) $(-15)^2$ (e) $-\dfrac{2^3}{5}$

   (f) $-\left(\dfrac{-3}{4}\right)^4$ (g) $(6)^{-2}$ (h) $\left(\dfrac{-7}{5}\right)^{-3}$

2. Find the value of each of the following expressions.

   (a) $\left(3^2 + 4^3\right)^2$ (b) $\left(\dfrac{2^2}{3} + (-3)^{-3}\right)^0$ (c) $\left(-\left(\dfrac{1}{4}\right)^{-2} + \left(\dfrac{3^2}{7^3} - 4\right)^0\right)$

3. Evaluate the radicals. Round your answers to the nearest thousandth.

   (a) $\sqrt{59}$ (b) $\sqrt[3]{68}$ (c) $-3\sqrt[4]{132}$ (d) $(64)^{\frac{1}{2}}$ (e) $\left(\dfrac{8}{27}\right)^{\frac{2}{3}}$

4. Perform the indicated operations. Round the answers to the nearest thousandth.

   (a) $-3\sqrt[4]{5} + 2\sqrt[4]{5} - 6\sqrt[4]{5}$ (b) $\dfrac{\sqrt{11}}{2} + \dfrac{\sqrt{11}}{3}$ (c) $\left(\sqrt{23}\,\sqrt[3]{3}\right)$

   (d) $\dfrac{5}{\sqrt{12}}$ (e) $\dfrac{\sqrt{4}+3}{\sqrt{6}-2} + \dfrac{(-2)^{-2}}{3}$

5. Solve the following equations for x.

   (a) $\sqrt{x-13} = 23$ (b) $\sqrt{x+3} - \sqrt{x-1} = 1$ (c) $\sqrt[3]{3x-4} = \sqrt[3]{x+10}$

6. Evaluate the following expressions when $x = 3$; $y = 1$; $z = \sqrt{3}$; $a = -1$; $b = 2$; and $c = -\dfrac{1}{2}$.

   (a) $\dfrac{a^2 + b}{z}$ (b) $\dfrac{z^3 - x^2}{y^3}$ (c) $\sqrt[3]{\dfrac{a^2 - b^2}{c^2}}$

155

LABORATORY 12: RADICALS AND EXPONENTS

7. The solution to the radical equation $\sqrt{x+a} - \sqrt{x} = 2$ is $x = 36$. Find the value of a.

8. A tree stump is to be removed from a backyard using two tow trucks. If two forces C and D pull at right angles (90 degrees) to each other, the resulting force F is given by the formula
$$F = \sqrt{C^2 + D^2}.$$
If tow truck C is exerting a force of 475 pounds and the resulting force is 900 pounds, find the force exerted by tow truck D (to the nearest thousandth of a pound).

9. The velocity v, in feet per second, of an object after it has fallen h feet accelerated by gravity g, in feet per second squared, is given by the formula
$$v = \sqrt{2gh}.$$
Find how far an object has fallen if its velocity is 100 feet per second, given that g is approximately 32 feet per second squared (round your answer to the nearest thousandth).

10. Use the formula for the radius r of a sphere given its surface area A,
$$r = \sqrt{\frac{A}{4\pi}}$$
to find the surface area of the Moon. The radius of the Moon is 1080 miles. Round your answer to the nearest square mile.

LABORATORY 13: INVERSE FUNCTIONS

Purpose:

The purpose of this laboratory is to look at the characteristics of both the inverse of a relation and the inverse of a function using the TI-83/82 and TI-85/86. A review of **Laboratory 8: Functions** would be appropriate before continuing with this laboratory.

Analytic Approach:

In **Laboratory 8: Functions**, a relation was defined as a set of ordered pairs (x, y). The **inverse of a relation** is the set of all ordered pairs of the form (y, x), where (x, y) belongs to the relation. The inverse of a relation is found by interchanging the first and second coordinates in the original relation. If R represents the relation, then R^{-1} represents the *inverse of the relation*. The symbol R^{-1} is the single symbol used to denote the inverse of the relation R. It is read as "R inverse". This symbol *does not mean* $\frac{1}{R}$.

When a relation is defined by an equation, interchanging x and y produces an equation of the inverse relation. The graph of the inverse of a relation such as $y = x + 4$ can be obtained algebraically by changing the position of the x and y variables in the equation $y = x + 4$. The equation that is obtained from switching the variables is $x = y + 4$. The new equation, $x = y + 4$, must now be solved for y. The resulting equation solved for y is $y = x - 4$. The new equation which is denoted by y^{-1}, is the inverse of the relation y.

REMEMBER: Always start a problem with the standard setting for the viewing rectangle.

Problem 1: Graph the relation $y = x + 4$ and show that $x = y + 4$ is its inverse, y^{-1}.

Remember to first solve the equation $x = y + 4$ for y. Then graph the result, $y = x - 4$.

TI-83 Graphing Calculator Solution

TI-85 Graphing Calculator Solution

Y=	CLEAR	\Y₁=	X,T,θ,n
+	4	ENTER	\Y₂=	X,T,θ,n
−	4	GRAPH

GRAPH	F1 [y(x) =]	y1=	CLEAR
x-VAR	+	4	ENTER	y2=	x-VAR
−	4	2nd	M5 [GRAPH]

LABORATORY 13: INVERSE FUNCTIONS

TI-83 Graphing Calculator Screen   TI-85 Graphing Calculator Screen

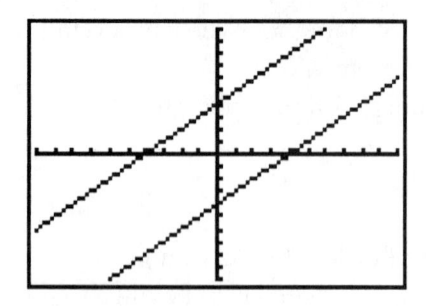

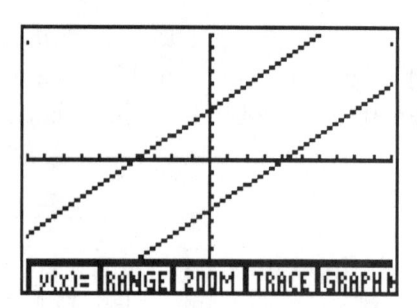

We will now show that the two lines are mirror images of each other by determining if the equations are indeed inverses. This can be done by showing that there exists a point (x_1, y_1) on the line $y = x + 4$, that corresponds to the point (y_1, x_1) on the line $y = x - 4$.

Examine a point such as $(0, 4)$ on the line $y = x + 4$ and see if there exists a corresponding point $(4, 0)$ on the line $y = x - 4$.

TI-83 Graphing Calculator Solution   TI-85 Graphing Calculator Solution

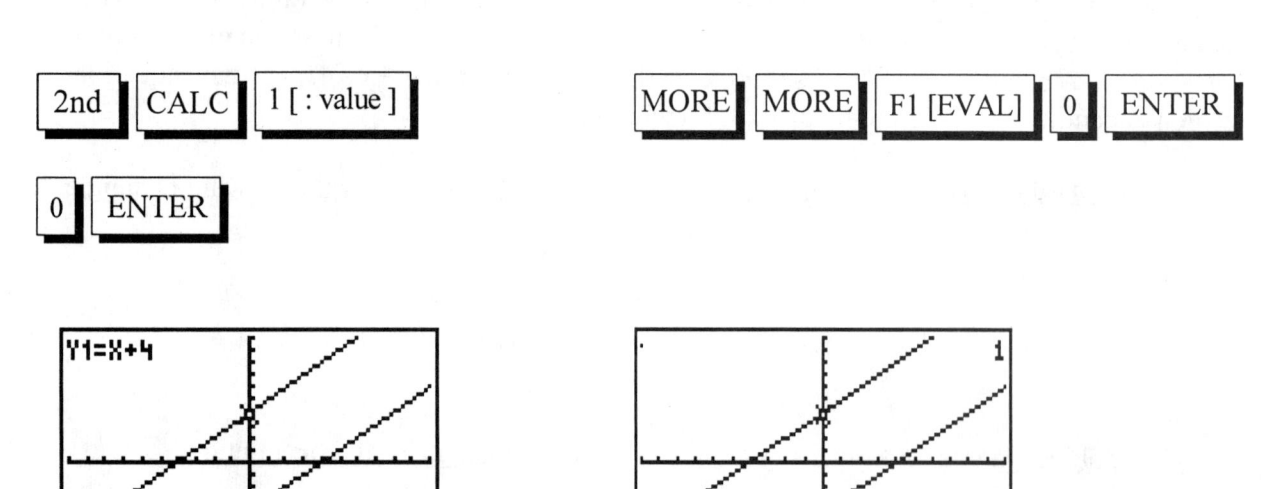

| 2nd | CALC | 1 [: value] |

| 0 | ENTER |

| MORE | MORE | F1 [EVAL] | 0 | ENTER |

Remember that you must use the up or down arrow keys to move between the lines.

| 2nd | CALC | 1 [: value] |

| 4 | ENTER | ▼ |

| EXIT | F1 [EVAL] | 4 | ENTER | ▼ |

158

LABORATORY 13: INVERSE FUNCTIONS

TI-83 Graphing Calculator Screen

TI-85 Graphing Calculator Screen

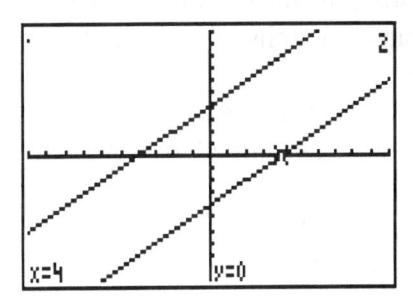

Check out at least two more pairs of points (x, y) from the graph of $y1 = x + 4$ and the corresponding two points (y, x) from the graph of $y2 = x - 4$.

Now include the graph of $y = x$, by pressing the following keys.

TI-83 Graphing Calculator Solution

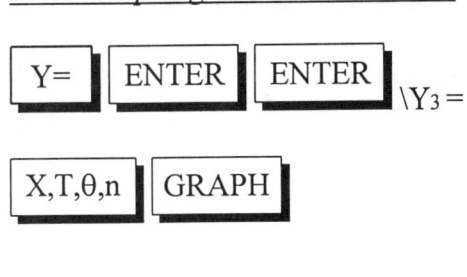

| Y= | ENTER | ENTER | \Y3 =
| X,T,θ,n | GRAPH |

TI-85 Graphing Calculator Solution

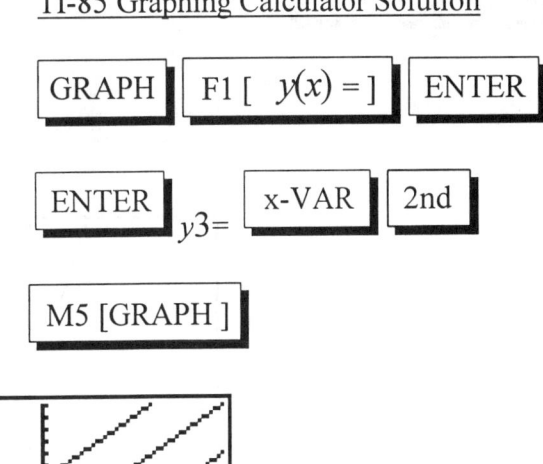

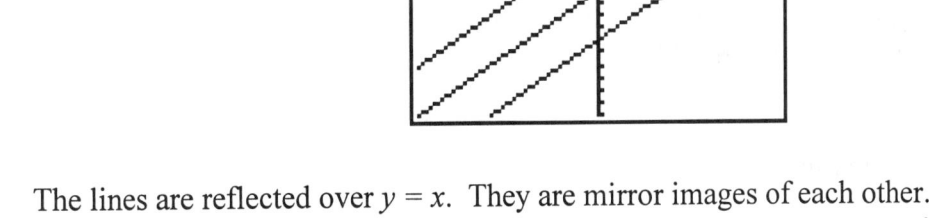

| GRAPH | F1 [$y(x) =$] | ENTER |
| ENTER | y3= | x-VAR | 2nd |
| M5 [GRAPH] |

The lines are reflected over $y = x$. They are mirror images of each other. Therefore $y = x + 4$ and $y = x - 4$ are inverses. The relationship between the two equations can be stated as $y^{-1} = x - 4$, is the inverse of $y = x + 4$.

We could also show that the equations y and y^{-1} are both functions by showing that a Vertical line intersects each graph *only once*. Refer to **Laboratory 8: Functions** for the keystrokes needed to show the vertical line test.

INVERSES ARE NOT NECESSARILY FUNCTIONS. The inverse of a relation may or may not be a function.

LABORATORY 13: INVERSE FUNCTIONS

The graph of the inverse of the function $y = x^2 + 1$ can be found from the graph of $y = x^2 + 1$ by locating the mirror image of each point with respect to the line $y = x$. For this reason, we say that the graphs of $y(x)$ and its inverse $y^{-1}(x)$ are symmetric about the line $y = x$.

Once $y = x^2 + 1$ and its inverse $x = y^2 + 1$ have been displayed on the calculator screen, a vertical line test can be performed to show that the inverse $x = y^2 + 1$ is not a function.

Problem 2: Determine if $y = x^2 + 1$ and its inverse $x = y^2 + 1$ are both functions. Clear all the y selections from *Problem 1*.

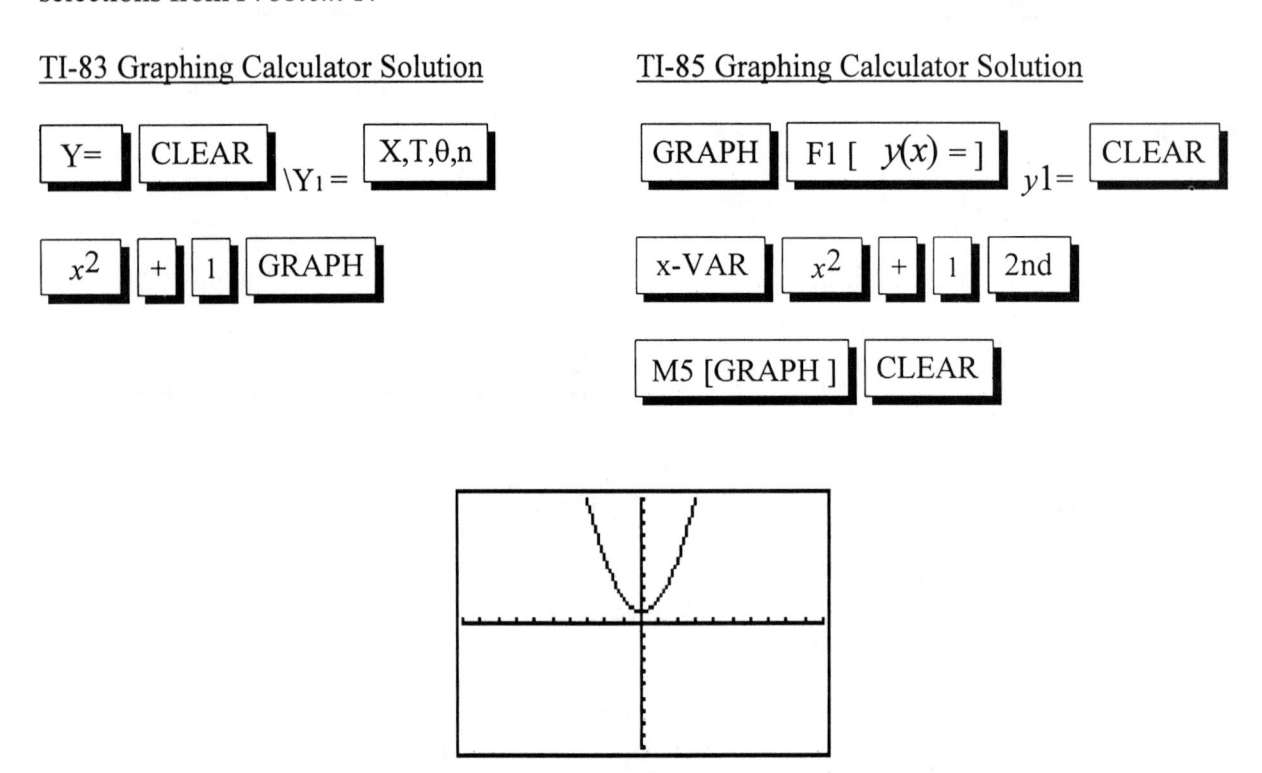

TI-83 Graphing Calculator Solution

| Y= | CLEAR | \Y₁= | X,T,θ,n |

| x^2 | + | 1 | GRAPH |

TI-85 Graphing Calculator Solution

| GRAPH | F1 [y(x) =] | y1= | CLEAR |

| x-VAR | x^2 | + | 1 | 2nd |

| M5 [GRAPH] | CLEAR |

Now that we have graphed $y = x^2 + 1$, show that it is a function by performing the vertical line test on your own. Refer to **Laboratory 8** for the proper keystrokes needed *to draw a vertical line* on the graph.

Next, graph $y = x$ by pressing the following keys.

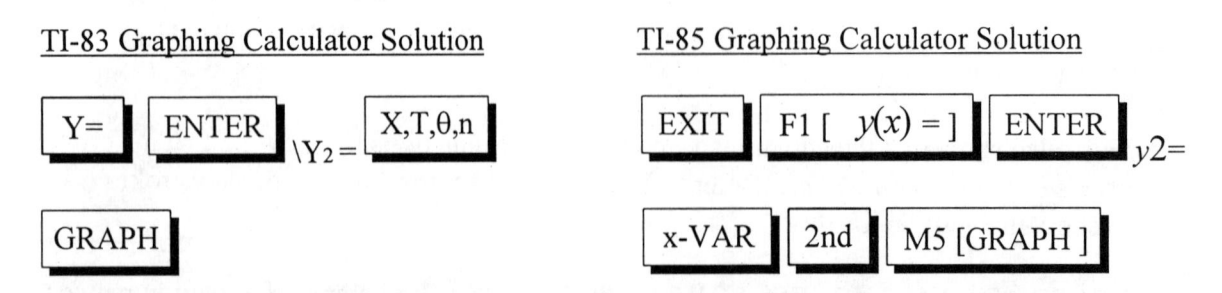

TI-83 Graphing Calculator Solution

| Y= | ENTER | \Y₂= | X,T,θ,n |

| GRAPH |

TI-85 Graphing Calculator Solution

| EXIT | F1 [y(x) =] | ENTER | y2= |

| x-VAR | 2nd | M5 [GRAPH] |

160

LABORATORY 13: INVERSE FUNCTIONS

Finally, graph y^{-1}. *Remember* that $x = y^2 + 1$ must first be solved for y. The resulting equation, $y = \pm\sqrt{x-1}$, is then graphed. *Recall* that you must graph each part of the equation separately. Graph $y = \sqrt{x-1}$, *then* graph $y = -\sqrt{x-1}$. The final graph is shown below.

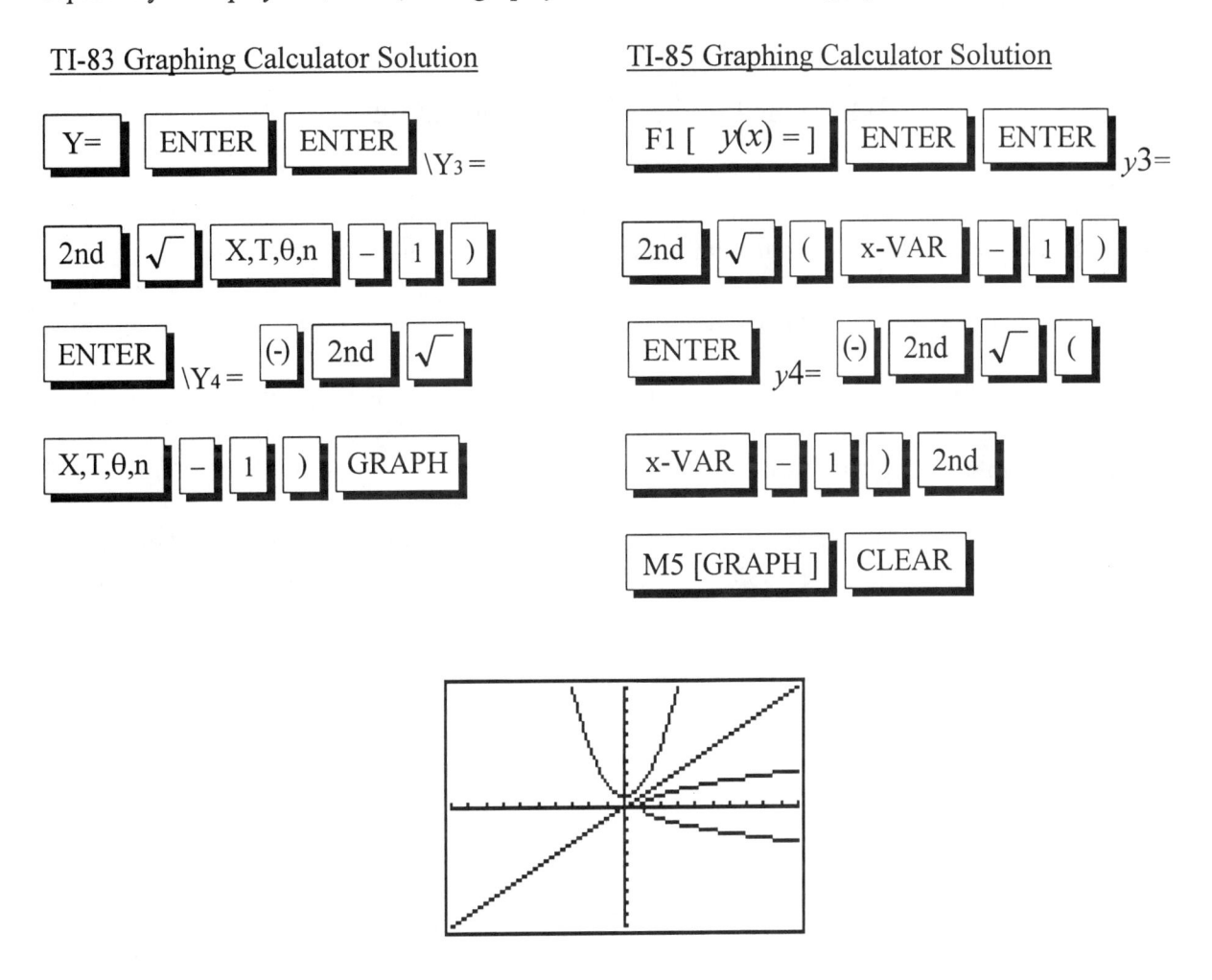

TI-83 Graphing Calculator Solution   TI-85 Graphing Calculator Solution

| Y= | ENTER | ENTER | \Y3=

| 2nd | √ | X,T,θ,n | − | 1 |)

| ENTER | \Y4= | (-) | 2nd | √ |

| X,T,θ,n | − | 1 |) | GRAPH |

| F1 [y(x) =] | ENTER | ENTER | y3=

| 2nd | √ | (| x-VAR | − | 1 |)

| ENTER | y4= | (-) | 2nd | √ | (

| x-VAR | − | 1 |) | 2nd |

| M5 [GRAPH] | CLEAR |

Remember to insert a left parenthesis after pressing the square root key when using the **TI-82**.

Perform a vertical line test on the graph of y^{-1}, which is the two parts of the equation, $y = \pm\sqrt{x^2-1}$. Why is y^{-1} not a function?

Identifying ONE-TO-ONE Functions

A function is a one-to-one function if each member of the range of f is paired with *one and only one* member of the *domain* of f. In a one-to-one function, no two different ordered pairs have the same second component.

LABORATORY 13: INVERSE FUNCTIONS

For a function to have an inverse that is a function, no two different numbers in the domain can correspond to the same number in the range. If the inverse relation of a function is also a function, it is called the **inverse function** and is denoted by $f^{-1}(x)$. A function f has an inverse f^{-1} if and only if f is **one-to-one**.

Horizontal Line Test For Inverse Functions

A function is not one-to-one if any horizontal line intersects its graph more than once. The inverse of a one-to-one function is also a function that passes a horizontal line test.

Problem 3:   Is $y = x^2 + 1$ a one-to-one function?

To determine if y is one-to-one, draw a horizontal line through the graph of y. This can be done either by "eyeing" the graph or by performing the following keystrokes to determine if the horizontal line intersects the graph only once.

On the **TI-85**, it is necessary to graph a constant value for y as the **TI-85** *does not have a horizontal key* feature as do the TI-83/82 and TI-86. **Use** the constant function $y = 4$ as your horizontal line on the **TI-85**.

Clear the functions $y = x$ and the two parts of the equation $y = \pm\sqrt{x-1}$ from the y selections that were entered in *Problem 2*. Use the following keystrokes to eliminate the unwanted functions.

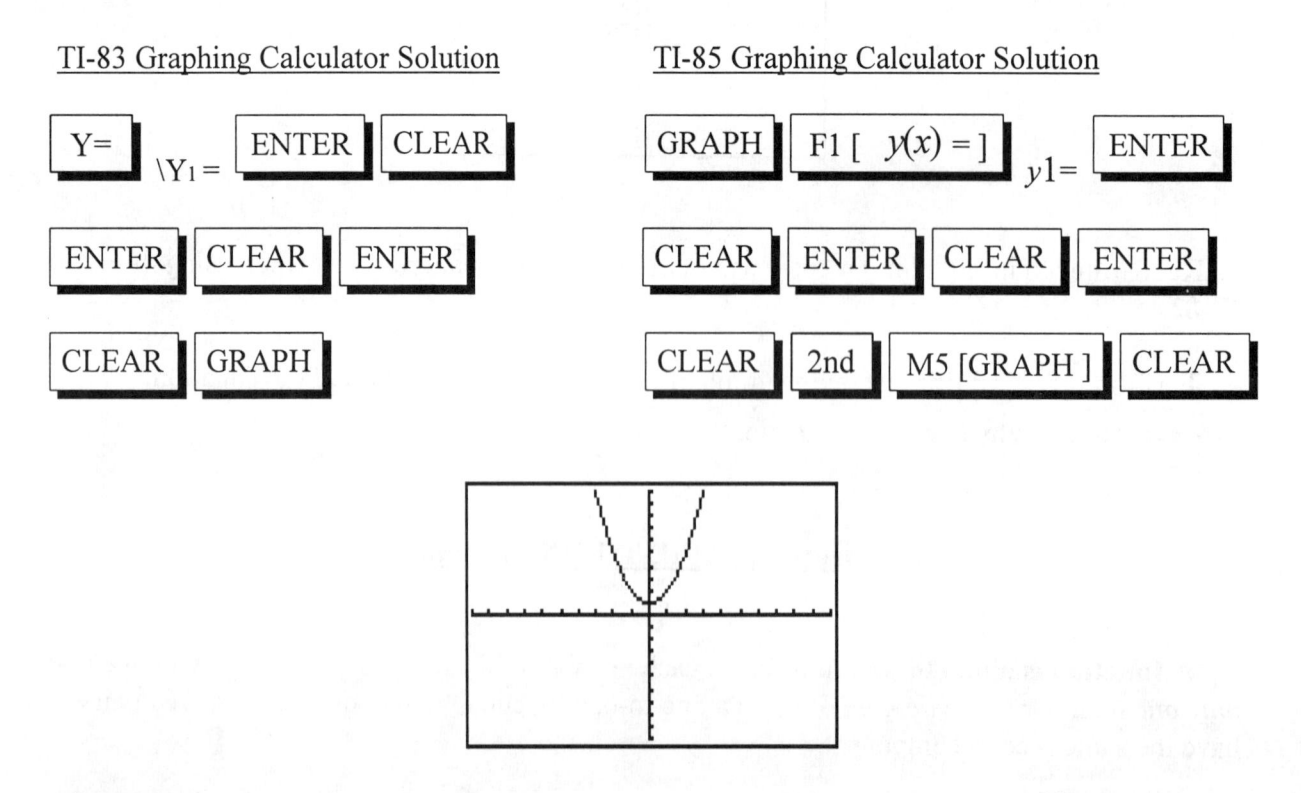

TI-83 Graphing Calculator Solution

| Y= | \Y₁= | ENTER | CLEAR |

| ENTER | CLEAR | ENTER |

| CLEAR | GRAPH |

TI-85 Graphing Calculator Solution

| GRAPH | F1 [y(x) =] | y1= | ENTER |

| CLEAR | ENTER | CLEAR | ENTER |

| CLEAR | 2nd | M5 [GRAPH] | CLEAR |

LABORATORY 13: INVERSE FUNCTIONS

Now enter a horizontal line on the **TI-83/82** and **TI-86**, or enter the function $y = 4$ on the **TI-85**, by pressing the keys below.

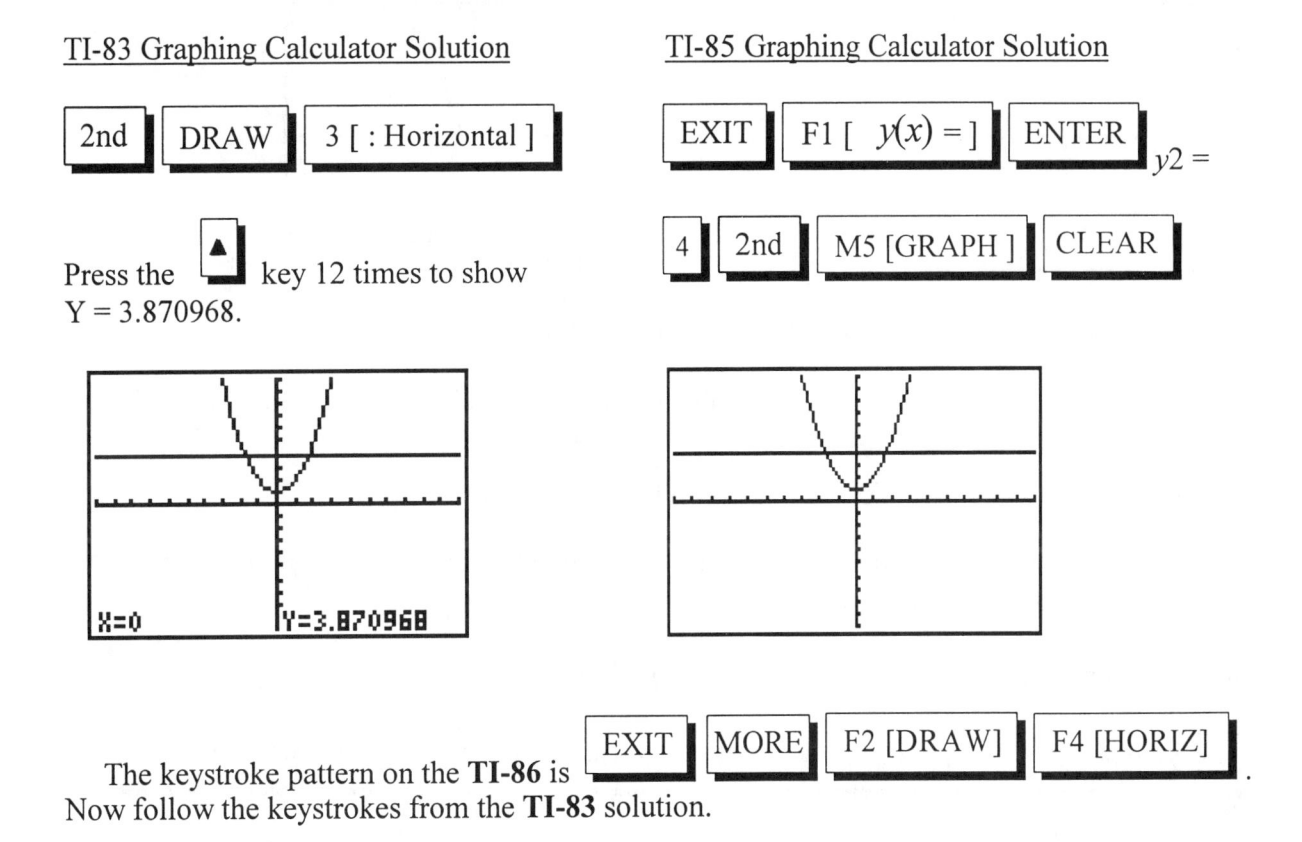

TI-83 Graphing Calculator Solution

| 2nd | DRAW | 3 [: Horizontal] |

Press the ▲ key 12 times to show Y = 3.870968.

TI-85 Graphing Calculator Solution

| EXIT | F1 [y(x) =] | ENTER | y2 =

| 4 | 2nd | M5 [GRAPH] | CLEAR |

The keystroke pattern on the **TI-86** is | EXIT | MORE | F2 [DRAW] | F4 [HORIZ] |.
Now follow the keystrokes from the **TI-83** solution.

It is evident from the screens above that the horizontal line intersects the graph more than once. This indicates that the function $y = x^2 + 1$ is **not** a one-to-one function. Since $y = x^2 + 1$ is not a one-to-one function, the inverse of $y = x^2 + 1$, y^{-1}, will not be a *function*.

Remember to **CLEAR** all graphs before starting this problem.

Problem 4: Given the function $f(x) = x^3$, determine if its inverse f^{-1} is a function.

First graph $f(x)$ using the following keystrokes.

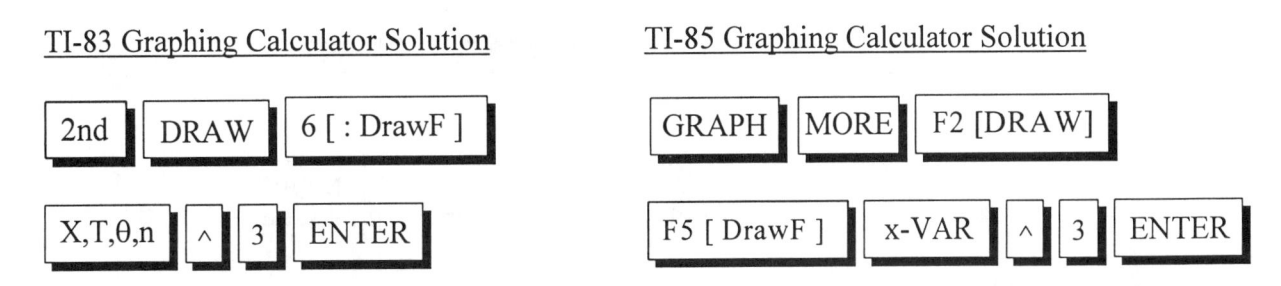

TI-83 Graphing Calculator Solution

| 2nd | DRAW | 6 [: DrawF] |

| X,T,θ,n | ^ | 3 | ENTER |

TI-85 Graphing Calculator Solution

| GRAPH | MORE | F2 [DRAW] |

| F5 [DrawF] | x-VAR | ^ | 3 | ENTER |

LABORATORY 13: INVERSE FUNCTIONS

TI-83 Graphing Calculator Screen   TI-85 Graphing Calculator Screen

Is the function a one-to-one function? State the reason(s) for your answer.

Now graph the inverse of $f(x) = x^3$. The algebra is performed by interchanging the x and y variables, obtaining $x = y^3$. Solve for y, to obtain y^{-1}, which is $y = \sqrt[3]{x}$. Now graph y^{-1} to determine if it is not only an inverse but also a function.

Perform **both** a *horizontal line test* to show that the function is one-to-one **and** a *vertical line test* to show that it is also a function.

TI-83 Graphing Calculator Solution

| 2nd | DRAW | 8[: DrawInv] |

| X,T,θ,n | ^ | 3 | ENTER |

TI-85 Graphing Calculator Solution

| MORE | F2 [DRAW] | MORE | MORE |

| F2[Drinv] | x-VAR | ^ | 3 | ENTER |

Users of the **TI-86** should note that **DrawF** is accessed by pressing [MORE] F1 of the **DRAW** menu. The **DrInv** key is, then, accessed by pressing [MORE] [MORE] F3.

164

LABORATORY 13: INVERSE FUNCTIONS

When graphs have been entered in the **DRAW** menu, they are cleared in the **DRAW** menu. Press $\boxed{1 \ [\ : \text{ClrDraw}]}$ on the **TI-83/82**, $\boxed{\text{F5 [CLDRW]}}$ on the **TI-85**, or $\boxed{\text{F1 [CLDRW]}}$ on the **TI-86**.

Is the inverse a function? What makes you sure that it is or is not a function? Are both f and f^{-1} one-to-one?

Both of the tests, a vertical line test to show that the given relations are functions and a horizontal line test to show that they are one-to-one, have been satisfied. Therefore, $f(x) = x^3$ has a function $f^{-1}(x) = \sqrt[3]{x}$ as its inverse.

From the examples presented, it is safe to say that not every function has an *inverse* that is a *function*. In general, the inverse $f^{-1}(x)$ of a function $f(x)$ is a function that "*reverses*" or "*undoes*" f. We can use *Problem 4* to help explain the statement. Since $y = f(x) = x^3$, then $x = \sqrt[3]{y} = f^{-1}(y)$ is the inverse function. Finding the cube root of a number "*undoes*" or "*reverses*" the cubing necessary to obtain the cube root.

LABORATORY 13: INVERSE FUNCTIONS

EXERCISES

1. Determine the inverse of each of the following relations by switching the coordinates (x_1, y_1) to its inverse (y_1, x_1).

(a) {(1, 1), (-1, -1), (0, 3), (3, 0)}

(b) {(10, 9), (5, 6), (6, 5), (12, 13)}

2. Label the following examples as true or false.

(a) The inverse of {(2, 5), (13, -2), (4, 0)} is {(-2, 13), (0, 4), (5, 2)}.

(b) The function $y(x) = -4$ is one-to-one.

(c) Any graph that passes the horizontal line test must be the graph of a function.

(d) All linear functions are one-to-one.

3. For each relation, write an equation for the inverse and then sketch the graph of the relation and its inverse.

(a) $y = 3$ (b) $y = x + 3$ (c) $y = x^2 - 4$ (d) $y = \sqrt{x}$

4. Are the following functions *one-to-one*? Sketch the graph showing *why* it is or is not *one-to-one*.

(a) $f(x) = x^2 + x + 1$ (b) $f(x) = x^3 + 2$ (c) $f(x) = \sqrt{x - 3}$

5. Determine if the inverse of each of the following functions is also a function. Show the graphs of both the function and its inverse. Are the functions *one-to-one*? Show a horizontal line to test your result.

(a) $f(x) = x^3 - 5x^2 + 8$ (b) $f(x) = \sqrt{x + 1}$

LABORATORY 14: EXPONENTIAL AND LOGARITHMIC FUNCTIONS

Purpose:

In this laboratory, the student will learn how to define and graph both the exponential and logarithmic function.

Analytic Approach:

An **exponential function** is a function in which the *independent variable x* appears in the exponent. An exponential function has the form $y = f(x) = b^x$; where *b is positive* and *b does not equal 1*. The exponent *x* can be any real number. Every *exponential function* is a *one-to-one function*.

Exponential Growth Functions

In the exponential function $f(x) = b^x$, when *b* is greater than 1, the *exponential function* is an *increasing function*. An increasing exponential function is referred to as to as an **exponential growth function**. Compound interest, future valuation of homes, wages, population, and many scientific problems such as a.c. voltage across a capacitor are described as exponential growth problems.

At the beginning of every problem in this laboratory, set the viewing rectangle to the standard setting.

Problem 1: Show that $f(x) = 4^x$ is an increasing exponential function and, therefore, an exponential growth function. State the domain and range of $f(x) = 4^x$.

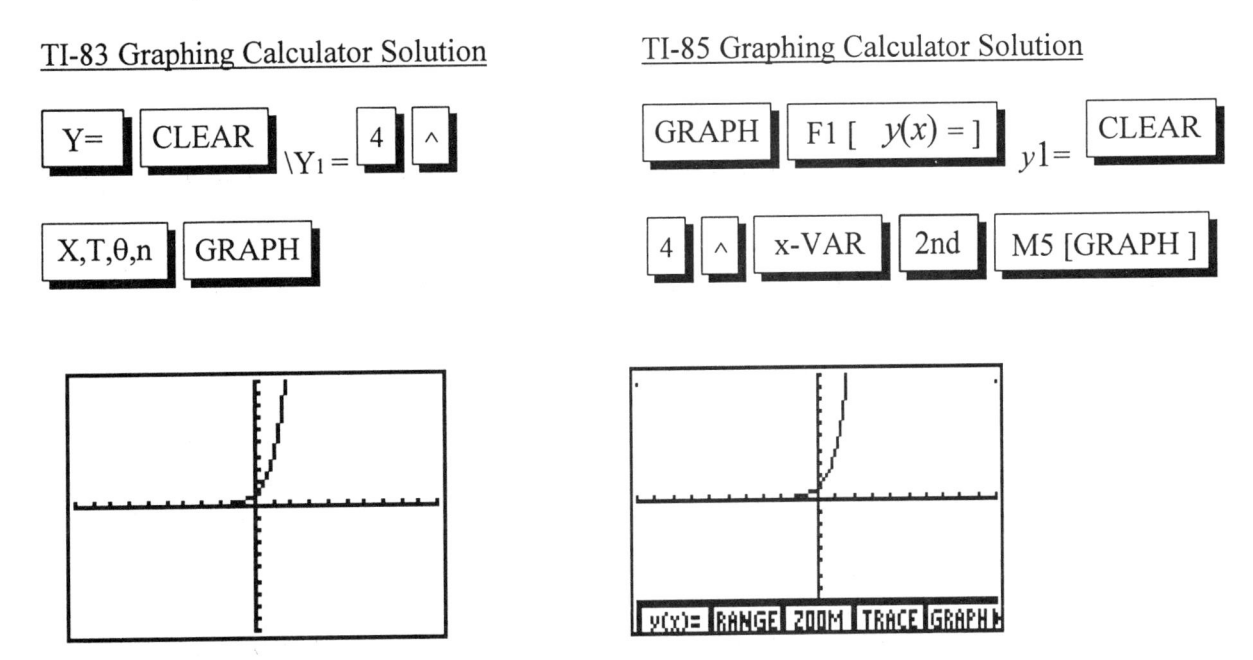

TI-83 Graphing Calculator Solution

| Y= | CLEAR | \Y₁= | 4 | ^ |
| X,T,θ,n | GRAPH |

TI-85 Graphing Calculator Solution

| GRAPH | F1 [y(x) =] | y1= | CLEAR |
| 4 | ^ | x-VAR | 2nd | M5 [GRAPH] |

167

LABORATORY 14: EXPONENTIAL AND LOGARITHMIC FUNCTIONS

As x increases from left to right, so does the value of y. Note that the *y-intercept is 1* and that the graph of $f(x) = 4^x$ does not cross the *x*-axis. The *x*-axis acts as a horizontal *asymptote*. The *domain* (*x*-values) of the function is the set of all real numbers. Because the function does not intercept the *x*-axis, the *range* (*y*-values) of $f(x) = 4^x$ must be greater than zero.

Is it always true that the *y*-intercept is 1 when the exponential function is a growth function? Is it also true that a growing exponential function will **not** cross the *x*-axis? Why?

Many exponential growth functions can be explained using the compound interest formula, $y = P(1 + r)^t$. In the function, $y = P(1 + r)^t$, y is the *value* of the function, P is the *principal*, or original value of the investment, r is the *rate* at which the principal is to grow for any given length of *time t*.

In the following problems, the **ZFIT** feature will be used. This feature is not found on the **TI-82**. It is important that the user of the **TI-82** realize that a graph may not appear on the viewing rectangle. It may be necessary to reset the values in **WINDOW** when using the **TI-82**, so that a graph will appear in the viewing rectangle. An algebraic approach to solving the function is suggested to determine the suitable values to use when setting the **WINDOW** on the **TI-82**. When using the **ZFIT** feature on the **TI-83, 85** and **TI-86**, it is important to check the **WINDOW** on the **TI-83/86** and **RANGE** on the **TI-85** to view the new settings for the viewing rectangle.

Problem 2: In 1992, you bought a painting for $250. For each year *x*, the value *y* of the painting has increased by 7%. Assuming that this trend continues, what is the value of the painting in 1997? What will the value of the painting be in 1999 and 2002?

First, determine the exponential function using the formula $y = P(1 + r)^t$. Let $P = \$250$, $r = .07$, and $t = x$. If *x* represents the number of years since 1992, for the year 1992 let $x = 0$. The formula to be graphed is $y = 250(1 + .07)^x$. Simplify inside the parentheses obtaining $y = 250(1.07)^x$.

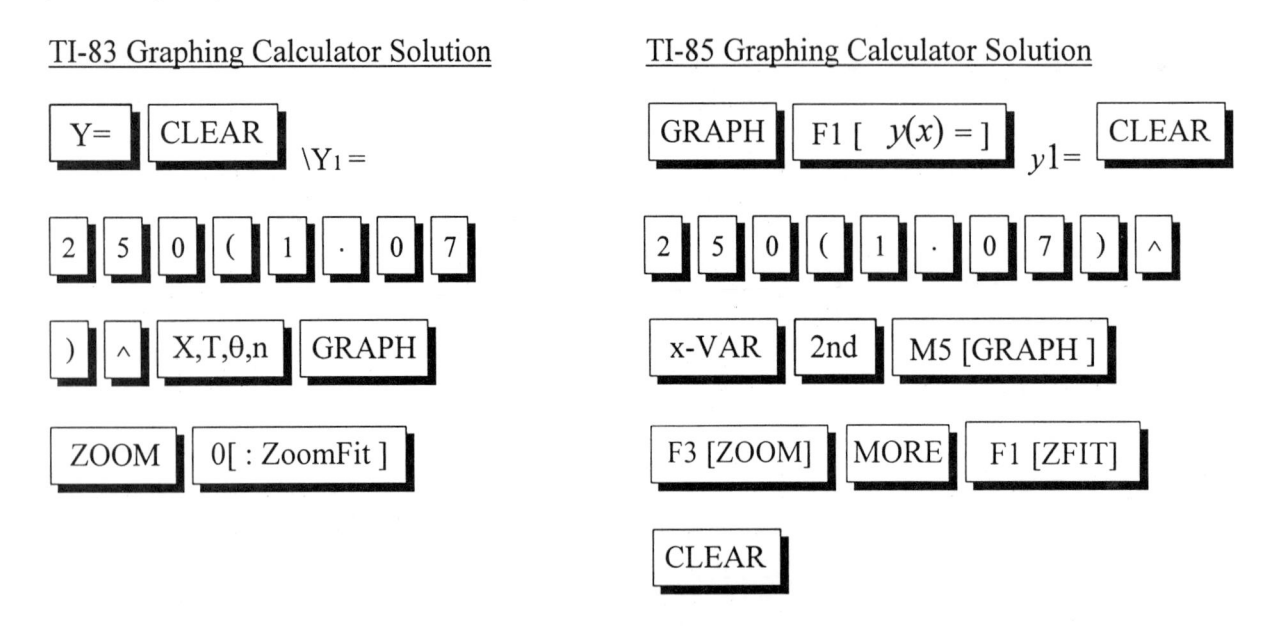

TI-83 Graphing Calculator Solution

| Y= | CLEAR | \Y₁= |

| 2 | 5 | 0 | (| 1 | . | 0 | 7 |

|) | ^ | X,T,θ,n | GRAPH |

| ZOOM | 0[: ZoomFit] |

TI-85 Graphing Calculator Solution

| GRAPH | F1 [y(x) =] | y1= | CLEAR |

| 2 | 5 | 0 | (| 1 | . | 0 | 7 |) | ^ |

| x-VAR | 2nd | M5 [GRAPH] |

| F3 [ZOOM] | MORE | F1 [ZFIT] |

| CLEAR |

LABORATORY 14: EXPONENTIAL AND LOGARITHMIC FUNCTIONS

Graphing Calculator Screen

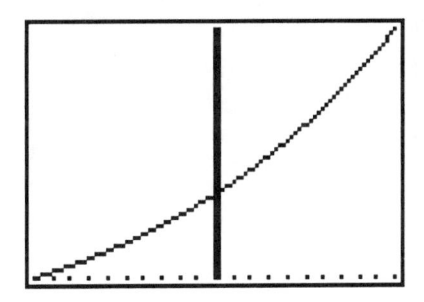

Notice that when **ZFIT** is used **"dots"** appear where the *x*-axis should be. The **"dots"** indicate that the *x*-axis is **not** the horizontal line at the bottom of the screen.

Find the value of the function $y = 250 \, (1.07)^x$, when $x = 5, 7$, and 10; these *x* values correspond to the years 1997, 1999 and 2002, respectively.

TI-83 Graphing Calculator Solution   TI-85 Graphing Calculator Solution

| 2nd | CALC | 1 [: value] |

| 5 | ENTER |

| EXIT | MORE | MORE | F1 [EVAL] |

| 5 | ENTER |

| 2nd | CALC | 1 [: value] |

| 7 | ENTER |

| EXIT | F1 [EVAL] | 7 | ENTER |

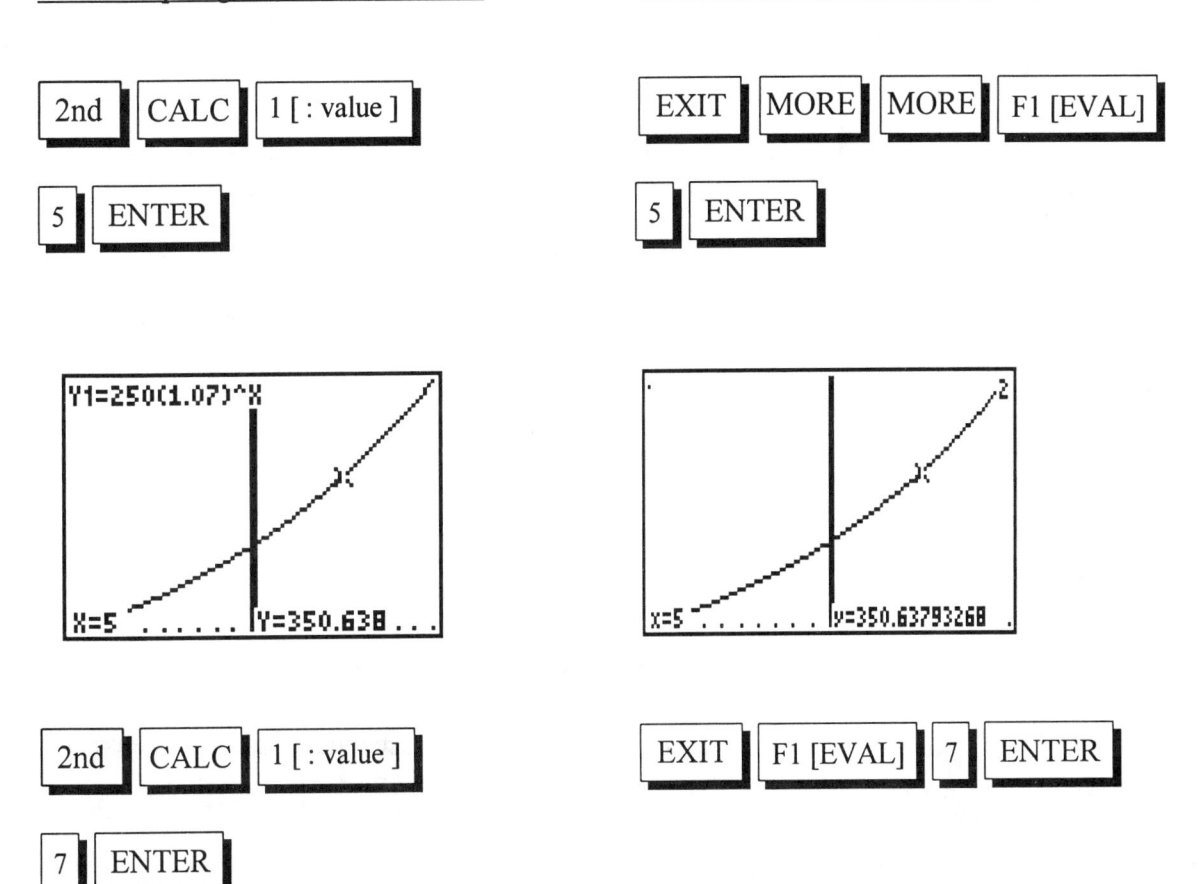

LABORATORY 14: EXPONENTIAL AND LOGARITHMIC FUNCTIONS

TI-83 Graphing Calculator Screen   TI-85 Graphing Calculator Screen

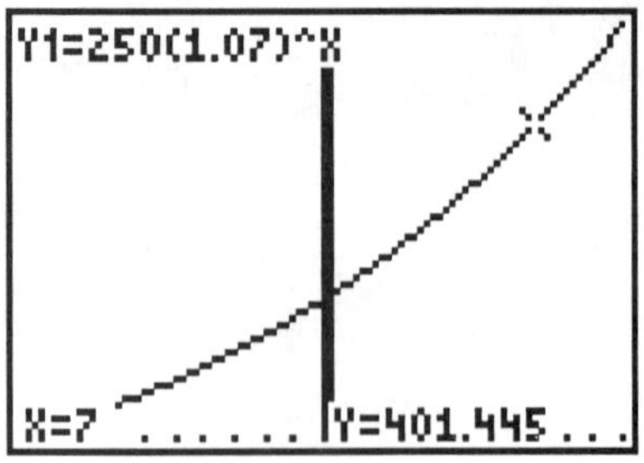

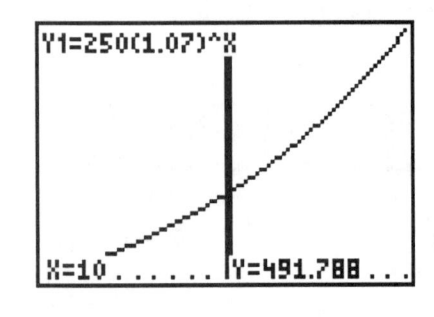

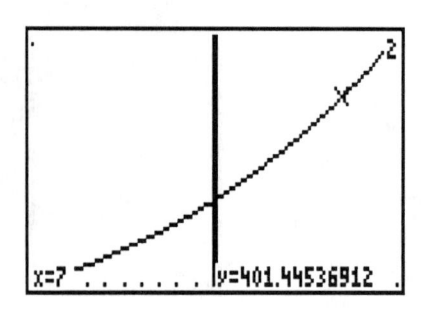

[2nd] [CALC] [1 [: value]]   [EXIT] [F1 [EVAL]] [1] [0] [ENTER]

[1] [0] [ENTER]

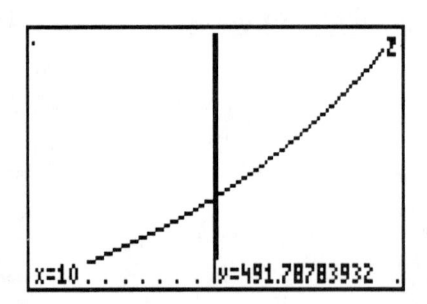

The value of the painting in 1997 is $350.64. In 1999, the painting's value is $401.45, and in the year 2002, the painting is worth $491.79.

Decaying Functions

Functions of the form;   i. $f(x) = b^{-x}$, where $b > 1$, and
ii $f(x) = b^x$, where $0 < b < 1$,

are exponential functions described as *decreasing* or *decaying functions*.

In the exponential function $f(x) = 3^{-x}$, the **negative exponent** causes the graph to decrease exponentially as x increases, and the curve approaches the positive x-axis. In $f(x) = \left(\frac{1}{3}\right)^x$, the fractional base has the same effect on the graph as the negative exponent in $f(x) = 3^{-x}$.

The function $y = 3^{-x}$ and the function $y = \left(\frac{1}{3}\right)^x$ are equivalent, why?

LABORATORY 14: EXPONENTIAL AND LOGARITHMIC FUNCTIONS

Problem 3: Graph $f(x) = 3^{-x}$ and state the domain and range of the function. Reset the viewing rectangle to the standard setting.

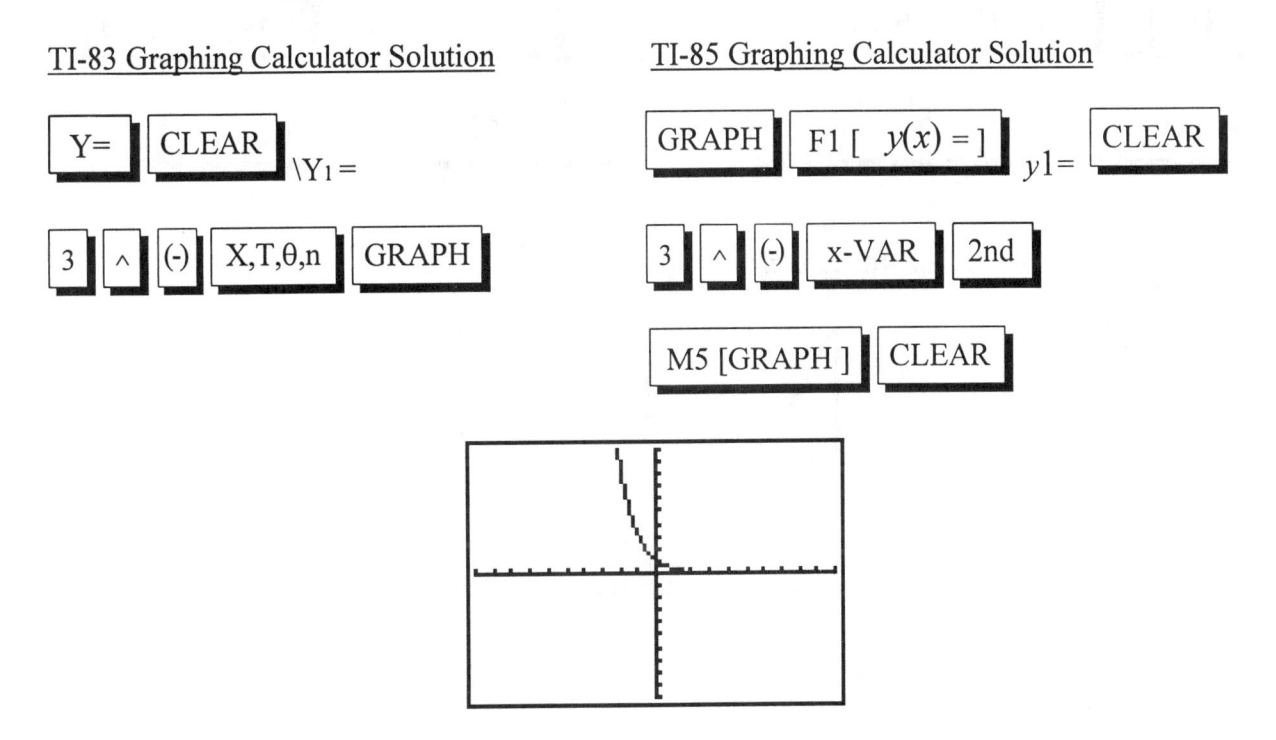

TI-83 Graphing Calculator Solution

| Y= | CLEAR | \Y₁=

| 3 | ^ | (-) | X,T,θ,n | GRAPH |

TI-85 Graphing Calculator Solution

| GRAPH | F1 [y(x) =] | y1= | CLEAR |

| 3 | ^ | (-) | x-VAR | 2nd |

| M5 [GRAPH] | CLEAR |

There are no apparent vertical asymptotes. The domain of the function is all real numbers. Since the *x*-axis acts as a horizontal asymptote, the range (*y*-values) must be greater than 0. As the value of *x* increases from left to right, the value of *y* decreases. The exponential function $f(x) = 3^{-x}$ is a decaying function.

Problem 4: If you bought a new car in 1994 for $18,000 and it depreciates at a rate of 13% a year, how much is the car worth in 1997? In 1999? In 2001? (The depreciation does not include any unusual wear and tear, etc.)

Set up the exponential function needed to find the depreciated value of the car using $y = P(1 - r)^t$, where *P* is the *initial cost* of the car, *r* is the *rate* at which the car is depreciating, and *t* is the *length of time* for depreciation. Note that the subtraction sign indicates that the car is losing its value. Using the stated values of the variables in *Problem 4*, and letting $t = x$, the function to be graphed is $y = 18000 (1 - .13)^x$. *x* has a value of 0 for the year 1994.

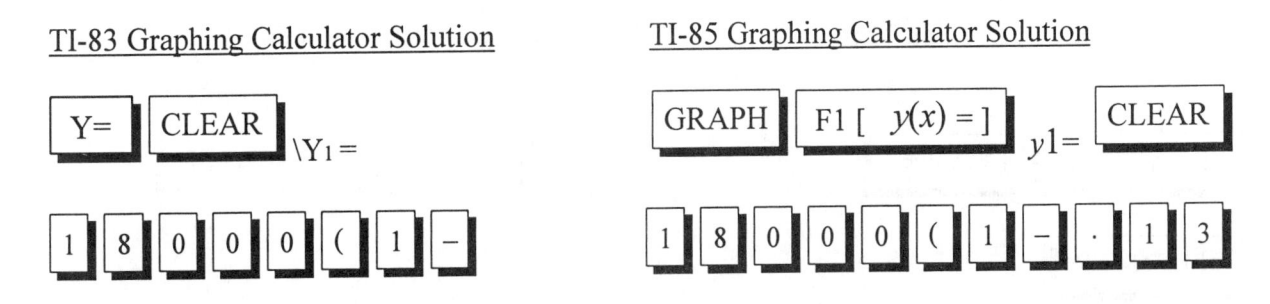

TI-83 Graphing Calculator Solution

| Y= | CLEAR | \Y₁=

| 1 | 8 | 0 | 0 | 0 | (| 1 | - |

TI-85 Graphing Calculator Solution

| GRAPH | F1 [y(x) =] | y1= | CLEAR |

| 1 | 8 | 0 | 0 | 0 | (| 1 | - | . | 1 | 3 |

LABORATORY 14: EXPONENTIAL AND LOGARITHMIC FUNCTIONS

TI-83 Graphing Calculator Solution

| . | 1 | 3 |) | ^ | X,T,θ,n |

| GRAPH | ZOOM | 0[: ZoomFit] |

The TI-82 does not have a Z-fit feature.

TI-85 Graphing Calculator Solution

|) | ^ | x-VAR | 2nd | M5 [GRAPH] |

| F3 [ZOOM] | MORE | F1 [ZFIT] |

| CLEAR |

Let $x = 3$ represent 1997, $x = 5$ represent 1999, and $x = 7$ represent 2001.

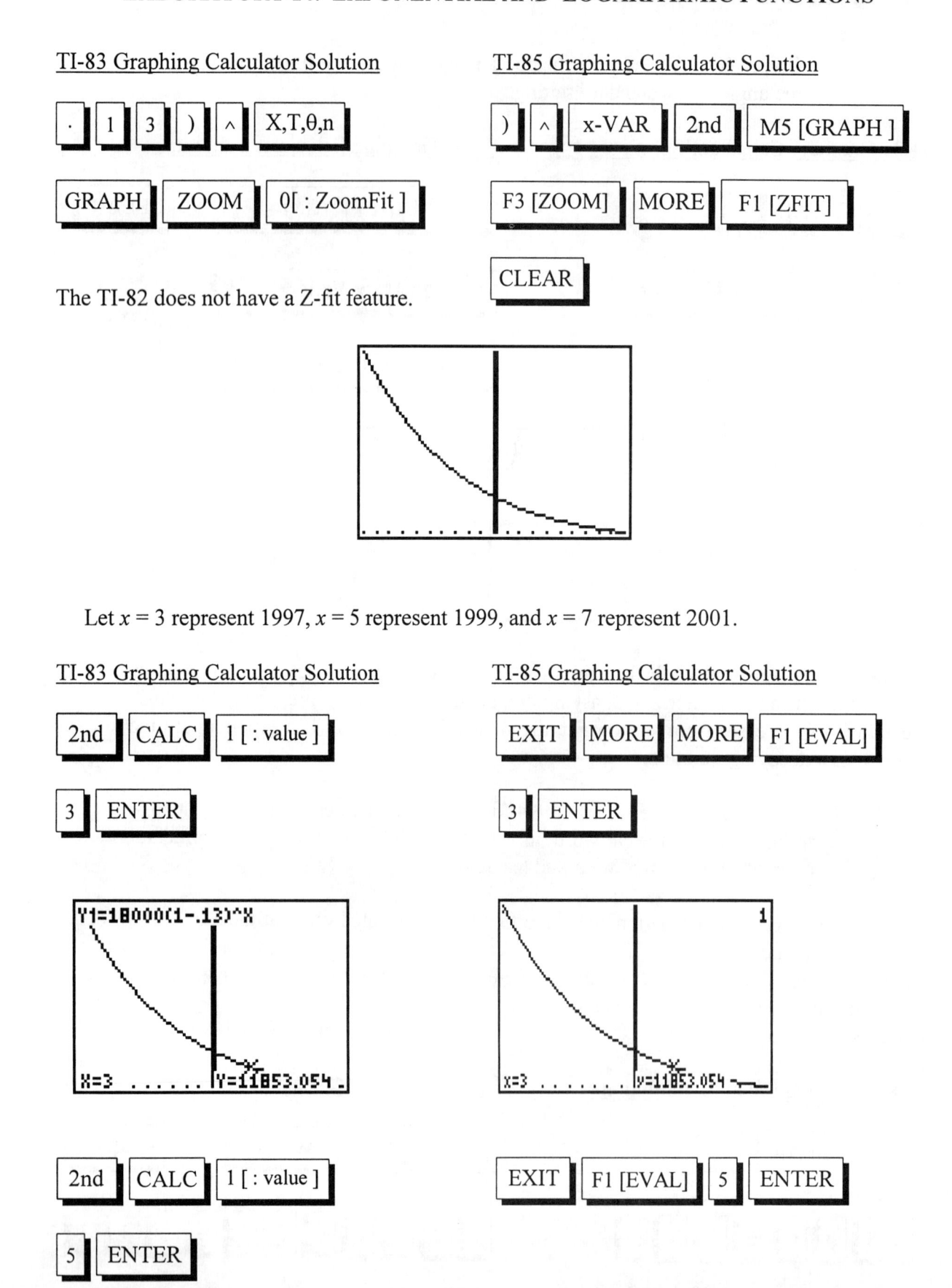

TI-83 Graphing Calculator Solution

| 2nd | CALC | 1 [: value] |

| 3 | ENTER |

Y1=18000(1-.13)^X
X=3   Y=11853.054

| 2nd | CALC | 1 [: value] |

| 5 | ENTER |

TI-85 Graphing Calculator Solution

| EXIT | MORE | MORE | F1 [EVAL] |

| 3 | ENTER |

x=3   y=11853.054

| EXIT | F1 [EVAL] | 5 | ENTER |

LABORATORY 14: EXPONENTIAL AND LOGARITHMIC FUNCTIONS

TI-83 Graphing Calculator Screen   TI-85 Graphing Calculator Screen

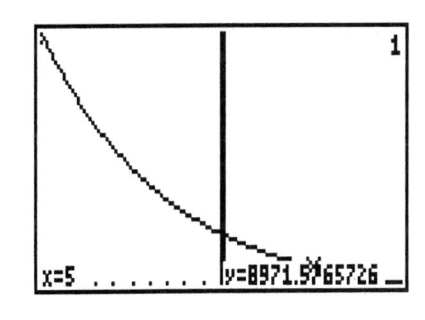

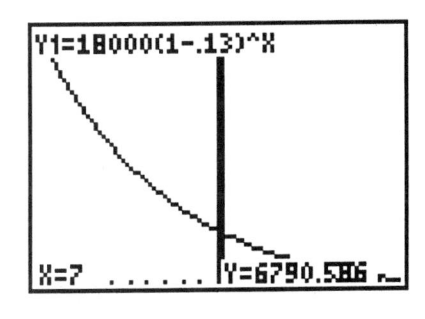

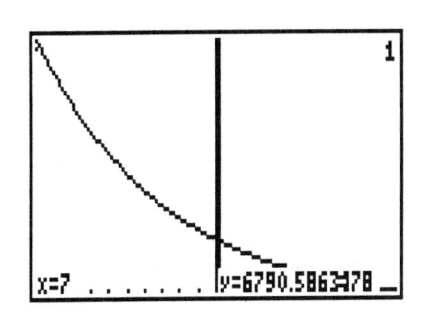

[Y1=18000(1-.13)^X X=5 Y=8971.577]   [X=5 y=8971.5765726]

| 2nd | CALC | 1 [: value] | | EXIT | F1 [EVAL] | 7 | ENTER |

| 7 | ENTER |

[Y1=18000(1-.13)^X X=7 Y=6790.586]   [X=7 y=6790.586378]

The depreciated value of the car in 1997 is $11853.05. How much of the original value of the car was lost in the first three years?

In 1999, the value of the car is $8971.58 and in 2001, the value is $6790.59. Why do you think the value of the car is depreciating at a slower rate as the car gets older?

The Irrational Number *e*

Many problems involving exponential functions use the irrational number *e*. The value of the irrational number *e* is approximately 2.718281828. This number can be easily verified by pressing the proper keys on the calculator: | 2nd | e^x | 1 | ENTER |

LABORATORY 14: EXPONENTIAL AND LOGARITHMIC FUNCTIONS

Problem 5: Graph $f(x) = e^x$. Reset the viewing rectangle to the standard setting.

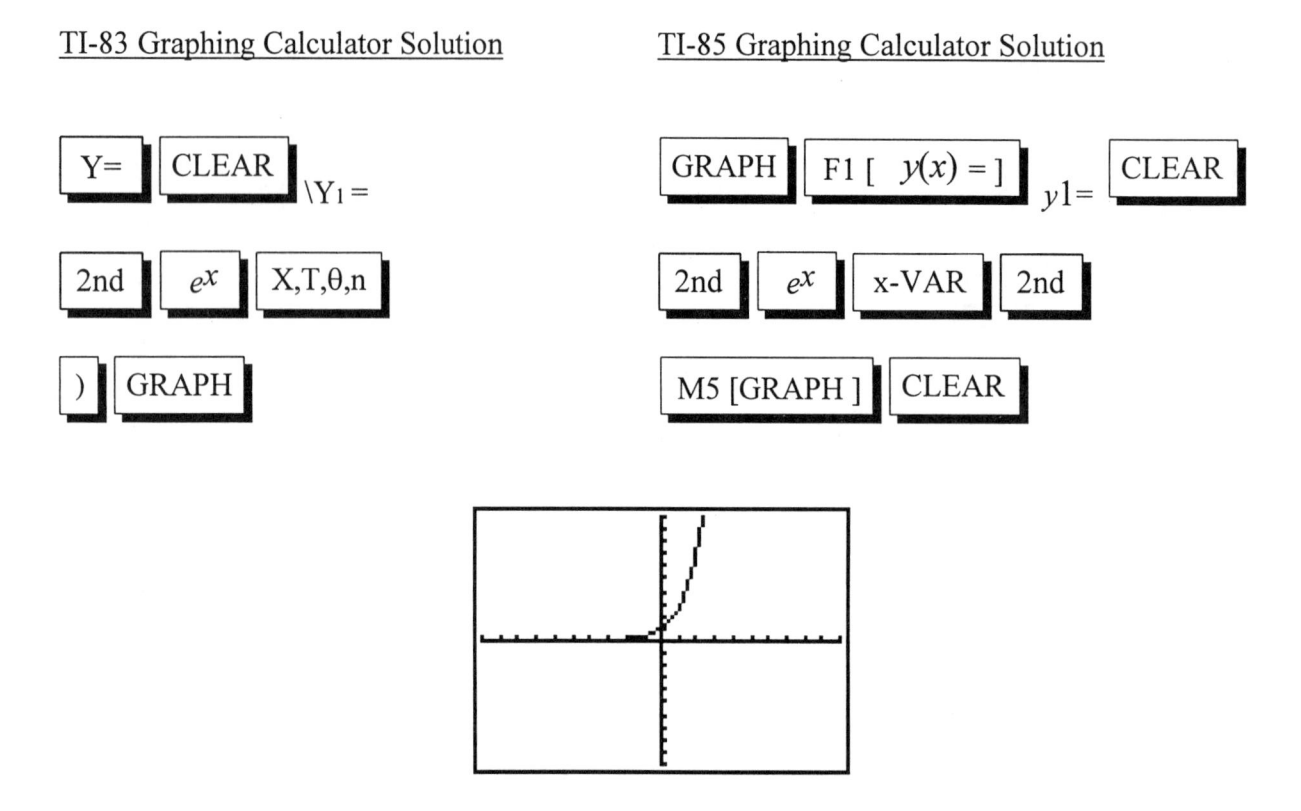

TI-83 Graphing Calculator Solution   TI-85 Graphing Calculator Solution

| Y= | CLEAR | \Y₁ = | GRAPH | F1 [y(x) =] | y1 = | CLEAR |

| 2nd | eˣ | X,T,θ,n | 2nd | eˣ | x-VAR | 2nd |

|) | GRAPH | M5 [GRAPH] | CLEAR |

Is the function an increasing (growth function) or a decreasing (decaying) function?

The **exponential growth function** using the irrational number e can be defined as $y = ae^{nt}$, for a and n greater than 0 and t is time. This formula can be adapted to fit any situation where the exponent is non-negative and the function is *increasing*, such as in continuous compounding of interest.

Problem 6: Find the interest if $550 is invested at 4% interest compounded continuously for t number of years. How much will the investment be worth in 5 years? How much of that amount is the interest?

Using the function $y = ae^{nt}$; y is the *worth of the investment*; a is the *principal*; n is the *interest rate* (in decimal form); and t is the *time period*. Substitute $550 for a, .04 for n, and x for t, to obtain the function $y = 550\, e^{.04x}$.

On all four TI calculators, set the viewing rectangle to **xMin:** 0, **xMax:** 12, **xScl:** 1, **yMin:** 550, **yMax:** 888, and **yScl:** 20.

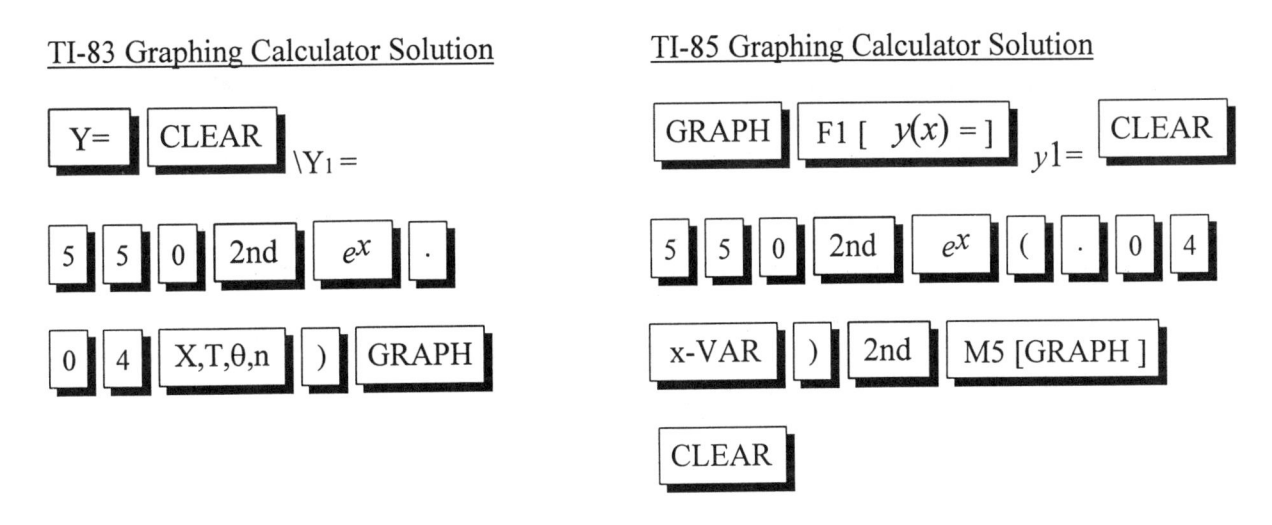

LABORATORY 14: EXPONENTIAL AND LOGARITHMIC FUNCTIONS

TI-83 Graphing Calculator Solution

[Y=] [CLEAR] \Y₁=

[5] [5] [0] [2nd] [eˣ] [.]

[0] [4] [X,T,θ,n] [)] [GRAPH]

TI-85 Graphing Calculator Solution

[GRAPH] [F1 [y(x) =]] y1= [CLEAR]

[5] [5] [0] [2nd] [eˣ] [(] [.] [0] [4]

[x-VAR] [)] [2nd] [M5 [GRAPH]]

[CLEAR]

Users of the **TI-82** need to *insert* a left parenthesis after the e^x key has been pushed.

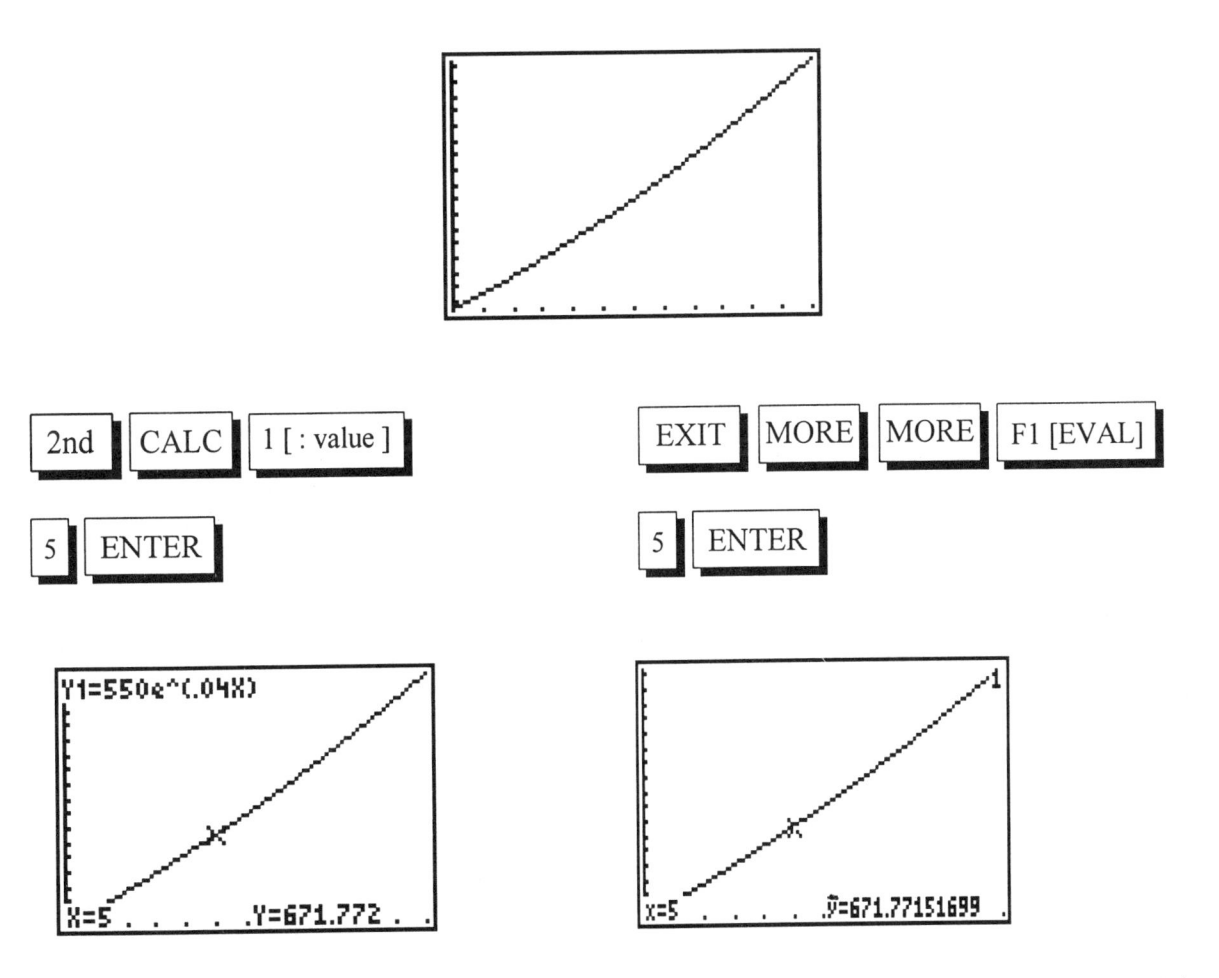

[2nd] [CALC] [1 [: value]]

[5] [ENTER]

[EXIT] [MORE] [MORE] [F1 [EVAL]]

[5] [ENTER]

Y1=550e^(.04X)
X=5   Y=671.772

x=5   y=671.77151699

The investment will be worth $671.77 in five years. To find the interest that was earned for five years *subtract* the principal, $550, from the worth of the investment in five years, $671.77. The interest earned for the five year period was $121.77.

LABORATORY 14: EXPONENTIAL AND LOGARITHMIC FUNCTIONS

The **exponential decay function** using the irrational number e can be defined as $y = ae^{-nt}$. The formula can be adapted to any *decreasing* function, such as radioactive decay in Chemistry and many problems found in Physics as well.

Problem 7: The atmospheric pressure p decreases exponentially with the height h (**in miles**) above the earth according to the function $p = 29.92\, e^{-(h/5)}$ inches of mercury. Find the pressure at a height of 20,000 feet.

In the function $p = 29.92\, e^{-(h/5)}$, let x represent h (height in miles). Remember to *convert* 20,000 feet to miles by dividing 20,000 by 5,280, obtaining 3.79 miles.

For those who graph with the **TI-82**, the only difference in the keystroke pattern is the addition of a left parentheses before pressing the negation key in the exponent. Set the viewing rectangle for **xMin**: 0, **xMax**: 15, **xScl**: 1, **yMin**: 0, **yMax**: 35, and **yScl**: 5.

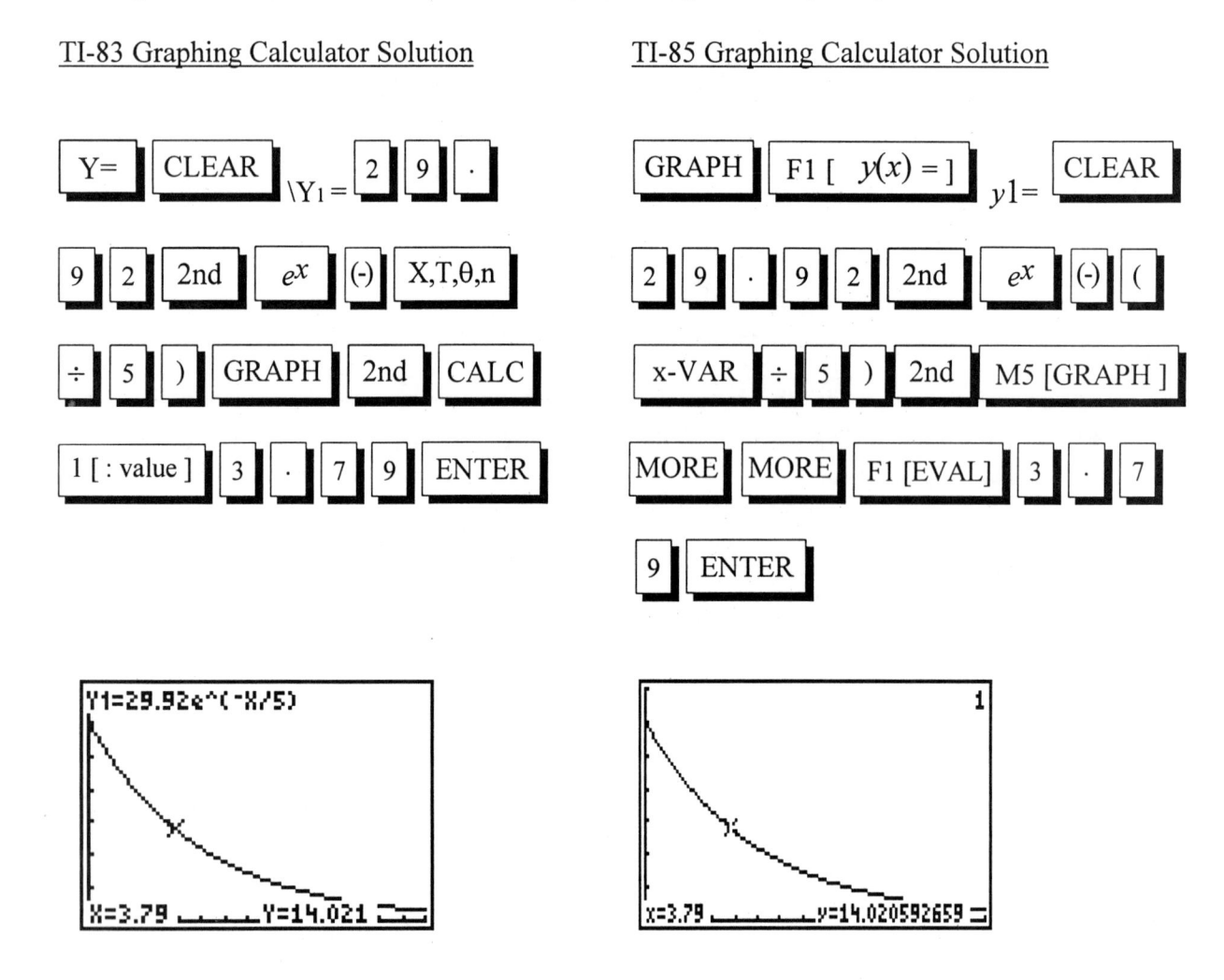

TI-83 Graphing Calculator Solution   TI-85 Graphing Calculator Solution

| Y= | CLEAR | \Y₁= | 2 | 9 | . |

| 9 | 2 | 2nd | eˣ | (-) | X,T,θ,n |

| ÷ | 5 |) | GRAPH | 2nd | CALC |

| 1 [: value] | 3 | . | 7 | 9 | ENTER |

| GRAPH | F1 [y(x) =] | y1= | CLEAR |

| 2 | 9 | . | 9 | 2 | 2nd | eˣ | (-) | (|

| x-VAR | ÷ | 5 |) | 2nd | M5 [GRAPH] |

| MORE | MORE | F1 [EVAL] | 3 | . | 7 |

| 9 | ENTER |

```
Y1=29.92e^(-X/5)




       X

X=3.79       Y=14.021
```

```
                              1



       X

x=3.79        y=14.020592659
```

The atmospheric pressure at a height of 3.79 miles is approximately 14.02 inches of mercury.

LABORATORY 14: EXPONENTIAL AND LOGARITHMIC FUNCTIONS

Logarithmic Functions

The *value* of $\log_b x$ is called the **logarithm of x with base b**. A logarithm is an exponent. The logarithm is commonly referred to as a "log".

The **logarithmic function** is the *inverse* of the **exponential function**. The inverse of the exponential function $y = b^x$ is called the **logarithmic function with base b** and is denoted by $y = \log_b x$.

The function $y = \log_{10} x$ is called the **common logarithmic function**, while the function $y = \log_e x$ is called the **natural logarithmic function**. Another form for writing the natural logarithmic function is $y = \ln x$.

However, when using the TI-83/82 and TI-85/86 calculators to evaluate logs, the only keys to use involve base 10 or base e. Therefore, the **Change of Base Theorem** must be used before evaluating a logarithm in any base other than 10 or e on a TI graphing calculator.

> The *Change of Base Theorem* states that if $y = \log_b x$; then $y = \dfrac{\log x}{\log b}$.

Using the Change of Base Theorem, the value of $y = \log_4 x$ can be found in base 10 by dividing $\log x$ by $\log 4$.

Problem 8: Evaluate the following logs:
  (a) $\log_5 30$
  (b) $\log_{12} 234$
  (c) $\ln 3.5$

Set the **Float** to reflect three decimal places. **TI-82/86** users should follow the keystroke patterns from the **TI-85** solution.

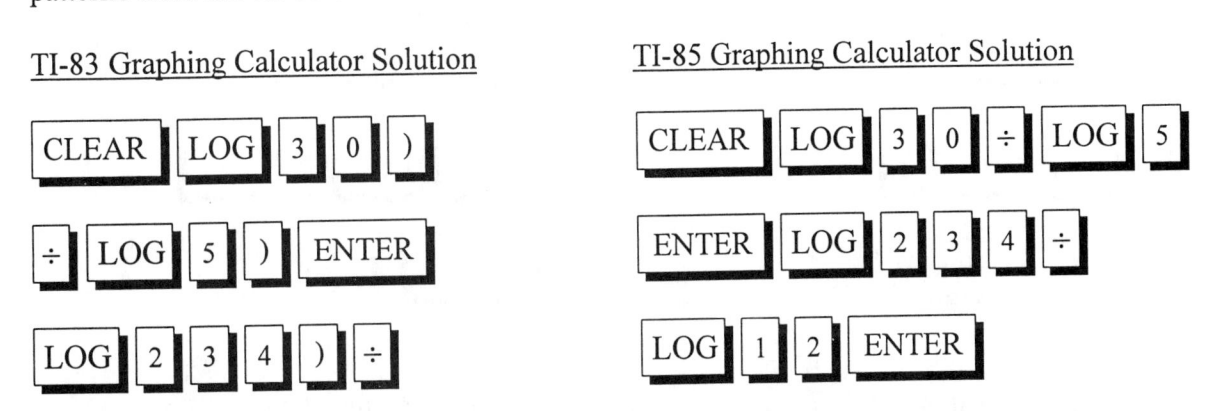

TI-83 Graphing Calculator Solution   TI-85 Graphing Calculator Solution

| CLEAR | LOG | 3 | 0 |) |

| ÷ | LOG | 5 |) | ENTER |

| LOG | 2 | 3 | 4 |) | ÷ |

| CLEAR | LOG | 3 | 0 | ÷ | LOG | 5 |

| ENTER | LOG | 2 | 3 | 4 | ÷ |

| LOG | 1 | 2 | ENTER |

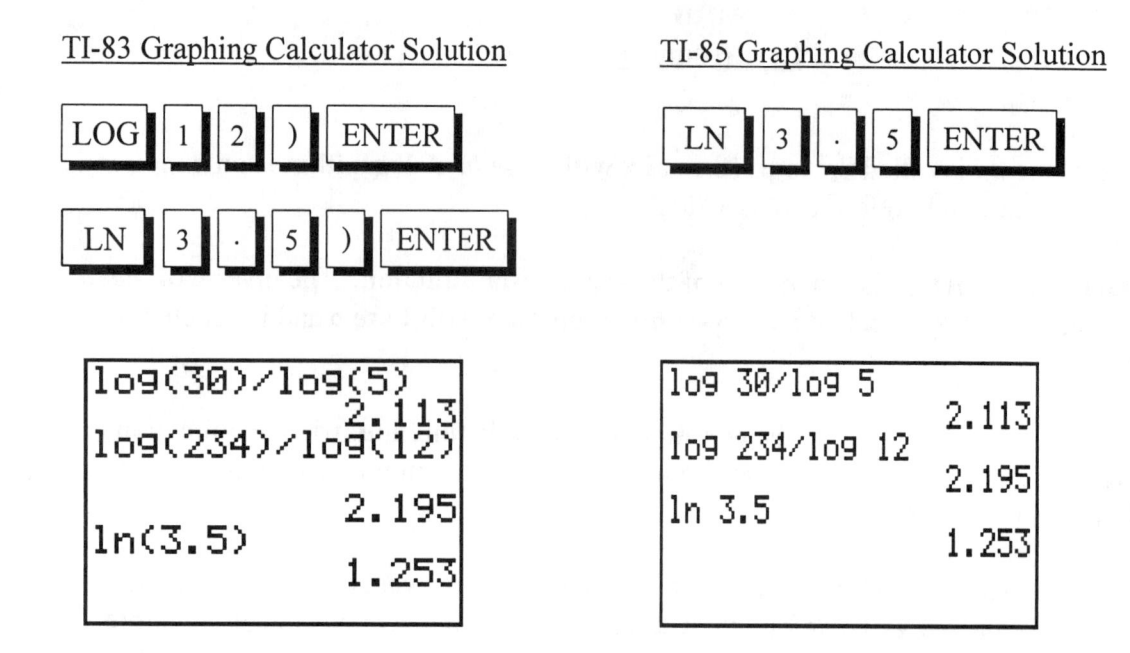

LABORATORY 14: EXPONENTIAL AND LOGARITHMIC FUNCTIONS

TI-83 Graphing Calculator Solution

[LOG] [1] [2] [)] [ENTER]

[LN] [3] [.] [5] [)] [ENTER]

```
log(30)/log(5)
              2.113
log(234)/log(12)
              2.195
ln(3.5)
              1.253
```

TI-85 Graphing Calculator Solution

[LN] [3] [.] [5] [ENTER]

```
log 30/log 5
              2.113
log 234/log 12
              2.195
ln 3.5
              1.253
```

The value of (a) $\log_5 30$ is 2.113 and the value of (b) $\log_{12} 234$ is 2.195. The natural log of 3.5 from part (c) is 1.253.

Problem 9: Graph the exponential function $y = 2^x$ and the logarithmic function $y = \log_2 x$ on the same set of axes. Reset the viewing rectangle to the standard setting.

First, use the Change of Base Theorem to change $y = \log_2 x$ to $y = \dfrac{\log x}{\log 2}$.

TI-83 Graphing Calculator Solution

[Y=] [CLEAR] \Y₁= [2] [^]

[X,T,θ,n] [ENTER] \Y₂=

[LOG] [X,T,θ,n] [)] [÷]

[LOG] [2] [)] [GRAPH]

TI-85 Graphing Calculator Solution

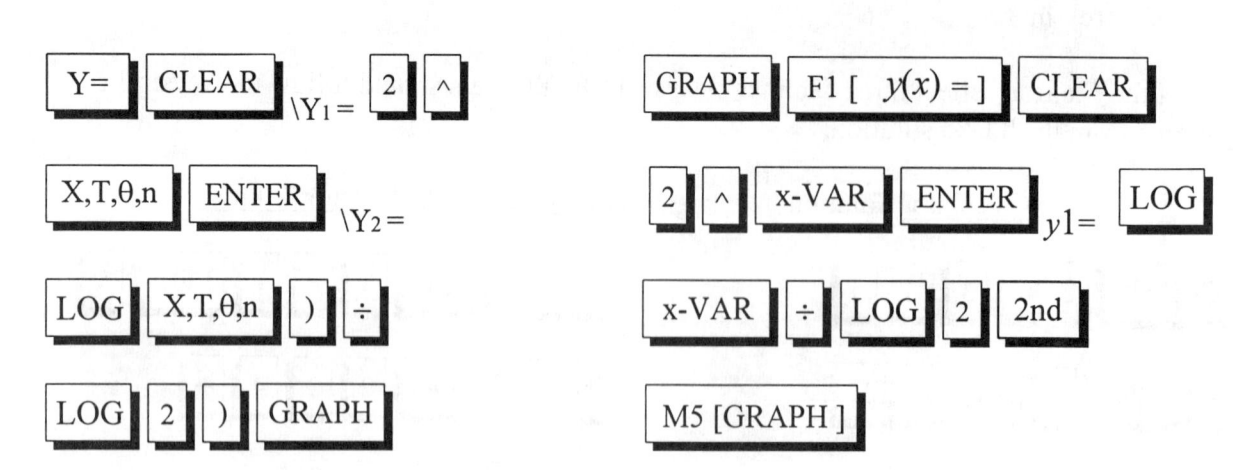

[GRAPH] [F1 [y(x) =]] [CLEAR]

[2] [^] [x-VAR] [ENTER] y1= [LOG]

[x-VAR] [÷] [LOG] [2] [2nd]

[M5 [GRAPH]]

TI-82 users may in some cases need a left parenthesis after pressing **LOG**.

178

LABORATORY 14: EXPONENTIAL AND LOGARITHMIC FUNCTIONS

TI-83 Graphing Calculator Screen

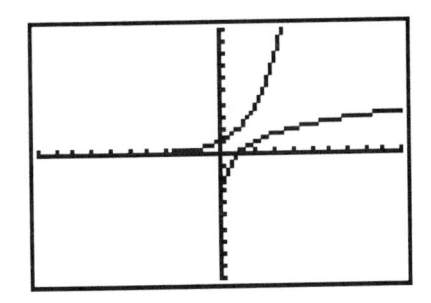

TI-85 Graphing Calculator Screen

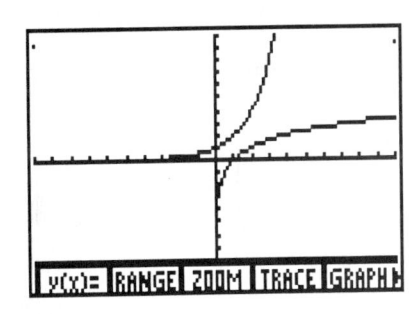

Why is $y = \log_2 x$ the inverse function of $y = 2^x$? Refer to **Laboratory 13: Inverse Functions**. What type of functions are $y = 2^x$ and $y = \log_2 x$, *growing* or *decaying*?

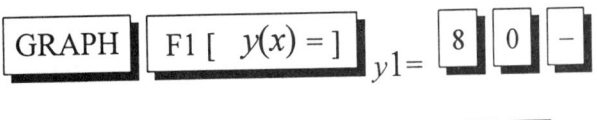

It is a good idea at this point to review the *Properties of Logarithms* which can be found in your textbook.

Before beginning the next problem, **CLEAR** all existing graphs from the calculators.

Problem 10: Ninth grade students in an Italian class were given a vocabulary test. The same students were retested several times over the next two years. The average score, *S*, of the class can be represented by the model

$$S = 80 - \log(x+1)^{14}$$

where *x* represents the time in months since the original exam. Graph this model.

Set the viewing rectangle to reflect **xMin**: 0, **xMax**: 30, **xScl**: 5, **yMin**: 0, **yMax**: 85, **yScl**: 5.

TI-83 Graphing Calculator Solution

| Y= | \Y₁= | 8 | 0 | − |

| LOG | (| X,T,θ,n | + | 1 |

|) | ^ | 1 | 4 |) | GRAPH |

TI-85 Graphing Calculator Solution

| GRAPH | F1 [y(x) =] | y1= | 8 | 0 | − |

| LOG | (| x-VAR | + | 1 |) | ^ | 1 |

| 4 | 2nd | M5 [GRAPH] |

LABORATORY 14: EXPONENTIAL AND LOGARITHMIC FUNCTIONS

TI-83 Graphing Calculator Screen

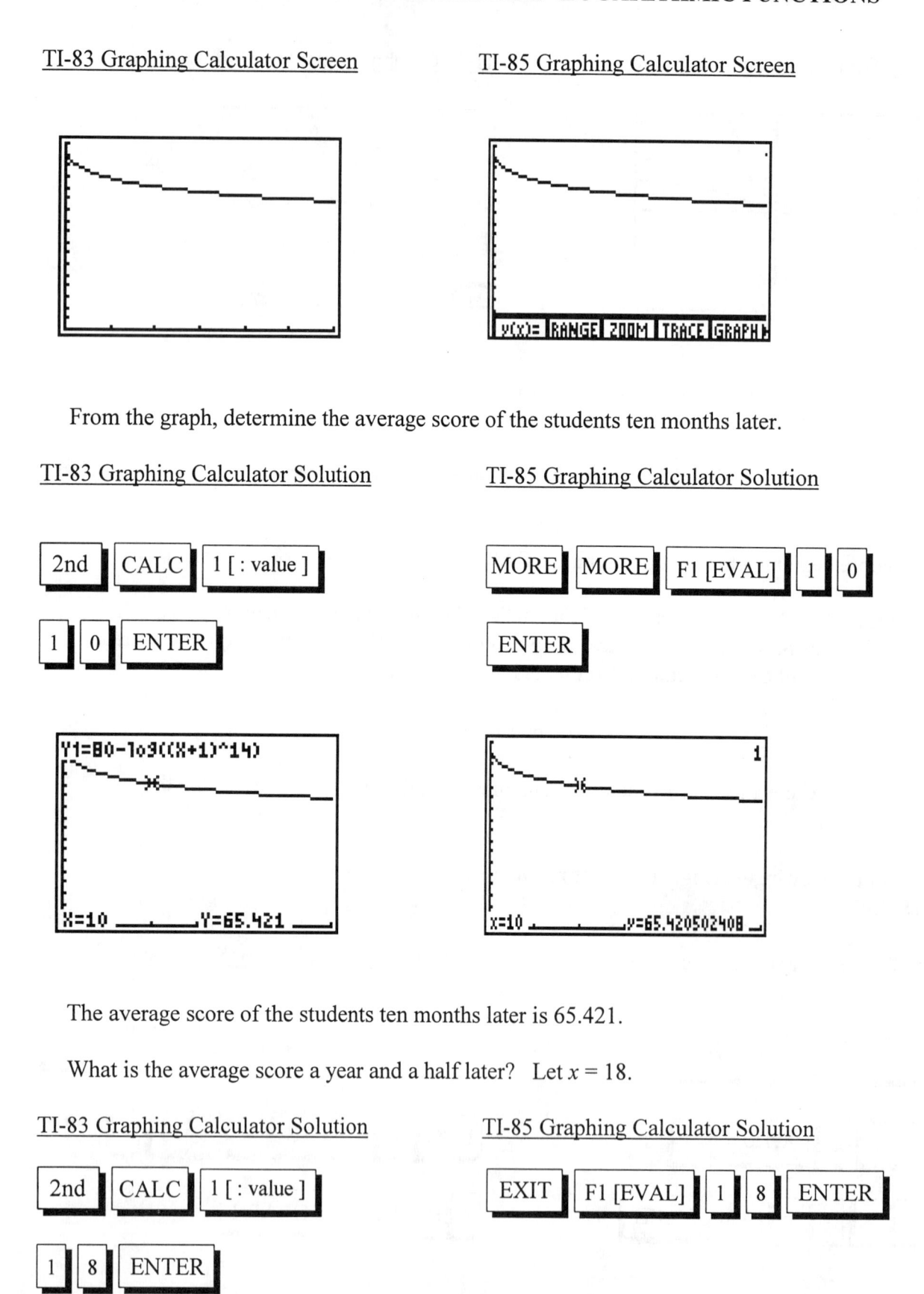

TI-85 Graphing Calculator Screen

From the graph, determine the average score of the students ten months later.

TI-83 Graphing Calculator Solution

| 2nd | CALC | 1 [: value] |

| 1 | 0 | ENTER |

TI-85 Graphing Calculator Solution

| MORE | MORE | F1 [EVAL] | 1 | 0 |

| ENTER |

Y1=80-log((X+1)^14)

X=10   Y=65.421

x=10   y=65.420502408

The average score of the students ten months later is 65.421.

What is the average score a year and a half later? Let $x = 18$.

TI-83 Graphing Calculator Solution

| 2nd | CALC | 1 [: value] |

| 1 | 8 | ENTER |

TI-85 Graphing Calculator Solution

| EXIT | F1 [EVAL] | 1 | 8 | ENTER |

LABORATORY 14: EXPONENTIAL AND LOGARITHMIC FUNCTIONS

TI-83 Graphing Calculator Screen

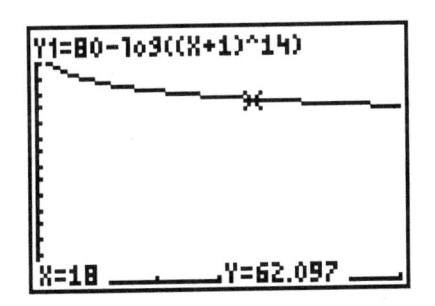

TI-85 Graphing Calculator Screen

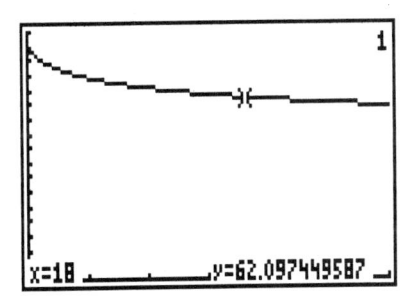

Y1=80-log((X+1)^14)
X=18   Y=62.097

x=18   y=62.097449587

The average score after 18 months is 62.097. What happens to the student's score as two or more years go by?

Problem 10 is an example of a **decaying** function. The more time that passes (*x* increases), the less the average student remembers (*S* decreases). The students' scores **dropped** as the months passed and approached two years. What do you think happens to the students' scores after two years?

LABORATORY 14: EXPONENTIAL AND LOGARITHMIC FUNCTIONS

EXERCISES

1. Find the value of each of the following.

   (a) 3^{-4} (b) $\left(\dfrac{2}{3}\right)^{-3}$ (c) $\log_4 34$ (d) $\log 97$ (e) $\ln 23$

   (f) $e^{-2.4}$ (g) $\log_e 5$ (h) $\log_7 0.735$

2. Evaluate for x. Use **SOLVE** on the **TI-82** or **SOLVER** on the **TI-83, TI-85** and **TI-86**.

   (a) $x^{5.2} = 352$ (b) $3^{x-2} = \left(\dfrac{1}{2}\right)$ (c) $2(3^{-x}) = 5$ (d) $e^{2x+1} = 72.4$

   (e) $3 \log x + 2 = 2 \log x$ (f) $\log(2x+1) = 5$

3. Graph (a) $y = \left(\dfrac{1}{2}\right)^{-x}$ and (b) $y = 2^x$ on the same set of axes. Which function, (a) or (b), is a **growth** function and which one is a **decaying** function? Are these inverse functions? Is the y intercept of both functions the same, and should it be the same?

4. A deposit of $2500 is made into a savings account at a compound interest rate of 5.5% quarterly. Write an equation showing how much would be in the account after x number of years. How much will be in the account in 4 years? In 7 years?

5. If the power input to a device is P_1 and the power output is P_2, the amount of decibels gained or lost in the device can be stated using the logarithmic function

$$G = 10 \log\left(\dfrac{P_2}{P_1}\right) \text{ dB.}$$

   If a certain amplifier gives a power output of 1500 W for an input of 70 W, find the dB gain.

6. A population of frogs in a pond is described by the model $f(x) = 12 \log_5(2x - 5)$, where x represents the number of months after the frogs are placed in the pond. Find:

   (a) $f(10)$ (b) $f(60)$ (c) $f(215)$

7. An isotope of a nuclear substance is very unstable. It is decaying at a rate of 21% each second. Find how much isotope remains 15 seconds after 6 grams of the isotope is created. Use $y = 6(2.7)^{-0.21t}$ as the model.

LABORATORY 15: CONIC SECTIONS

Purpose:

The purpose of this laboratory is to learn more about the graphs of the different types of conic sections using the TI-83/82 and TI-85/86 calculators.

Analytic Approach:

The intersection of a cone and a plane results in what is called a **conic section**. The parabola, ellipse, circle and hyperbola are the conic sections that result from the intersection of a cone and a plane.

The Parabola

In **Laboratory 8: Functions**, we looked at parabolas that were functions. Remember that a vertical line drawn through the graph can intersect the curve, at one and only one, point.

The graph of $y = a(x - h)^2 + k$ is a parabola that is a function.
- Its vertex is (h, k),
- Its axis of symmetry is the line $x = h$,
- It opens up if $a > 0$ and down if $a < 0$, and
- The points one unit to the left and right of the vertex on the parabola have a y-coordinate of $k + a$.

In this laboratory, we will look at the graph of a parabola that is not a function.

The graph of $x = a(y - k)^2 + h$ is a parabola.
- Its vertex is (h, k),
- Its axis of symmetry is the line $y = k$,
- It opens right if $a > 0$ and left if $a < 0$, and
- The points one unit above and below the vertex on the parabola have an x-coordinate of $h + a$.

Set the viewing rectangle on all four TI calculators to **xMin**: -5, **xMax**: 10, **yMin**: -5, and **yMax**: 5. Make sure to **CLEAR** all existing graphs from the screen before beginning each problem in this laboratory.

Problem 1: Graph the parabola whose equation is $x = 3y^2$.

Written in standard form, the equation $x = 3y^2$ is $x = 3(y - 0)^2 + 0$ with $a = 3$, $h = 0$, and $k = 0$. Its graph is a parabola with vertex $(0, 0)$ and its axis of symmetry is the line $y = 0$.

LABORATORY 15: CONIC SECTIONS

Before the graph can be entered on the TI calculators, solve the equation for y. Then enter the two equations, $y = \sqrt{\dfrac{x}{3}}$ and $y = -\sqrt{\dfrac{x}{3}}$.

Remember to add a left parenthesis after pressing the root key when using the **TI-82**.

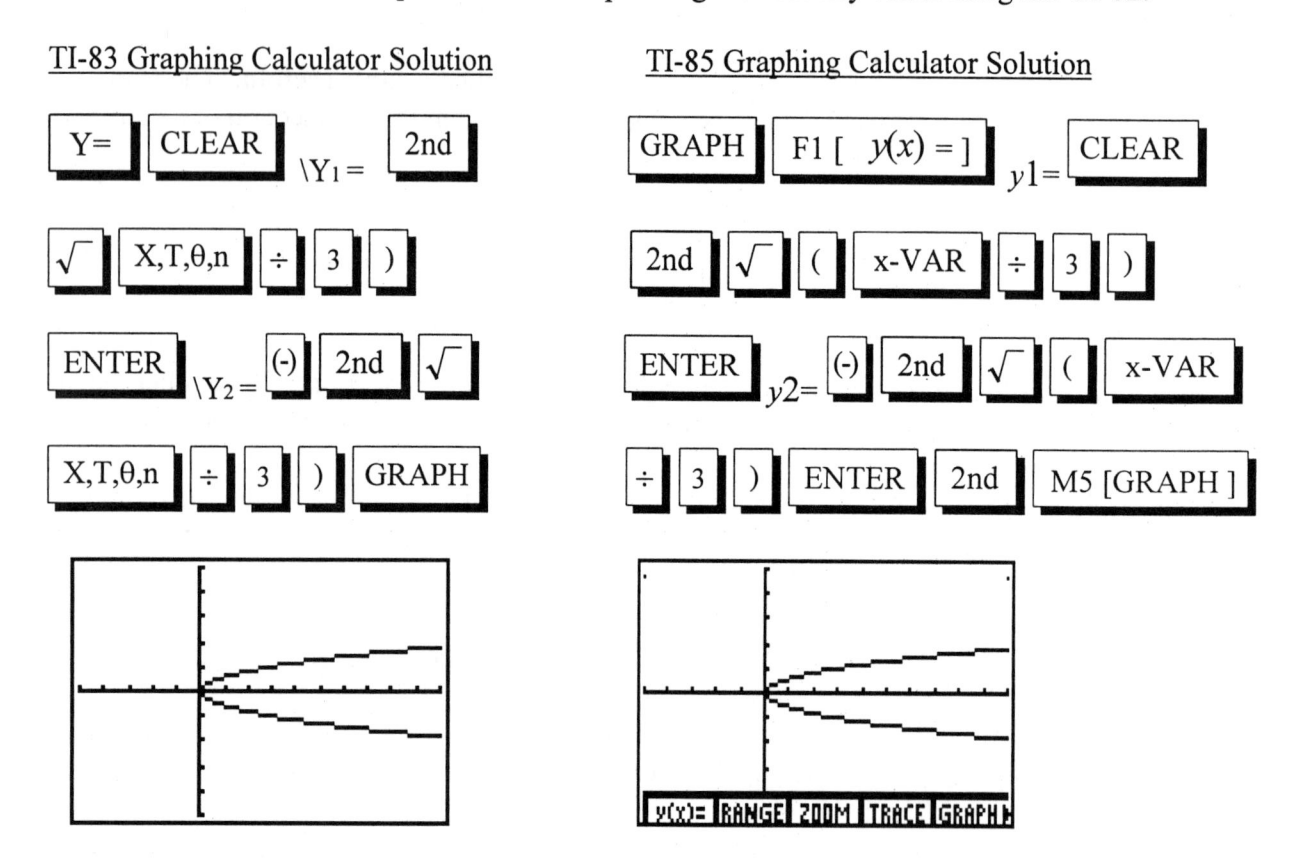

TI-83 Graphing Calculator Solution

[Y=] [CLEAR] \Y₁= [2nd]

[√] [X,T,θ,n] [÷] [3] [)]

[ENTER] \Y₂= [(-)] [2nd] [√]

[X,T,θ,n] [÷] [3] [)] [GRAPH]

TI-85 Graphing Calculator Solution

[GRAPH] [F1 [y(x) =]] y1= [CLEAR]

[2nd] [√] [(] [x-VAR] [÷] [3] [)]

[ENTER] y2= [(-)] [2nd] [√] [(] [x-VAR]

[÷] [3] [)] [ENTER] [2nd] [M5 [GRAPH]]

Since $a > 0$, this parabola opens to the right as you can clearly see from the graph. Why is this graph not a function?

The Ellipse

Another type of conic section is the **ellipse**. An **ellipse** is the set of points in a plane such that the sum of the distances of those points from two fixed points, called the **foci**, is constant. Each fixed point is called a **focus**. The point midway between the foci is called the **center**.

The graph of an equation of the form $\dfrac{x^2}{a^2} + \dfrac{y^2}{b^2} = 1$ is an ellipse with its center at (0, 0). The x-intercepts are a and $-a$ and the y-intercepts are b and $-b$.

LABORATORY 15: CONIC SECTIONS

Problem 2: Graph $\dfrac{x^2}{4} + \dfrac{y^2}{25} = 1$.

The equation is of the form $\dfrac{x^2}{a^2} + \dfrac{y^2}{b^2} = 1$, with $a = 2$ and $b = 5$, so its graph is an ellipse with center $(0, 0)$, x-intercepts 2 and -2, and y-intercepts 5 and -5.

Remember, before the equation can be entered on the TI calculators, to clear the denominators first, obtaining $25x^2 + 4y^2 = 100$. The equation then must be solved for y. The two equations to be graphed are $y = \sqrt{25 - \dfrac{25x^2}{4}}$ and $y = -\sqrt{25 - \dfrac{25x^2}{4}}$.

Set the viewing rectangle on all four TI calculators to **xMin: -7, xMax: 7, yMin: -6,** and **yMax: 6**. **CLEAR** all existing graphs from the screen.

Remember to add a left parenthesis after pressing the root key when using the **TI-82**.

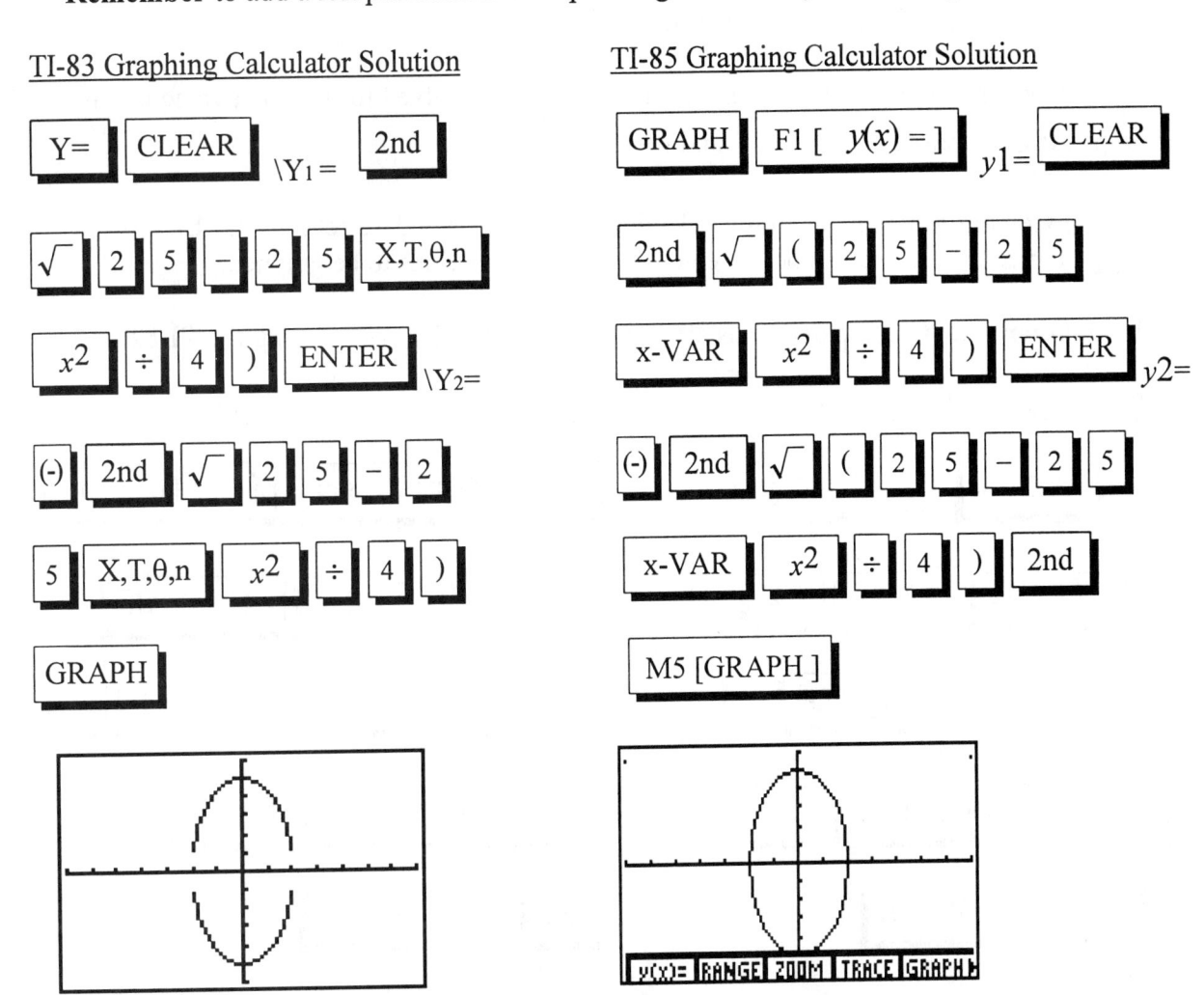

TI-83 Graphing Calculator Solution   TI-85 Graphing Calculator Solution

| Y= | CLEAR | \Y₁= | 2nd | | GRAPH | F1 [y(x) =] | y1= | CLEAR |

| √ | 2 | 5 | − | 2 | 5 | X,T,θ,n | | 2nd | √ | (| 2 | 5 | − | 2 | 5 |

| x² | ÷ | 4 |) | ENTER | \Y₂= | | x-VAR | x² | ÷ | 4 |) | ENTER | y2= |

| (-) | 2nd | √ | 2 | 5 | − | 2 | | (-) | 2nd | √ | (| 2 | 5 | − | 2 | 5 |

| 5 | X,T,θ,n | x² | ÷ | 4 |) | | x-VAR | x² | ÷ | 4 |) | 2nd |

| GRAPH | | M5 [GRAPH] |

Use the graph and draw a vertical line to state why the ellipse is not a function.

LABORATORY 15: CONIC SECTIONS

The Circle

The **circle** is another type of conic section. A **circle** is the set of all points in a plane that are the same distance from a fixed point known as the **center** of the circle. This distance is called the **radius** of the circle. The *distance formula* is used to determine the standard form of the equation of the circle.

The graph of $(x - h)^2 + (y - k)^2 = r^2$ is a circle with center (h, k) and radius r. The form $(x - h)^2 + (y - k)^2 = r^2$ is called the **standard form** of a circle. If an equation can be written in the standard form, then its graph is a circle which can be sketched by graphing the center (h, k) and using its radius r.

Problem 3: Graph $x^2 + y^2 = 16$.

The equation can be written in the standard form as $(x - 0)^2 + (y - 0)^2 = 4^2$. The center of the circle is the origin $(0, 0)$ and the radius is 4.

Remember that to graph the circle, the equation must be solved for y. The two equations to be graphed are therefore $y = \sqrt{16 - x^2}$ and $y = -\sqrt{16 - x^2}$.

Set the viewing rectangle on all four TI calculators to the standard setting. **CLEAR** all existing graphs from the screen before entering the new equations to be graphed.

Remember to add a left parenthesis after pressing the root key when using the **TI-82**.

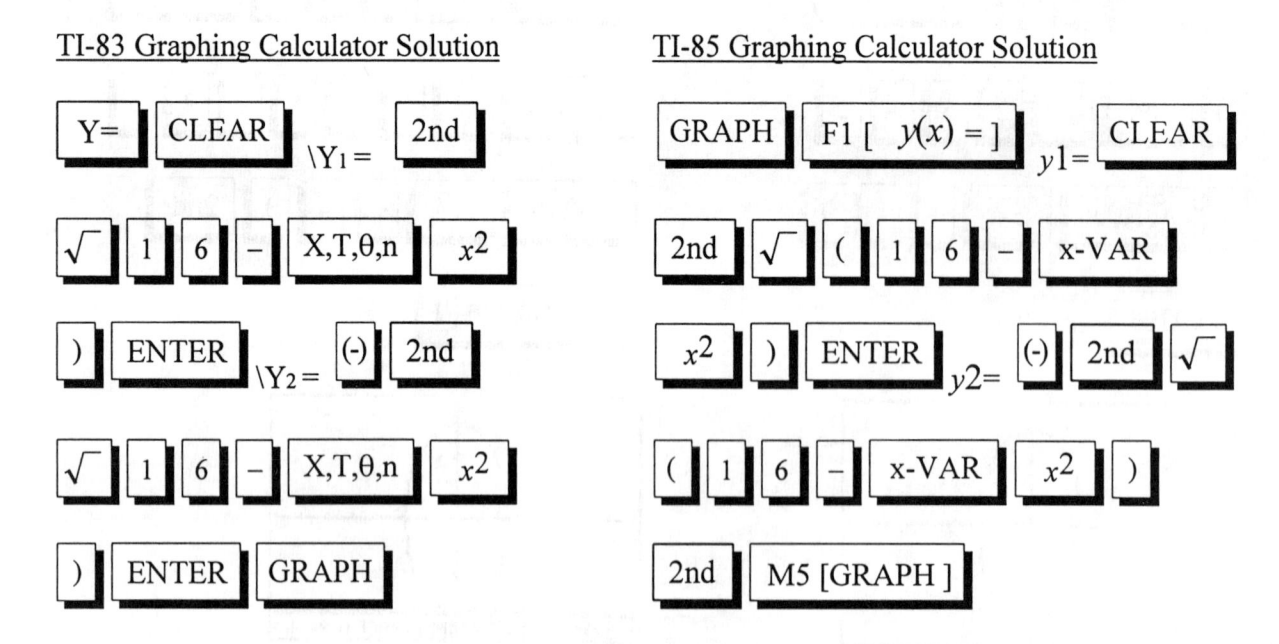

TI-83 Graphing Calculator Solution

| Y= | CLEAR | \Y₁= | 2nd |

| √ | 1 | 6 | − | X,T,θ,n | x² |

|) | ENTER | \Y₂= | (-) | 2nd |

| √ | 1 | 6 | − | X,T,θ,n | x² |

|) | ENTER | GRAPH |

TI-85 Graphing Calculator Solution

| GRAPH | F1 [y(x) =] | y1= | CLEAR |

| 2nd | √ | (| 1 | 6 | − | x-VAR |

| x² |) | ENTER | y2= | (-) | 2nd | √ |

| (| 1 | 6 | − | x-VAR | x² |) |

| 2nd | M5 [GRAPH] |

LABORATORY 15: CONIC SECTIONS

TI-83 Graphing Calculator Screen

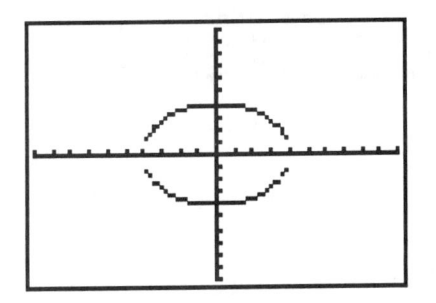

TI-85 Graphing Calculator Screen

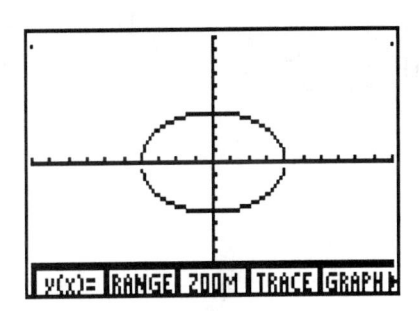

Notice that the graph of $x^2 + y^2 = 16$ looks just like an ellipse. There is a feature of **ZOOM** that will help to make the graph look more like a circle. The **ZOOM** feature will *square* the viewing rectangle making a circle look more like a circle than an ellipse. Press the following keys to see that this is indeed a circle.

TI-83 Graphing Calculator Solution

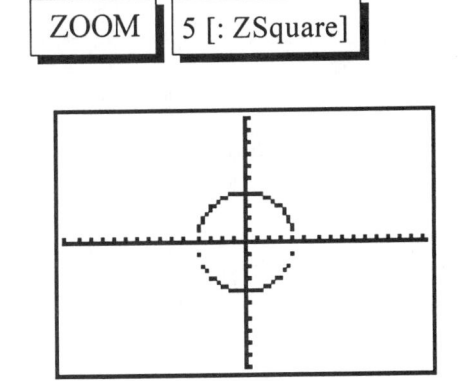

| ZOOM | 5 [: ZSquare] |

TI-85 Graphing Calculator Solution

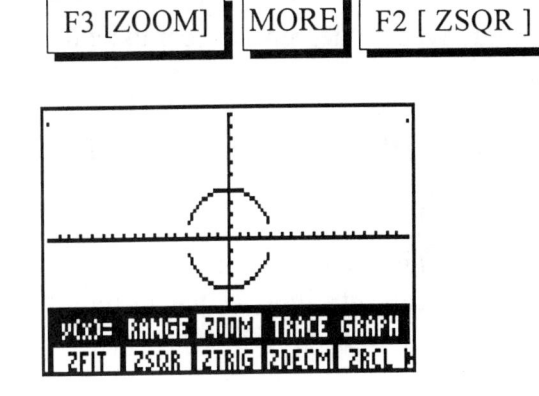

| F3 [ZOOM] | MORE | F2 [ZSQR] |

Even though the graph looks more like a circle, the graph still is not connected. Why? Press the following keys to determine that the x-intercepts of 4 and -4 exist on the graph.

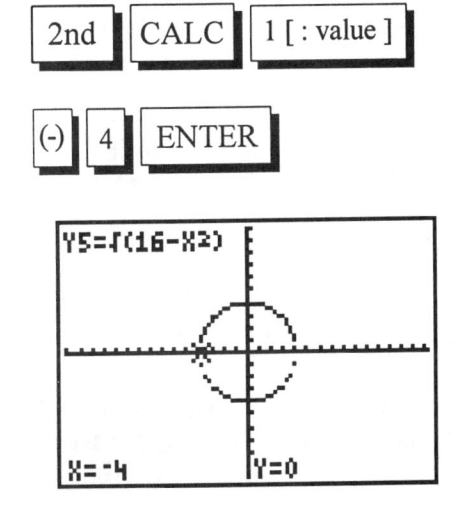

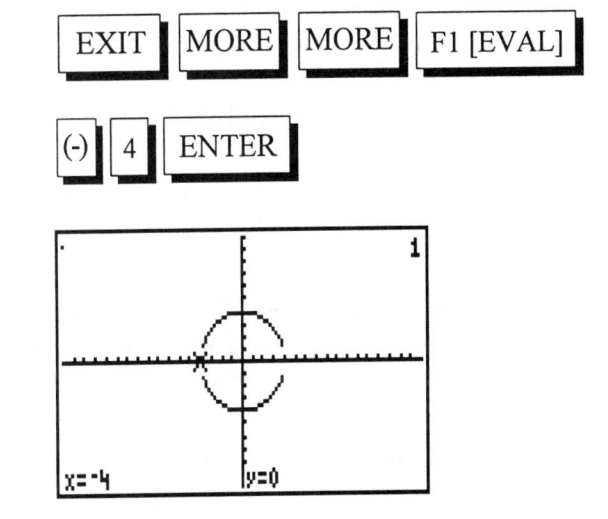

| 2nd | CALC | 1 [: value] |

| (-) | 4 | ENTER |

| EXIT | MORE | MORE | F1 [EVAL] |

| (-) | 4 | ENTER |

187

LABORATORY 15: CONIC SECTIONS

TI-83 Graphing Calculator Solution

[2nd] [CALC] [1 [: value]]

[4] [ENTER]

TI-85 Graphing Calculator Solution

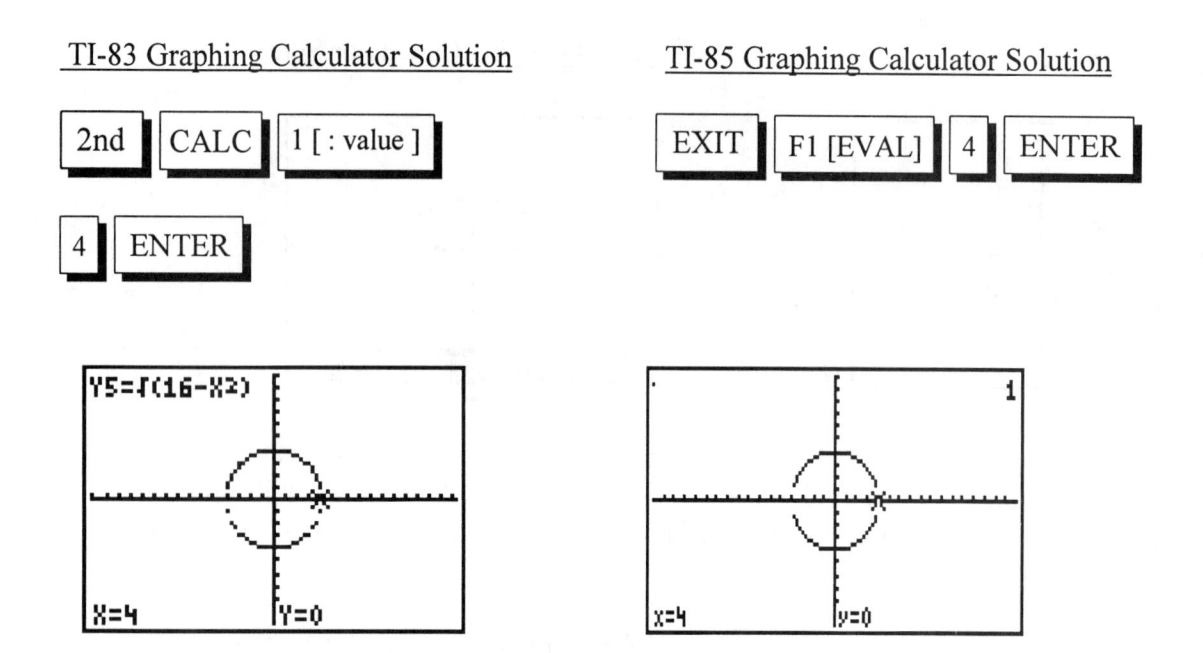

[EXIT] [F1 [EVAL]] [4] [ENTER]

Even though the x-intercepts appear to be missing, they indeed are located at the points (-4, 0) and (4, 0).

Problem 4: Graph the circle with a radius of 6 and with a center at (-5, 7). State the equation of the circle.

The value of h is given as -5, the value of k is given as 7, and the radius r is given as 6. Using these values, we can substitute the values into the equation $(x - h)^2 + (y - k)^2 = r^2$ obtaining the equation for the circle as $(x + 5)^2 + (y - 7)^2 = 6^2$. The equation would then be solved for y and graphed.

Besides using the **GRAPH** menu to sketch circles, we can also use the **DRAW** menu found on all four TI calculators. There is also a **Circle** feature found in the **CATALOG** on the **TI-83 TI-85** and **TI-86**. The **Circle** feature found in the catalog draws the circle given the center and radius. When using the **CIRCL** feature in the **DRAW** menu on the **TI-85** and **TI-86**, position the cursor at the center of the circle you want to draw. Press [ENTER]. Move the cursor to a point on the circumference using the radius of the circle to locate the point. Press [ENTER]. The circle is drawn on the graph.

Set the viewing rectangle on all four TI calculators to **xMin**: -15, **xMax**: 10, **yMin**: -1, and **yMax**: 15. Remove all existing graphs from the screen. Start with a blank home screen.

There isn't any keystroke adjustment necessary for the **TI-82**. The keystroke adjustment for the **TI-86** is as follows. Use the first two keystrokes from the **TI-85** solution. Then press **F1** to

LABORATORY 15: CONIC SECTIONS

access the catalog. Move the arrow key to locate the Circle feature. Then follow the **TI-85** solution.

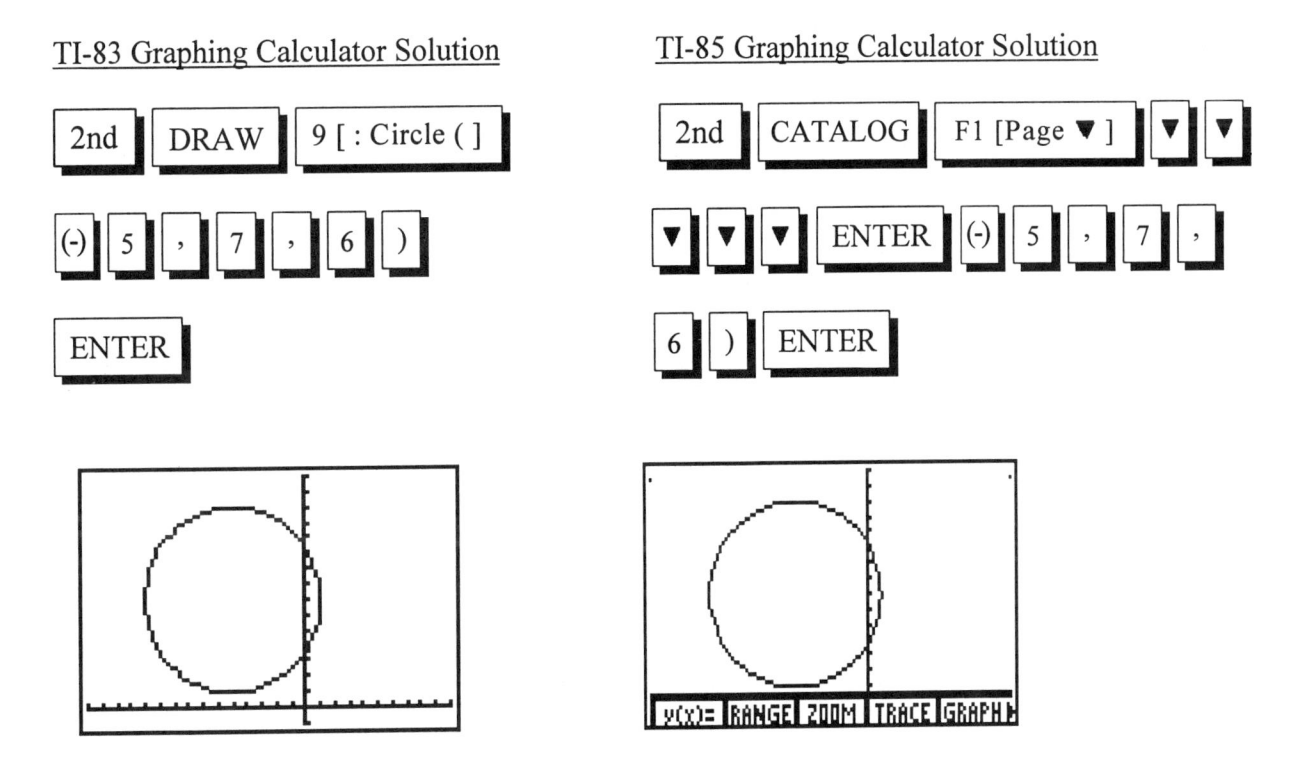

TI-83 Graphing Calculator Solution

| 2nd | DRAW | 9 [: Circle ( ] |

| (-) | 5 | , | 7 | , | 6 |) |

| ENTER |

TI-85 Graphing Calculator Solution

| 2nd | CATALOG | F1 [Page ▼] | ▼ | ▼ |

| ▼ | ▼ | ▼ | ENTER | (-) | 5 | , | 7 | , |

| 6 |) | ENTER |

Note that you must clear the screen in the **DRAW** menu before going on to the next problem. Use 1 [: ClrDraw] on the **TI-83/82** or F5 [CLDRW] on the **TI-85**. Use the **F1** key of the **DRAW** menu when **CLDRW** is needed with the **TI-86**. Remember to clear the home screen as well before moving on to the next problem.

The Hyperbola

The last of the conic sections to be examined in this laboratory is known as the **hyperbola**. A *hyperbola* is the set of points in a plane such that the absolute value of the difference of the distance from two fixed points, called the foci, is constant. The **center** is midway between the foci.

If the **center** of the hyperbola is at the origin (0, 0), the equation of a *horizontal* hyperbola is $\frac{x^2}{a^2} - \frac{y^2}{b^2} = 1$. Its *x*-intercepts are *a* and *-a*.

LABORATORY 15: CONIC SECTIONS

If the **center** of the hyperbola is at the origin (0, 0), the equation of a vertical hyperbola is $\frac{y^2}{b^2} - \frac{x^2}{a^2} = 1$. Its y-intercepts are b and -b.

The asymptotes of the hyperbolas $\frac{x^2}{a^2} - \frac{y^2}{b^2} = 1$ and $\frac{y^2}{b^2} - \frac{x^2}{a^2} = 1$ are defined in the same manner, $y = \pm \frac{b}{a} x$.

Problem 5: Graph the equation $\frac{x^2}{25} - \frac{y^2}{16} = 1$.

The graph of this equation has its center at the origin (0, 0), and x-intercepts -5 and 5. **Remember** to clear the denominators first and then solve the equation for y. The two equations that must be graphed are $y = \sqrt{-16 + \frac{16x^2}{25}}$ and $y = -\sqrt{-16 + \frac{16x^2}{25}}$.

Set the viewing rectangle on all four TI calculators to **xMin: -8, xMax: 8, yMin: -9,** and **yMax: 9**. **CLEAR** all existing graphs from the screen before entering the new equations.

Remember to add a left parenthesis after pressing the root key when using the **TI-82**.

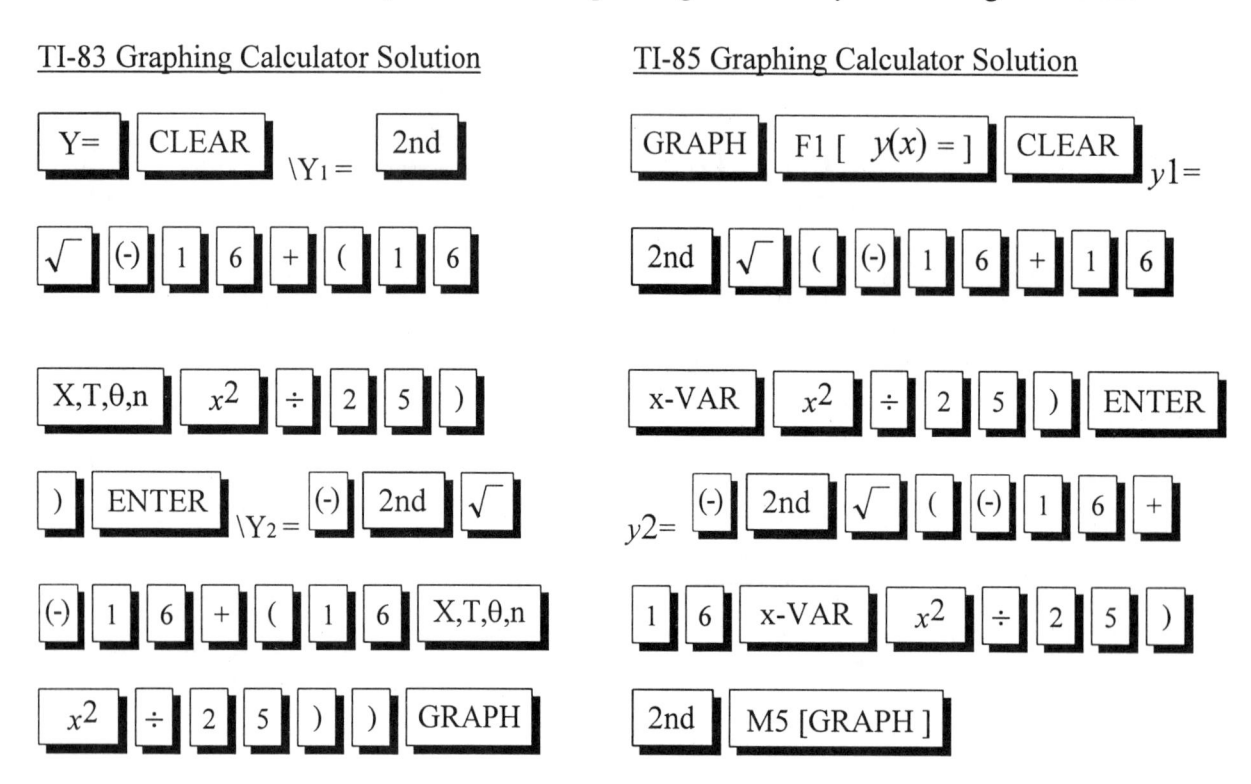

TI-83 Graphing Calculator Solution

[Y=] [CLEAR] \Y₁= [2nd]

[√] [(-)] [1] [6] [+] [(] [1] [6]

[X,T,θ,n] [x²] [÷] [2] [5] [)]

[)] [ENTER] \Y₂= [(-)] [2nd] [√]

[(-)] [1] [6] [+] [(] [1] [6] [X,T,θ,n]

[x²] [÷] [2] [5] [)] [)] [GRAPH]

TI-85 Graphing Calculator Solution

[GRAPH] [F1 [y(x) =]] [CLEAR] y1=

[2nd] [√] [(] [(-)] [1] [6] [+] [1] [6]

[x-VAR] [x²] [÷] [2] [5] [)] [ENTER]

y2= [(-)] [2nd] [√] [(] [(-)] [1] [6] [+]

[1] [6] [x-VAR] [x²] [÷] [2] [5] [)]

[2nd] [M5 [GRAPH]]

LABORATORY 15: CONIC SECTIONS

TI-83 Graphing Calculator Screen   TI-85 Graphing Calculator Screen

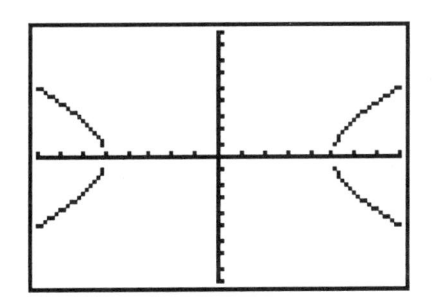

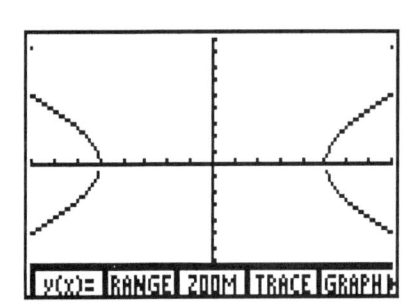

The asymptotes can be found by drawing lines through opposite corners of the rectangle whose vertices are (-5, 4), (-5, -4), (5, -4) and (5, 4). The diagonals of the rectangles are the asymptotes of the hyperbola.

Is the hyperbola defined by the equation $\dfrac{x^2}{25} - \dfrac{y^2}{16} = 1$ a function? Perform a vertical line test.

Problem 6:   Graph $4y^2 - 7x^2 = 28$. Determine the vertices of the rectangle which the asymptotes pass through.

To find the find the vertices and asymptotes of $4y^2 - 7x^2 = 28$, rewrite the equation in standard form. The equation in standard form is $\dfrac{y^2}{7} - \dfrac{x^2}{4} = 1$. We see that this is the standard form of a vertical hyperbola whose center is at (0, 0). Since $b^2 = 7$, $b = \sqrt{7}$; $a^2 = 4$, then $a = 2$.

Remember that to graph $4y^2 - 7x^2 = 28$, one must solve the equation for *y*. The two equations that must be graphed are $y = \sqrt{7 + \dfrac{7x^2}{4}}$ and $y = -\sqrt{7 + \dfrac{7x^2}{4}}$.

Note that when using a **TI-82** a left parenthesis must be inserted after the root key.

TI-83 Graphing Calculator Solution   TI-85 Graphing Calculator Solution

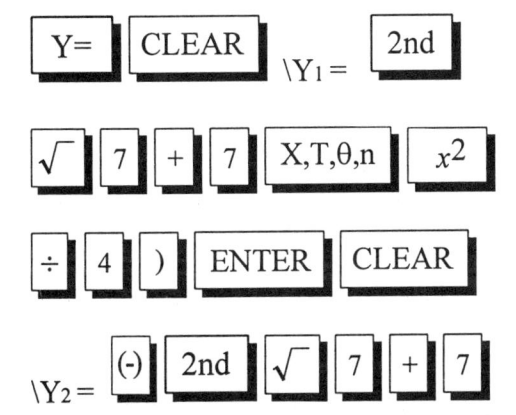

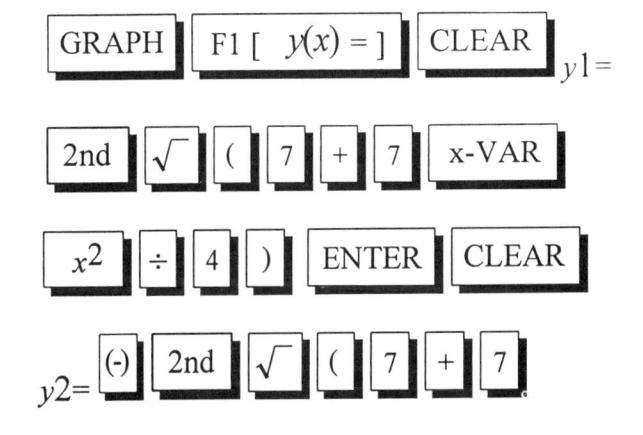

Y=	CLEAR	\Y₁=	2nd

√	7	+	7	X,T,θ,n	x²

÷	4	)	ENTER	CLEAR

\Y₂=	(-)	2nd	√	7	+	7

GRAPH	F1 [y(x) =]	CLEAR	y1=

2nd	√	(	7	+	7	x-VAR

x²	÷	4	)	ENTER	CLEAR

y2=	(-)	2nd	√	(	7	+	7

191

LABORATORY 15: CONIC SECTIONS

TI-83 Graphing Calculator Solution

| X,T,θ,n | x^2 | ÷ | 4 |) |

| GRAPH |

TI-85 Graphing Calculator Solution

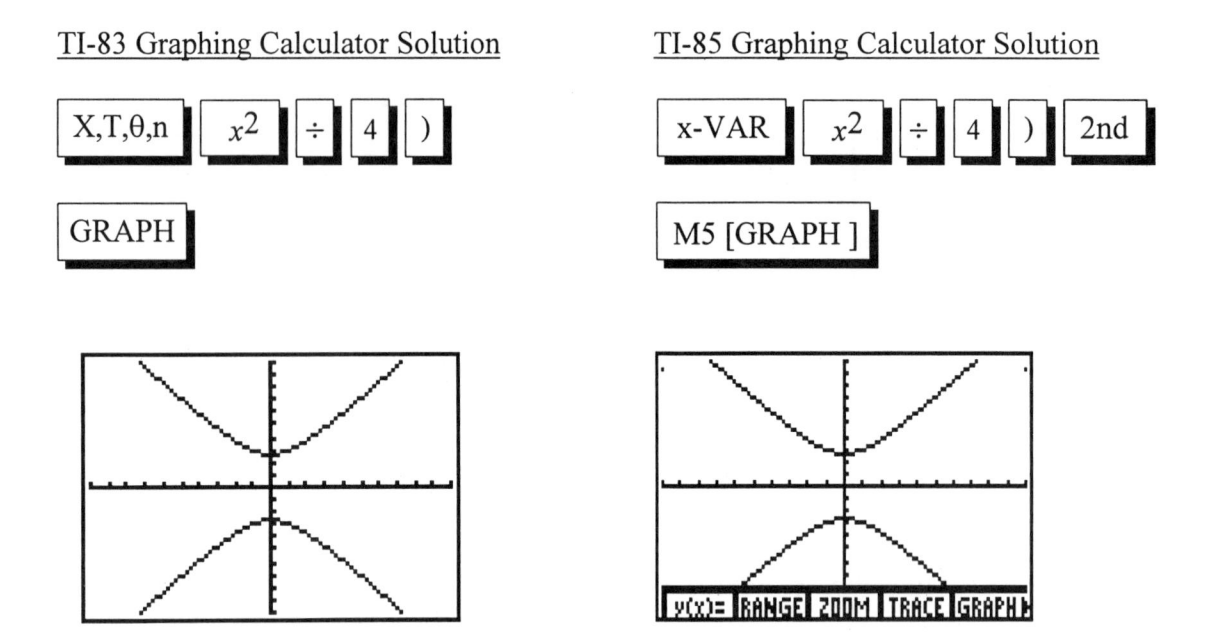

| x-VAR | x^2 | ÷ | 4 |) | 2nd |

| M5 [GRAPH] |

The vertices of the rectangle whose diagonals form the asymptotes of $4y^2 - 7x^2 = 28$ are $(2, \sqrt{7})$, $(2, -\sqrt{7})$, $(-2, \sqrt{7})$, and $(-2, -\sqrt{7})$.

Problem 7 demonstrates what a horizontal hyperbola would look like if the center of the hyperbola was *not* located at (0, 0).

Problem 7: Graph the equation $9x^2 - 4y^2 - 54x - 16y + 29 = 0$. Where is the center of the hyperbola? Name the vertices of the rectangle that can be used to draw the asymptotes.

The standard form of the equation is $\dfrac{(x-3)^2}{4} - \dfrac{(y+2)^2}{9} = 1$. The algebra necessary to write the equation $9x^2 - 4y^2 - 54x - 16y + 29 = 0$ in standard form is left to the student.

Remember that to graph the equation you must solve it for y obtaining the two equations $y = \sqrt{-9 + \dfrac{9(x-3)^2}{4}} - 2$ and $y = -\sqrt{-9 + \dfrac{9(x-3)^2}{4}} - 2$.

Set the viewing rectangle on all four TI calculators to **xMin**: -5, **xMax**: 8, **yMin**: -8, and **yMax**: 5. **CLEAR** all existing graphs from the screen before entering the new equations.

Remember to add a left parenthesis after pressing the root key when using the **TI-82**.

LABORATORY 15: CONIC SECTIONS

TI-83 Graphing Calculator Solution

[Y=] [CLEAR] \Y₁= [2nd]
[√] [(-)] [9] [+] [(] [9] [(]
[X,T,θ,n] [−] [3] [)] [x²] [÷]
[4] [)] [)] [−] [2] [ENTER] \Y₂=
[(-)] [2nd] [√] [(-)] [9] [+] [(] [9]
[(] [X,T,θ,n] [−] [3] [)] [x²]
[÷] [4] [)] [)] [−] [2] [GRAPH]

TI-85 Graphing Calculator Solution

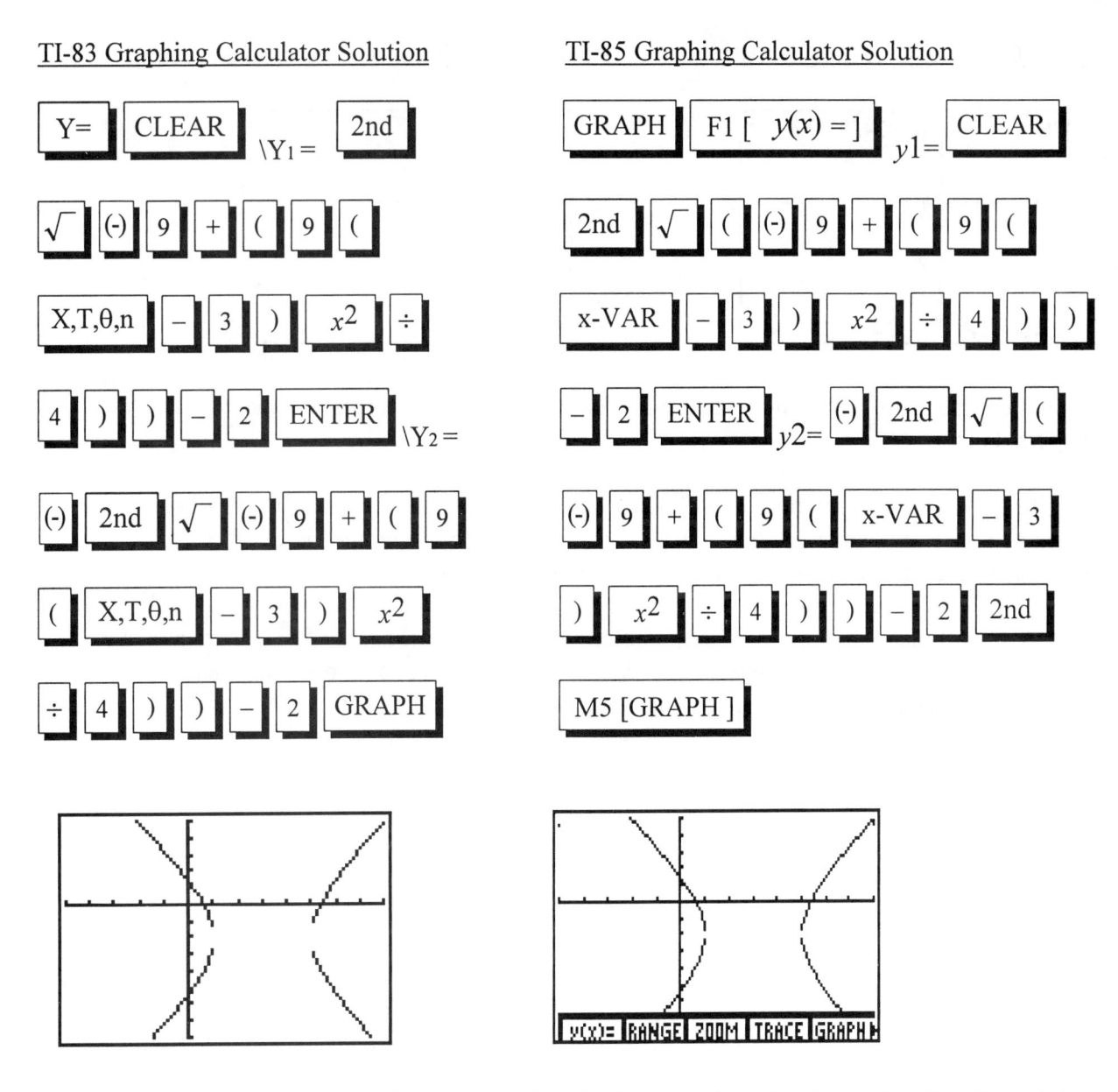

[GRAPH] [F1 [y(x) =]] y1= [CLEAR]
[2nd] [√] [(] [(-)] [9] [+] [(] [9] [(]
[x-VAR] [−] [3] [)] [x²] [÷] [4] [)] [)]
[−] [2] [ENTER] y2= [(-)] [2nd] [√] [(]
[(-)] [9] [+] [(] [9] [(] [x-VAR] [−] [3]
[)] [x²] [÷] [4] [)] [)] [−] [2] [2nd]
[M5 [GRAPH]]

Note that the graph looks as though the point (1, -2) is missing. To determine that the point (1, -2) is indeed on the graph press the following keys.

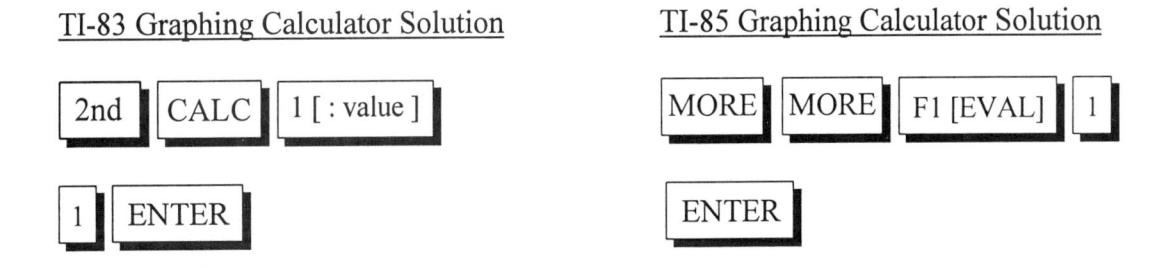

TI-83 Graphing Calculator Solution

[2nd] [CALC] [1 [: value]]
[1] [ENTER]

TI-85 Graphing Calculator Solution

[MORE] [MORE] [F1 [EVAL]] [1]
[ENTER]

193

LABORATORY 15: CONIC SECTIONS

TI-83 Graphing Calculator Screen   TI-85 Graphing Calculator Screen

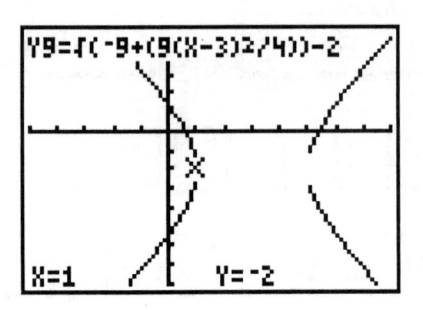

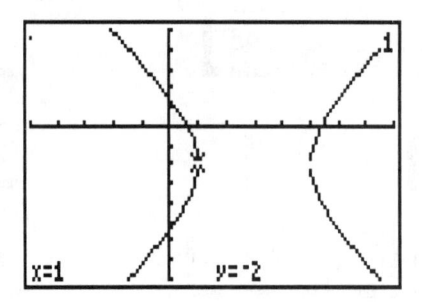

Follow the same procedure to determine that the point (5, -2) is also on the graph of the hyperbola.

The center of the hyperbola is at (3, -2). Because $a^2 = 4$, the hyperbola passes through points 2 units to the right and left of the center. The vertices of the rectangle (1, 1), (5, 1), (1, -5), and (5, -5). The asymptotes are the lines determined by the diagonals of the rectangle.

Are hyperbolas functions? Are hyperbolas one-to-one? **Laboratory 8: Functions** and **Laboratory 13: Inverse Functions** are two laboratories you should review to help you in determining the answers to those two questions.

LABORATORY 15: CONIC SECTIONS

EXERCISES

Set the viewing rectangle to the standard setting before beginning each of the exercises below. Use a **ZOOM** feature to help determine a better viewing rectangle if the need arises. You may also decide to use the **DRAW** menu to draw any of circles.

1. Graph $y = x^2 + 4$ and determine what type of conic section it represents.

2. Graph $x^2 + y^2 = 25$ and determine if it is a circle.

3. Graph $\dfrac{(y+4)^2}{4} - \dfrac{x^2}{16} = 1$ and determine if it is a hyperbola or an ellipse. Find the coordinates of the center. If it is a hyperbola, find the vertices of the rectangle that define the asymptotes.

4. Graph $\dfrac{x^2}{4} + \dfrac{y^2}{25} = 1$ and determine which type of conic section it is. Find the center of the section.

5. Graph $9x^2 + 4y^2 = 36$. Which type of conic section does this represent and why?

6. Graph $8y^2 - 8x^2 = 24$. Determine what type of conic section this represents.